Balancing River Health and Hydropower Requirements in the Lancang River Basin

Xuezhong Yu • Daming He • Phouvin Phousavanh

Balancing River Health and Hydropower Requirements in the Lancang River Basin

Springer

Xuezhong Yu
Ecofish Research Ltd.
Vancouver, BC, Canada

Phouvin Phousavanh
Faculty of Agriculture
National University of Laos
Vientiane, Lao People's Democratic Republic

Daming He
Asian International Rivers Center
Yunnan University
Kunming, Yunnan, China

ISBN 978-981-13-1564-0 ISBN 978-981-13-1565-7 (eBook)
https://doi.org/10.1007/978-981-13-1565-7

Library of Congress Control Number: 2018955289

This Springer imprint is published by the registered company Springer Nature Singapore Pte Ltd.
The registered company address is: 152 Beach Road, #21-01/04 Gateway East, Singapore 189721, Singapore

Preface

Hydropower development is the primary influence on river health in the Lancang River Basin, and it also affects the ecosystem of Mekong River. However, hydropower developers are short of knowledge on the state of river health and the impacts of hydropower projects in general and how best to mitigate and compensate for these impacts at a basin level. Meanwhile, institutions and communities in downstream countries do not have adequate information and opportunity to understand transboundary environmental effects of hydropower projects on the Lancang River.

From 2015 to 2017, an international team including scientists and engineers from Canada, China, and Laos was established to explore approaches for balancing river health and hydropower requirements in the Lancang River Basin. With a vision of integrating ecological values and social values, we assessed river health of the Lancang River Basin with respect to the impacts of hydropower projects. A state and impact assessment framework was developed and applied to evaluate river health and the impacts of hydropower projects. Based on the evaluation and comparison to international best practices, the environmental mitigation and compensation measures of hydropower projects were studied at a basin level to fill existing knowledge gaps in the Lancang River Basin. This work can help to improve knowledge, skills, and the practice of sustainable hydropower development, including those in the Lancang River Basin. We studied key components of transboundary effects including hydrology, sediment transport, water temperature, and fish community. The research work and communication activities strengthened dialog and communication on transboundary effects of hydropower projects on the Lancang River. The team also studied the specific needs of women, the specific impacts on women's lives and livelihood, and the factors that influence women's participation in river health management. The gender study is helpful for enhancing awareness and involvement of women in river health assessment and management. Based on the policy and practice analysis, the experience and lessons of environmental protection of hydropower development in the Lancang River Basin were shared with hydropower developers and regulators in Laos for improving their knowledge and skills of environmental protection.

Team members are from Ecofish Research Ltd. (Canada); Asian International Rivers Center, Yunnan University (China); and the Faculty of Agriculture, National University of Laos (Laos). The authors of each chapter are Xuezhong Yu and Daming He (Chap. 1); Xuezhong Yu, Andrew Harwood, and Heidi Regehr (Chap. 2); Xuezhong Yu and Andrew Harwood (Chap. 3); Yungang Li, Daming He, Xian Luo, and Xuan Ji (Chap. 4); Ying Lu and Daming He (Chap. 5); Kaidao Fu (Chap. 6); Chengzhi Ding, Chao Zhang, and Liqiang Chen (Chap. 7); Yanbo Li and Wenling Wang (Chap. 8); Phousavanh Phouvin and Xuezhong Yu (Chap. 9); and Xuezhong Yu and Heidi Regehr (Chap. 10). The constructive and thoughtful reviews and comments of Susan Johnson, Autumn Cousins, Deborah Lacroix, and Todd Hatfield are highly appreciated.

This work was funded by the CGIAR (4500025375). We would like to thank all donors who supported this work through their contributions to the CGIAR Fund.

Vancouver, BC, Canada — Xuezhong Yu
Kunming, Yunnan, China — Daming He
Vientiane, Lao People's Democratic Republic — Phouvin Phousavanh

Contents

Chapter 1
Introduction

Hydropower development is the primary issue affecting river health and aquatic ecosystems in the Lancang River Basin and potentially more broadly within the Lancang-Mekong River (Grumbine et al. 2012; Fan et al. 2015). Nevertheless, hydropower developers lack adequate knowledge on the impacts of hydropower projects on river health in general and on how best to mitigate and compensate for adverse effects. Although numerous studies both in China and abroad have assessed the impacts of hydropower projects on the river ecosystem in the Lancang River Basin, few studies have applied a river health perspective that integrates ecological and human values. It is also apparent that institutions and communities in countries downstream of China, along the Lancang-Mekong River, do not have adequate knowledge to understand transboundary environmental effects of hydropower projects, the opportunities to influence hydropower development on the Lancang River, or the way in which water regulation is managed. Another substantial knowledge gap is the impacts of hydropower development on women, which may be greater than those on men; these impacts have not been considered in China. Protection of environmental resources and social values during hydropower development in the Lancang River Basin is most effectively achieved with a cooperative approach in which skills, experiences, and lessons learned are shared at local and international scales.

This book presents key components of a project ("the Project") funded by the Consultative Group for International Agricultural Research (CGIAR) Research Program on Water, Land and Ecosystems (WLE), which investigates the ecological and social effects of hydropower development on the Lancang River, within China and in countries downstream, and makes recommendations for improvement. Results are presented on five key Project topics related to hydropower development on the Lancang River, each of which is presented one or a number of chapters: (1) assessment of river health, (2) improving river health through mitigation and monitoring, (3) transboundary environmental effects, (4) effects of hydropower on women, and (5) cross-border experience sharing. The sections below introduce the purpose and objectives of the Project, describe the Project setting, summarize overall Project goals, and outline general approaches, some of which are accomplished through this book.

X. Yu et al., *Balancing River Health and Hydropower Requirements in the Lancang River Basin*, https://doi.org/10.1007/978-981-13-1565-7_1

1.1 Project Purpose and Objectives

The CGIAR Research Program on WLE is a global research-for-development program that promotes sustainable solutions for people and societies. Sustainable intensification is one of the concepts promoted by the program, in which a healthy functioning ecosystem is seen as a prerequisite to agricultural development, resilience of food systems, and human well-being. A number of programs have been developed within WLE, and among them is the Greater Mekong program (WLE Mekong). WLE Mekong has a focus on water governance and large-scale reservoir management and the goal of improving the governance and management of water resources and associated land and ecosystems in the Greater Mekong Region by generating and sharing knowledge and experience.

In October 2014, WLE Mekong released a Call for Expressions of Interest (EOI) to solicit projects on a number of topics. Among these was the "healthy rivers" topic, the objective of which was to inform and strengthen river management decisions within the Greater Mekong through regionally appropriate, equitable, river health frameworks, as well as data collection and monitoring systems. A consortium led by Ecofish Research Ltd. (Ecofish), partnered with Asian International River Center (AIRC) of Yunnan University and Faculty of Agriculture (FOA) of National University of Laos, submitted a proposal on the "healthy rivers" topic for a project focused on the Lancang River Basin, which was subsequently approved. An international multidisciplinary Project team was established in 2015 that included biologists, engineers, hydrologists, chemists, and sociologists from Canada, China, and Laos. The Project team collaborated for 3 years on literature reviews, field surveys, and modeling analyses to assess the status of river health, impacts of HEPs, transboundary environmental effects of hydropower, and women's involvement in river health management in the Lancang River Basin. In addition, the Project team organized a variety of activities in China, Laos, Cambodia, Thailand, and Myanmar for consultation and communication with target groups including government agencies, hydropower operators, and civil society. The overarching goal of the Project was to, through research and collaboration, identify key issues and provide recommendations that will maximize the positive and minimize the negative ecological and social effects of hydropower development.

This Project has many diverse objectives and, as such, has many components. General Project objectives addressed in this book include defining and assessing "river health" of the Lancang River in relation to hydropower development, identifying potential transboundary ecological effects of hydropower projects, investigating their effects on women, documenting the challenges and lessons learned in international collaboration, and sharing experience of environmental management associated with hydropower development.

To address these high-level objectives, this book includes:

- An assessment of river health from a perspective that integrates positive and negative impacts of hydropower development on ecological and human values, through the use of physical, chemical, biological, and social indicators

- Mitigation and monitoring strategies to improve river health in light of hydropower development
- An assessment of potential transboundary effects of Lancang hydropower development for key physical and biological components of the Lancang-Mekong River system
- An assessment of the impacts of hydropower development on women's livelihood
- A case study from Laos that presents an example of the cross-border experience sharing undertaken

The details, specific objectives, and approaches for each aforementioned objective are presented separately in the book chapters that follow. Related to these Project objectives are a variety of outreach activities conducted and products designed to achieve and support the Project goals. These goals and outputs are briefly discussed in Sect. 1.3 below.

1.2 Project Setting

1.2.1 Study Area

The Lancang River is the upper half of the 4880-km-long Lancang-Mekong River (Fig. 1.1). Its headwaters are located on Guozongmucha Mountain in Zaduo County, Qinghai Province, from where it drops 4853 m over 2153 km to the China-Myanmar border. At this point the river exits China and is no longer called the Lancang River. The drainage area of the Lancang River (up to the China-Myanmar border) is 167,487 km^2, and the mean annual discharge is 2180 m^3/s (Zhou and Guan 2001). Its mean annual runoff is approximately 640×10^8 m^3, which represents 13.5% of the total Lancang-Mekong runoff (4750×10^8 m^3) (Zhao et al. 2000; Adamson et al. 2009). The mainstem of the Lancang River is typically divided into three reaches: (1) upstream Tibet, (2) upstream Yunnan, and (3) the middle and lower reaches downstream of Gongguoqiao dam (Fig. 1.1). After leaving China, the Lancang-Mekong River flows through Myanmar, Laos, Thailand, Cambodia, and Vietnam before entering the South China Sea.

1.2.2 Hydropower Development

The mainstem of the Lancang River has substantial potential for the generation of hydropower. The 4th national survey of hydropower resources estimated that the total theoretical hydropower potential of the Lancang River is 35.9 GW, and the technically exploitable installed capacity is 34.8 GW (Li and Shi 2006; Yuan

Fig. 1.1 Overview map of the Lancang-Mekong River Basin

2010). Given this potential, there were originally 23 proposed hydropower developments with associated dams planned on the mainstem of the Lancang River, some of which are now in various stages of completion. As shown in Table 1.1 and Fig. 1.2, there are four proposed hydropower projects in the planning stage in the upstream Tibet reach, for which the total installed capacity would be 3358 MW, and one 155 MW project in the lower reach. Seven projects in the upstream Yunnan reach, for which the total installed capacity would be 8930 MW, are

Table 1.1 Characteristics and status of hydropower projects existing or planned for the mainstem Lancang River in 2016

Dam	Installed capacity (MW)	Dam height (m)	Total storage (km^3)	Regulating storage (km^3)	Regulation type	Status
Cege	129					Planned
Yuelong	129					Planned
Kagong	240					Suspended
Banda	1,000					Planned
Rumei	2100					Planned
Bangduo	680					Suspended
Guxue	1700					Suspended
Quzika	405					Suspended
Gushui	1900	245	1.54	0.07	Yearly	Site preparation
Wunonglong	990	133.5	0.65	0.15	Daily	Site preparation
Lidi	420	63	0.1	0.01	Daily	Site preparation
Tuoba	1,400	140	1.04	0.24	Seasonal	Site preparation
Huangdeng	1,900	189	1.51	0.41	Seasonal	Site preparation
Dahuaqiao	920	106	0.29	0.04	Weekly	Site preparation
Miaowei	1400	131.3	0.66	0.16	Weekly	Site preparation
Gongguoqiao	900	130	0.51	0.01	Daily	Completed (2012)
Xiaowan	4200	292	15.1	10	Multi-yearly	Completed (2010)
Manwan	1550	132	1.06	0.26	Seasonal	Completed (2007)
Dachaoshan	1350	120.5	0.88	0.37	Seasonal	Completed (2003)
Nuozhadu	5850	261.5	23.7	11.3	Multi-yearly	Completed (2012)
Jinghong	1750	110	1.14	0.31	Seasonal	Completed (2009)
Ganlanba	155	60.5			Run-of-river	Planned
Mengsong	600					Canceled

[a]Phase 1 was completed in 1995, phase 2 in 2007

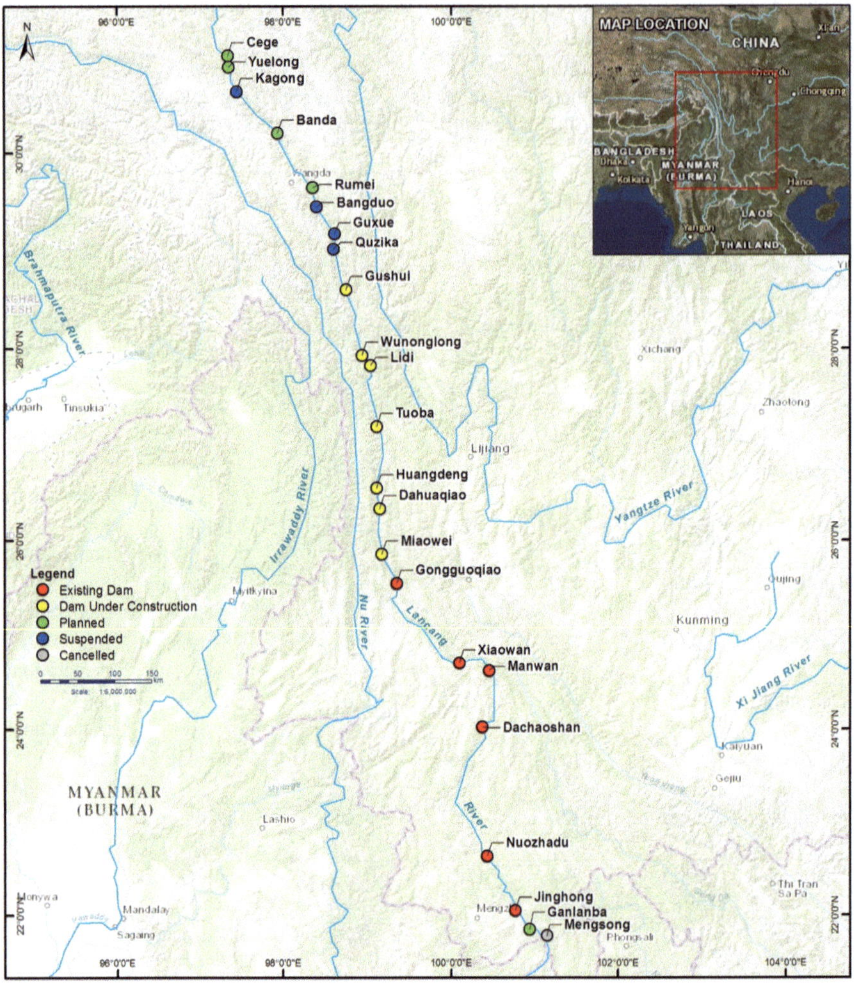

Fig. 1.2 Existing or planned hydropower projects on the mainstem Lancang River in 2016

currently under site preparation or under construction, and six projects in the middle and lower reaches that have a total installed capacity of 15,600 MW are now completed (Huang 2013; Wang 2015). Although 23 projects were originally planned, based on the strategic environmental assessment (SEA) for hydropower planning on the mainstem Lancang (Huang 2013), the total number of projects on the mainstem was reduced to 18, with 4 of the original 23 (Kagong, Bangduo, Guxue, Quzika) suspended and 1 (Mengsong) canceled.

Although hydropower development on the Lancang River was first planned in the 1950s (Zhao 2000), construction of the first dam of the Lancang cascade of hydropower projects, the Manwan dam, was not initiated until 1986. Over the last three decades, dam construction along the Lancang River advanced rapidly. As noted above, six large mainstem projects in the middle and lower reaches were in operation by 2015 (Manwan, Dachaoshan, Jinghong, Xiaowan, Gongguoqiao, and Nuozhadu in chronological order), and 12 dams are expected to be constructed in the next decades (Fan et al. 2015). Except for the Dachaoshan project, these projects are developed and operated by Huaneng Lancang River Hydropower Co., Ltd. (HydroLancang).

In addition to the large mainstem dams, there are also 782 small hydropower projects (installed capacity not exceeding 50 MW) planned on the tributaries of the Lancang River Basin. According to a survey conducted in 2008, 374 projects were completed or in construction (NSRH 2008). Most of these small projects are developed and operated by private enterprises.

1.3 Project Goals and Outputs

This book represents a key Project output that addresses, at least in part, a number of the high-level goals of the Project. Five high-level goals related to the Project objectives were identified:

1. Improve the knowledge, skills, and practices of management and technical personnel in HydroLancang that are required to improve environmental mitigation and compensation measures of hydropower projects with respect to river health and river ecosystem services and resilience.
2. Strengthen dialog and communication on transboundary effects of the Lancang River hydropower projects based on objective and comprehensive study outputs.
3. Enhance involvement of women in river health assessment and management in hydropower-affected areas in Yunnan and increase government (e.g., local government and resettlement bureau) awareness and attention to impacts from hydropower projects that may be specific to women.
4. Improve the knowledge and skills related to environmental protection of hydropower developers and regulators in Laos.
5. Improve the knowledge and skills of graduate students in the Asian International Rivers Center of Yunnan University regarding river health assessment and transboundary environmental effects through their involvement in investigation and research on river health assessment and transboundary effects.

1.4 General Approach

The methodologies and approaches of this project study were illustrated in Fig. 1.3. The project was comprised of three main stages:

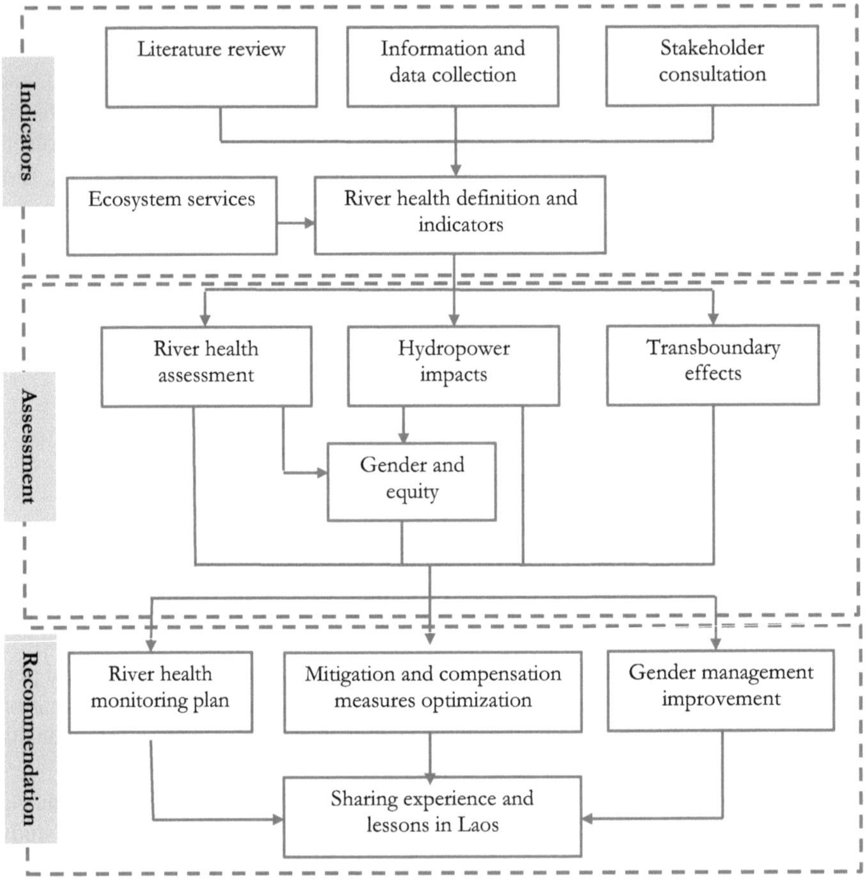

Fig. 1.3 Methodologies and approaches of the project study

1.4.1 *Identification of River Health Indicators*

Definition and indicators of river health are basis for river health assessment and analysis. Lancang River is a large river both for spatial characteristics and social implications. Thus, the definition of river health we adopted for this project incorporated both ecological and human values. The identification of river health indicators is based on literature review, information, and data collection in the study area and stakeholder consultation.

1.4.2 Assessment of River Health and Hydropower Impacts

This project assessed river health of the Lancang River with respect to the impacts of hydropower projects. A state and impact assessment framework was developed and applied to evaluate river health and the relationship between hydropower and river health based on the monitoring data and survey results. The river health assessment evaluated the healthy status of the Lancang River in a quantitative approach. The river health indicators were clarified into different categories by comparing with reference conditions. Meanwhile, the impacts of hydropower projects on river health were qualitatively assessed. Combining the assessment of river health and hydropower impact, critical indicators can be identified for the purpose of improving river health and mitigating hydropower impacts. For example, the river health indicators, which were assessed as critical for river health and significant negative for hydropower impacts, are critical components for river health and hydropower impact improvement.

In addition to the river health assessment of the Lancang River, we also studied the transboundary environmental effects of hydropower project on the Lancang River by analyzing key components of river health including hydrology, sediment transport, water temperature, fish community, flood, drought, and navigation.

1.4.3 Recommendations and Experience Sharing

Based on the analysis on critical components identified in the river health and hydropower impact assessment, a river health monitoring plan was developed for improving the existing monitoring scheme. We also proposed recommendations for optimizing the mitigation and compensation measures of hydropower projects on the Lancang River according to the assessment of river health and hydropower impact and international best practices. We proposed recommendations to government agencies to improve their awareness and understanding of specific impacts of hydropower projects on women. Thus, they will pay more attention in management and policy-making to promote women's participation in river health management to realize women's empowerment and rights.

Hydropower development remained rapid in the past three decades in China. And a variety of experience and lesson were gained in this period in terms of technology and management of environmental protection. Chinese experience will be beneficial for some Mekong countries that are seeking hydropower to power the economic and social development. The experience and lessons of environmental protection of hydropower development in the Lancang River Basin were shared with hydropower developers and regulators in Laos by distributing technical documents and organizing technical exchange and training workshops.

1.4.4 Gender Impact of Hydropower Projects

The gender study of this project is an extension of the assessment of river health and hydropower impact. Based on two case studies, we followed the long-term evolvement of local women's livelihood and social status in areas affected by hydropower development in the Lancang River Basin, and explored how larger institutional changes addressed impacts to women and their ability to adapt during resettlement. Policy recommendations were discussed regarding how to improve the livelihood of women resettlers, to empower women, to promote gender equality, and to strengthen river health management. The analysis was conducted based on community interview, key informants interview and literature reviews.

Given the broad scope of these goals, the investigations and analyses presented in this book generally form an important component of the tasks required to meet each goal. For example, this study contributes to the goal of improving the knowledge, skills, and practices of the management and technical personnel in HydroLancang by providing a river health assessment framework and assessment of river health and impacts of hydropower projects. Similarly, the goals of strengthening dialog and communication on transboundary effects and enhancing the involvement and engagement of women both required that these topics be investigated and analyzed. Our investigation and analysis identified key issues that communication and collaboration can now address. However, given the broad scope of the Project, effectively meeting its goals required a diversity of approaches. These were accomplished through various means in addition to producing this book. These included holding workshops, producing publications, developing a report card of river health, providing media coverage, and supporting graduate theses.

References

Adamson, P.T., I.D. Rutherfurd, M.C. Peel, and I.A. Conlan. 2009. Chapter 4 – The hydrology of the Mekong River. In *The Mekong-biophysical environment of an international river basin*, ed. I.C. Campbell, 53–76. San Diego: Academic.

Fan, H., D. He, and H. Wang. 2015. Environmental consequences of damming the mainstream Lancang-Mekong River: A review. *Earth-Science Reviews* 146: 77–91.

Grumbine, R.E., J. Dore, and J. Xu. 2012. Mekong hydropower: Drivers of change and governance challenges. *Frontiers in Ecology and the Environment* 10 (2): 91–98.

Huang, G. 2013. Practices in environmental protection of hydropower development in the Lancang River Basin. *Proceeding of the Hydropower 2013-CHINCOLD 2013 Annual Meeting and the 3rd International Symposium on Rockfill Dams*, Kunming, China (in Chinese).

Li, J., and L. Shi. 2006. Brief description of hydropower resources in China. *Water Power* 32 (1): 3–7 (in Chinese).

Steering Group of the National Survey and Assessment of Rural Hydropower (NSRH). 2008. *Survey and assessment report of rural hydropower in People's Republic of China*, General report. Beijing: Water & Power Press (in Chinese).

Wang, Y. 2015. *Sustainable hydropower development on the Lancang River*. http://www.hydropower.org.cn/showNewsDetail.asp?nsId=15588. Accessed 16 Feb 2016 (in Chinese).

Yuan, X. 2010. Thinking on speeding up the hydropower development of Lancang River in Tibet. *Water Power* 36 (11): 1–4 (in Chinese).

Zhao, A. 2000. Planning and development of hydropower resources on the middle and lower Lancang River. *Pearl River* 2: 5–8.

Zhao, C., Z. Zhu, and D. Zhou. 2000. *Worldwide rivers dams*. Beijing: China Water Power Press (in Chinese).

Zhou, C., and Z. Guan. 2001. The source of Lancangjiang (Mekong) River. *Geographical Research*, 20 (2): 184–190 (in Chinese).

Chapter 2
River Health Assessment

2.1 Background

To frame and guide our river health assessment, we collated and reviewed information and documents relevant to the study of river health and hydropower impacts on river health. Literature that was considered for review included Chinese and international scientific peer-reviewed publications, industry reports, publications by government and nongovernment organizations, and reports on technical standards and protocols. Relevant literature sources were compiled, reviewed, and analyzed to frame our assessment of river health, to assist with the selection of appropriate indicators, and to understand the current status of Lancang River health and how hydropower has influenced river health.

The following sections of this chapter define river health for the purpose of this project (Sect. 2.1.1), provide an overview of the literature on river health assessment (Sect. 2.1.2) and how this is affected by hydropower development (Sect. 2.1.3); we then describe the specific methods (Sect. 2.2) and results of our assessment (Sect. 2.3). Finally, we draw conclusions on the current state of health for the Lancang River and the influence of hydropower in Sect. 2.4.

2.1.1 Concept and Definition of River Health

River ecosystems encompass the full diversity of rivers, streams, and creeks, as well as riparian areas and groundwater systems that are linked to them. River ecosystems provide important ecological services, have substantial cultural heritage and scientific values, and support a rich diversity of plant and animal life. They also support a variety of human uses such as fisheries and recreation. The study of river health has become an important part of ecosystem health research, and the stresses on the structure and function of ecosystems from human activities have been recognized

X. Yu et al., *Balancing River Health and Hydropower Requirements in the Lancang River Basin*, https://doi.org/10.1007/978-981-13-1565-7_2

around the world. The degradation of aquatic ecosystems (river, lake, and wetland) is widespread and has resulted in an 81% decline in freshwater species populations since 1970, as measured by the Freshwater Living Planet Index (WWF 2016). As such, the maintenance and restoration of healthy rivers has become a critical objective of river management (Gore 1985; Karr 1991; Rapport 1991) and is included as a target within Sustainable Development Goal 6 that seeks to "ensure availability and sustainable management of water and sanitation for all" (UN DESA 2018). The study of river health has been ongoing for several decades, and numerous publications address components of river health such as definitions, indicators, assessment, monitoring, and management.

Many studies assessing river health have focused on physical, chemical, and some biological characteristics. Such assessments may be useful for protecting river ecosystems themselves or small rivers and creeks; however, they are inadequate for large-scale management of catchments (Norris and Thoms 1999). Some researchers have also argued that ecosystem health cannot be assessed on purely ecological grounds but that human values must also be considered (Karr 1996; Meyer 1997; Boulton 1999). Healthy rivers provide an array of ecological services that help meet social needs and expectations, and in China, it is widely accepted by river researchers and regulators that river health should incorporate both ecological and social concepts (Zhao and Yang 2005; Liu et al. 2006; Guo and Huang 2008; MEP 2013). Based on the substantial contribution that freshwater makes to human well-being, four ecosystem services were described by the Millennium Ecosystem Assessment categorization: (1) supporting services, e.g., nutrient cycling; (2) provisioning services for consumptive and non-consumptive uses; (3) regulating services, e.g., water purification; and (4) cultural services, e.g., aesthetic. This book explicitly incorporated both ecological integrity (maintaining structure and function) and human values (providing goods and services) into the definition of river health.

Following the work of Meyer (1997), Karr (1999), and Bunn et al. (2010), we define a healthy river as a river that is resilient to stress, *maintains its ecological structure and function* over time similar to the natural (undisturbed) ecosystems of the region, and has the ability to *recover from disturbance* while providing an array of unimpaired ecological services that continue to *meet social needs and expectations*. This definition incorporates both the ability of the river to provide ecosystem services and the resilience of the river ecosystem to meet social needs and expectations. Based on this, the impacts of hydropower projects on river health indicators included assessment of impacts to the four main types of ecosystem services listed above, which involved an analysis of the mechanisms and processes involved. It was recognized that hydropower projects may improve some of these services but have adverse effects on others. The changes in these services due to hydropower development are presented in Sect. 2.3 in a scientific and comprehensive way so that stakeholders are able to understand hydropower effects objectively.

2.1.2 Assessment of River Health

A variety of indicators and methods have been developed to assess the ecological conditions of river ecosystems around the world. The US Environmental Protection Agency (EPA) developed the Rapid Bio-assessment Protocols (RBP) to assess the health of streams and wadeable rivers (Barbour et al. 1999) and the Qualitative Habitat Evaluation Index (QHEI) for evaluating stream habitat quality (Rankin 1989, 1995). Karr (1981) developed the Index of Biotic Integrity (IBI) to describe and evaluate the condition of small warmwater streams in the USA. This index has been used throughout the USA and many countries internationally and has been proven to be a reliable means of assessing the effect of human disturbance on streams and watersheds. In Australia, the Australian River Assessment System (AUSRIVAS) was developed under the National River Health Program (NRHP) in response to growing concern in Australia for maintaining the ecological values of rivers. In addition to AUSRIVAS, the Index of Stream Condition (ISC) was also developed and applied in Australia to assess stream conditions in terms of hydrology, physical form, streamside zone, water quality, and aquatic life (Ladson and White 1999; Ladson et al. 1999). In 1994 the Department of Water Affairs and Forestry of South Africa initiated the River Health Programme (RHP) to monitor and evaluate the health of rivers in South Africa. The RHP primarily makes use of instream and riparian biological communities (e.g., fish, invertebrates, vegetation) to characterize the response of the aquatic environment to various disturbances. The River Eco-status Monitoring Programme (REMP), which is a component of the National Aquatic Ecosystem Health Monitoring Programme (NAEHMP), updated and replaced the RHP in 2016. The Index of Habitat Integrity (IHI) was developed to assess the integrity of instream and riparian zones based on hydrological, physicochemical, river bed, bank, and connectivity modifications (Kleynhans 1996; Kleynhans et al. 2008).

In China, progress in the assessment of river health has focused mainly on the Yellow, Yangtze, Pearl, Lancang, and Hai rivers. In 2004, the Yellow River Conservancy Commission proposed the vision "keeping the Yellow River healthy" as the target of water resources management in the Yellow River Basin (Li 2004). The targets for this vision of health were expressed as the total amount of water resources, flood discharging capacity, sediment carrying capacity, self-purification capacity, and the capacity to maintain ecosystems. To assess whether targets were being met, Liu et al. (2006) identified the following indicators of river health: minimum flow, maximum flood discharging capacity, bankfull discharge, floodplain transversal slope, water quality, wetland area, aquatic life, and water supply. In addition to these ecological indicators, social indicators were considered in the Yellow River health assessment. For the Yangtze River, Guo and Huang (2008) identified river morphology, chemical and ecological status, water resources utilization, and flood security as assessment indicators, whereas for the Pearl River, Lin et al. (2006) identified river morphology, water quality, aquatic life, human values, and monitoring capacity as assessment indicators. For the Lancang River, the functions of environment, ecology, service, flood protection, and utilization were considered to assess the status of health (Geng et al. 2006).

In collaboration with China's Ministry of Water Resources (MWR) and Ministry of Environmental Protection (MEP), the International Water Centre (IWC) of Australia assessed river health of the Yellow River, Pearl River, and Liao River from August 2009 until March 2012 (Speed et al. 2012). Variables of land use, hydrologic alteration, water quality, algae, macroinvertebrates, fish, riparian and channel condition, and riparian and instream vegetation were used in the assessment of the Pearl and Liao rivers. In 2012, the Chinese Research Academy of Environmental Sciences (CRAES 2012) released the Technical Regulation for Assessment of River Health in China. The river health indicators used in this Technical Regulation were hydrology, physical form (i.e., geomorphological processes), water quality, fish, benthic macroinvertebrates, algae, and riverine vegetation. The Regulation acknowledged that rivers provide goods and services that benefit society but focused on the physical and biological attributes of rivers for the river health assessment.

A comprehensive analysis of river health in China was conducted in 2012 by Feng et al. (2012) who reviewed approximately 150 research papers, documents, and standards that had been produced between 1972 and 2010 on the topic of river health. Through the use of statistics and correlation analysis, they evaluated 902 indicators used in 45 assessments internationally and in China and identified 8 indicators that were critical for assessing river health: riparian vegetation coverage, protection rate of wetlands, river continuity, flow alteration, water quality compliance rates, fish index of biotic integrity, utilization rate of water resources, and land use.

The various approaches described above share similarities, but not every approach emphasizes the same components of aquatic ecosystems, and hence the same indicators. These methods, and the relative importance of different indicators in the context of the Lancang River, were considered when developing the river health indicators used in this study (Sect. 2.2.2).

2.1.3 Hydropower Impacts on River Health

The damming of a river to develop a hydropower project results in a variety of changes to the physical, chemical, and biological properties of the river, all of which are fundamental components of ecological structure and function of the river ecosystem. Petts (1984) categorized these impacts into three orders, where the first-order impacts represent the physical properties, the second-order impacts represent some physical (channel morphology) and some biotic (primary production) properties, and third-order impacts represent the biotic properties of higher trophic levels. These impacts also have the potential to adversely affect the ecosystem goods and services that rivers provide. Given the focus of our assessment of river health on both ecological integrity and human values, our assessment of hydropower impacts also considers both ecological and social impacts.

An important tool for the assessment of hydropower impacts is the identification of Valued Environmental Components (VECs) (Beanlands and Duinker 1983; Hegmann et al. 1999; IFC 2012). These represent components of the natural and human environment that are considered to have scientific, ecological, economic, social, cultural, archaeological, historical, or other importance. The VECs that may be affected by a hydropower project include (IFC 2012; The World Bank 2012; Meynell and Nazia 2014):

- Water (e.g., flow regime, water temperature, water quality, greenhouse gases)
- Aquatic habitats (e.g., geomorphology and sediment transport, connectivity, fish habitat, aquatic life)
- Aquatic species (e.g., phytoplankton, macroinvertebrates, fish species)
- Terrestrial habitats and wildlife (e.g., vegetation, wildlife resources)
- Land use (e.g., agriculture, forests)
- Public health (e.g., vector-borne diseases, air quality, noise pollution)
- Protected areas (e.g., national parks, wild preservation areas)
- Cultural heritage (e.g., physical and nonphysical cultural resources)

Many studies have been conducted to describe the direction (positive or negative), magnitude, and possible mitigation measures of the impacts of hydropower on the VECs listed above (e.g., IEA 2000; UNEP 2007). Such studies have demonstrated that hydropower projects can impact a specific VEC in a number of different ways. For example, flow regulation associated with a storage hydropower project has both positive and negative impacts. Positive impacts include flood mitigation through limiting of flood flow impacts to human infrastructure (Hearnshaw et al. 2010); however, the lack of flood and high flows may have negative impacts on the river ecosystem (Postel and Richter 2003). Similarly, both positive and negative effects may be observed for fisheries and water quality (Hearnshaw et al. 2010).

In China, hydropower impacts on hydrological regimes, water quality, sediment and geomorphology, aquatic life, fish habitat, and riparian vegetation have been systematically studied (Wang 2004; Chang et al. 2006; Dong 2007). Specifically, the ecological effects of hydropower projects on the Lancang River have been a focus of recent research given the river's importance as a hydropower resource (He et al. 2006; Fu and He 2007; Fu et al. 2007; Gu et al. 2008; Yu et al. 2011; Huang 2013; Shi 2013).

In addition to the analysis of environmental effects of hydropower projects, the impacts on aquatic ecosystem services, such as provisioning, regulating, and supporting cultural services, have also been theoretically analyzed or economically evaluated (Xiao et al. 2007; Hearnshaw et al. 2010; Costanza et al. 2011; Fu et al. 2014). The results indicate that hydropower may have positive effects on electricity, water supply, natural hazard regulation, navigation, aquaculture, and recreation. However, these positive services were offset through negative impacts on biodiversity, sediment transport, nutrient cycling, and invasive species. For a number of services, including water purification and aesthetic values, the direction of the impact was found to vary spatially and temporally. This was demonstrated by Wei et al. (2008), who studied the impact mechanisms of hydropower development on river ecosystem services. Wei et al. (2008) developed assessment indicators and used these to assess impacts of the Manwan hydropower project on the Lancang River.

Because most projects on the Lancang River have dams that provide a substantial amount of storage and change the natural river flow (except Ganlanba), the reservoir/storage hydropower project was the focus of this review. Hydropower projects with large regulation reservoirs are usually multipurpose projects developed to meet social demands of water supply, flood control, electricity, navigation, aquaculture, and tourism. However, providing these benefits requires that dam operations change the natural flow regime of a river and this flow alteration can negatively impact river ecosystems (Poff et al. 1997; Postel and Richter 2003). These impacts of hydropower have led to high rates of endangerment among freshwater species and losses of productive fisheries from regulated rivers. On the other hand, hydropower projects can also positively impact river ecosystems. For example, a hydropower project's capacity to control floods, or alleviate drought, can in certain circumstances limit adverse effects to river geomorphology, vertebrates, and fish (Jones et al. 2013). The assimilative capacity of the reservoir or the reach downstream of the dam may also be increased and pollutant concentrations decreased based on water storage and higher water volume discharges at certain times (Segar 1998; Wei et al. 2009).

2.2 Methods

2.2.1 Overview

The study area for the river health assessment is the middle and lower reaches of the mainstem Lancang River (Gongguoqiao to China-Myanmar border). This reach was selected because most of the existing hydropower projects are located in this reach and there are data available to support the assessment.

The river health assessment component of this Project consisted of the following steps:

1. Identify river health indicators through literature review, information and data collection, and stakeholder consultation (Sect. 2.2.4).
2. Assess the current status of river health for the Lancang River (Sect. 2.3).
3. Assess how hydropower development impacts Lancang River health (Sect. 2.3).
4. Assess existing measures to mitigate hydropower impacts and their effectiveness, along with the current monitoring regime (Sect. 3.3).
5. Recommend improvements to existing mitigation measures, and develop a framework for a river health monitoring program to improve river health (Sect. 3.4).
6. Assess how hydropower impacts result in transboundary effects for a subset of river health indicators (Chaps. 4, 5, 6 and 7).

2.2.2 Identification of River Health Indicators

Identifying appropriate indicators by which river health could be evaluated was a critical step in our river health assessment. River health indicators were identified based on a literature review, compilation of information and data available from the study area (Fig. 1.1), and consultation with hydropower developers, government officials, and academic researchers in the region. Due to the vast spatial scale of the Lancang River and the social implications inherent in its management, the definition of river health for this Project incorporated both ecological and human values (Sect. 2.1.1). The concepts of river health and ecosystem services were intrinsically linked in this study; thus, consideration of the river ecosystem services described in Sect. 2.1.1 provided the theoretical basis for identifying appropriate river health indicators. The indicators most commonly used in the literature, especially those applicable to large Chinese rivers, were selected. Additionally, the Lancang River is the upstream portion of the Mekong River, and transboundary effects are critical concerns. Consequently, the indicators selected also reflect the concerns of downstream countries.

The indicators of river health selected for this project are categorized into ecological and social categories (Fig. 2.1). The ecological category includes physical (hydrology, river connectivity, and sediment and river geomorphology), chemical (water quality), and biological (fish community and river vegetation) sub-categories. The social indicators evaluated in this book include flood protection, water supply, navigation, electricity, and water recreation. Additional social elements, such as resettlement, project-affected communities and livelihoods, and indigenous peoples, are also important components potentially impacted by hydropower projects. However, a full assessment of the social impacts of hydropower development is complex and beyond the scope of this Project. Our evaluation of river health has focused on the ecosystem services provided from a social perspective, and we recommend specialized research be conducted to understand the influence of hydropower on factors associated with resettlement and project-affected communities.

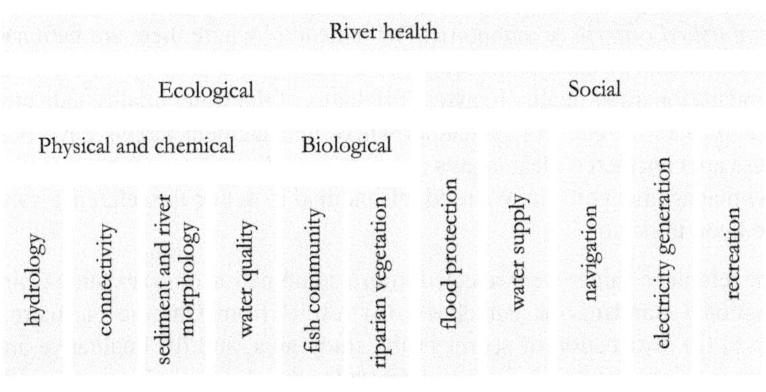

Fig. 2.1 Indicators of river health assessment of the Lancang River

Details of the selected indicators, their reference values, and scoring categories are provided in Sect. 2.2.4.

2.2.3 Reference Values and Scoring Assignment

2.2.3.1 Reference Values

The status of a river without disturbance from human activities was considered the ideal reference state for comparison with current river health conditions. However, the influence of human activities on rivers in China is so extensive that an undisturbed river is hard to find. Alternative approaches were therefore adopted to determine reference values. The selected reference values differed depending on the characteristic of the indicators. Five distinct cases were used to establish reference river health values:

1. *Natural condition*: For those indicators where undisturbed status can be quantitatively defined, the natural condition was used as the reference value. For example, the natural condition for river connectivity in systems where there are no natural barriers to fish passage is an unobstructed, free-flowing state. We used this method to define the reference values for hydrology, connectivity, and sediment and river geomorphology indicators.
2. *Pre-construction condition*: For indicators that are mainly impacted by hydropower projects (as opposed to those primarily influenced by non-hydropower impacts), the reference condition was considered the condition of the river ecosystem prior to the construction and operation of hydropower projects. We used this method to define the reference values for fish community and riparian vegetation indicators because there were no major disturbances to these two parameters prior to the construction of hydropower projects on the Lancang River.
3. *Best attainable condition*: This represents the expected condition if best management practices are used on the river. The indicators of water supply, navigation, and electricity were defined with this method.
4. *Established criteria or standards*: For indicators where there are national standards that can be used as the reference condition. We used the Chinese national standard for water quality to assess the status of the water quality indicator.
5. *Management target*: For indicators where management targets have been set; these are considered ideal targets to be attained through a variety of engineering and management efforts. We used this method to define the reference values for the flood indicator.

The reference values were selected from a combination of (a) existing Chinese or international standards or guidelines, (b) results from Chinese or international research, (c) distribution of scores in the study area, and (d) qualitative analysis. Reference values for each indicator were used to develop a scoring system, whereby each indicator could be assigned a score using the following categories: critical (1),

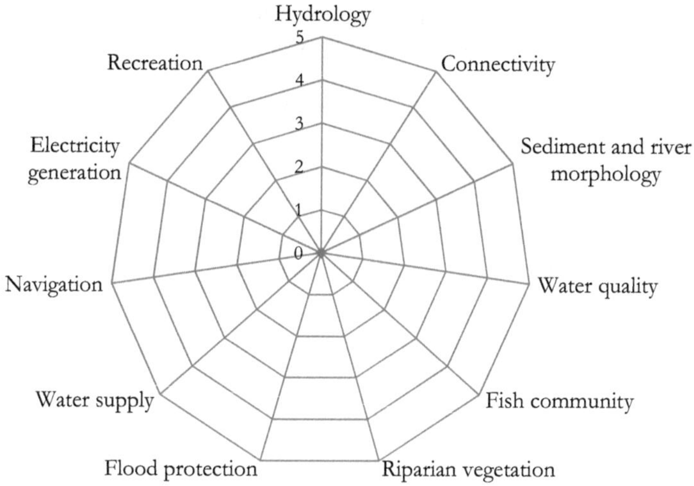

Fig. 2.2 Example spider diagram used to illustrate results of the river health assessment

poor (2), fair (3), good (4), and very good (5). In cases where multiple indicators were averaged to generate an aggregated score, the final score was categorized as follows: critical (0.0–1.0), poor (1.1–2.0), fair (2.1–3.0), good (3.1–4.0), and very good (4.1–5.0). Scores from the river health assessment were illustrated using a spider diagram presented in Fig. 2.2.

2.2.3.2 Assessment of Hydropower Impacts on River Health

A key objective of this Project was to assess the relationship between river health of the Lancang River and the impacts of hydropower projects. To accomplish this, a state and impact assessment framework was developed, and the relationship between hydropower and river health was evaluated using existing monitoring data and modeling results. The river health indicators were categorized relative to reference conditions, and the impacts of hydropower projects on river health were qualitatively assessed. As discussed in Sect. 2.1.3, hydropower development can impact river health both positively and negatively. However, there may also be cases when hydropower does not impact river health. Accordingly, hydropower impacts were categorized as positive, negative, or no impact. The magnitude of impact was classified into two categories, moderate and significant, based on the severity of the impacts relative to reference conditions for each indicator. Thus, the assessment of hydropower impacts on river health resulted in one of the five possible outcomes:

- *Significant negative*: The hydropower project impacts the river health indicator negatively, and the impact is significant compared to other key impact pathways.

- *Moderate negative*: The hydropower project impacts the river health indicator negatively, and the impact is moderate in severity compared to other key impact pathways.
- *No impact*: There is clear and strong evidence that the hydropower project has not affected the river health indicator.
- *Moderate positive*: The hydropower project impacts the river health indicator positively, and the impact is moderate compared to other key impact pathways.
- *Significant positive*: The hydropower project impacts the river health indicator positively, and the impact is significant compared to other key impact pathways.

The assessment of hydropower impacts on each river health indicator began with a review of relevant literature with a focus on storage hydropower (given that most hydropower projects on the Lancang River are storage projects; see Sect. 2.1.3). The literature review was followed by an analysis of the impacts of hydropower projects on the Lancang River on each indicator, as well as a comparison with other impact pathways that have the potential to affect the river health indicator.

2.2.3.3 River Health and Hydropower Impact Matrix

In this study, we developed a state-impact matrix to illustrate the status of river health and hydropower's impacts on river health, rather than combining scores from individual indicators into a summary river health score. The matrix allowed identification of indicators that are limiting river health and where mitigation of hydropower impacts through improved management should be focused. The categories for river health status (from critical to very good) were combined with the categories of hydropower impacts (defined by direction and severity) within a five-by-five matrix (Fig. 2.3).

To facilitate comparison of the significance of hydropower projects in maintaining river health, the 25 possible combinations within this matrix were divided into 4 overall categories of state and impact (Fig. 2.3):

(I) The river/reach has good river health and the impacts of hydropower are positive. This indicates that the hydropower project enhances the indicator of river health.

(II) The river/reach has good river health but there are negative impacts of hydropower. This indicates that the indicator of river health is in a good state despite negative hydropower impacts. Fair river health in combination with negative impacts of hydropower was also encompassed by this category due to similar implications.

(III) The river/reach has poor river health and there are negative impacts of hydropower. This indicates that hydropower project development and operation is likely a key stressor responsible for the observed degradation in river health.

(IV) The river/reach has poor river health but there are positive impacts of hydropower. This indicates that, although river health is poor, the hydropower proj-

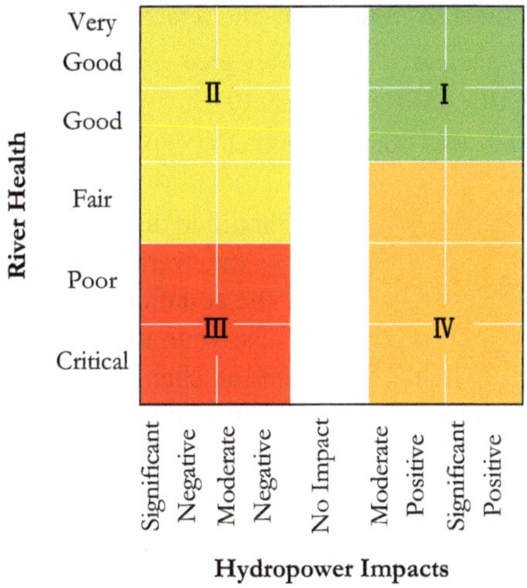

Fig. 2.3 Matrix of river health and hydropower impacts

ect enhances the performance of the indicator. Fair river health in combination with positive impacts of hydropower was also encompassed by this category due to similar implications.

The five grids indicating that hydropower was unlikely to have an impact on river health (Fig. 2.3, white cells) are not emphasized in this book given the focus of this Project on hydropower impacts.

2.2.4 Assessment Indicators

2.2.4.1 Physical and Chemical Indicators

Hydrology

The operation of reservoirs can modify the temporal and spatial processes of the natural flow regime, especially for large reservoirs with significant regulatory capacity (Poff et al. 1997; Postel and Richter 2003). The amended annual proportional flow deviation (AAPFD; Gehrke et al. 1995; Ladson and White 1999; Biemans et al. 2011) is a measure of deviation between affected and natural flow patterns. In this study, we use AAPFD to express the flow change (HF) as a proportion of the hydrology characteristic of river health, such that:

$$\mathrm{HF} = \left[\sum_{i=1}^{12}\left(\frac{r_i - r_{i0}}{\overline{r_{i0}}}\right)^2\right]^{\frac{1}{2}}$$

Table 2.1 Assessment criteria for flow change

Indicator	Very good	Good	Fair	Poor	Critical
HF (AAPFD)	≤0.10	0.11–0.30	0.31–1.0	1.10–3.0	>3.0

Modified from Ladson et al. (1999)

where r_i stands for modified discharge in month i, r_{i0} stands for the natural discharge in month i, and $\overline{r_{i0}}$ is the average natural discharge.

As indicated by the equation, the greater the AAPFD value, the more modified the flow regime is relative to natural conditions. The minimum value of AAPFD is zero which represents no change to natural flow. The value of AAPFD has no upper limit, and greater values are expected to result in more significant impacts to river ecosystems and poorer ecosystem health. Ladson et al. (1999) established the rating criteria for the AAPFD in the Index of Stream Condition (ISC) by categorizing the value of AAPFD from 0 to 5 into 11 levels. For this project, this rating system was modified by combining levels into five categories as shown in Table 2.1.

The AAPFD can be calculated at any arbitrary point on a river system if both modified and natural discharge data are available. In practice, natural discharge data are unavailable after the modification of the flow regime by one or multiple dams. Moreover, hydrological changes of rivers reflect a combination of the effects of climate change and the impacts of hydropower operation; thus to understand the impacts of hydropower, it is necessary to distinguish the relative contributions of climate variation and reservoir operation. An integrated model to simulate water movement through a river system and reservoir operation has been applied to study the impacts of hydropower on flow regimes (Shi 2013). This model simulates water movement in the river channel based on the Muskingum method, which is a hydraulic method for channel routing. This model uses a storage relationship to relate inflow and outflow in a channel reach:

$$\begin{cases} I - Q = \dfrac{\mathrm{d}W}{\mathrm{d}t} \\ W = K\left[xI + (1-x)Q\right] \end{cases}$$

in which I and Q are inflow and outflow rates (m^3/s), respectively, during the incremental time dt(s), and W is the channel reach storage (m^3), where K is a storage coefficient, and x is a weighting factor with a range of $0 \le x \le 0.5$. For each reach, inflow from large tributaries and small streams are also incorporated by collecting data from gauging stations or calculating inflow for sub-basin areas.

An operation scheme exists for each hydropower project, in which information regarding inflow and a number of water use components including electricity output, flood control requirements, irrigation, and downstream flow requirements are used to optimize reservoir outflow, depending on the purposes and features (storage, stage-storage curve) of the project. The operation scheme of the reservoir is programmed so that the reservoir outflow, storage, and water level can be calculated in

Table 2.2 Assessment criteria for changes to hydraulic conditions

Indicator	Very good	Good	Fair	Poor	Critical
HC (hydraulic index)	0.90–1.00	0.80–0.89	0.60–0.79	0.30–0.59	HC < 0.30

a discrete time-step mode, and then the reservoir operation model is integrated with the Muskingum model to simulate the river flow under the impacts of hydropower operation.

The numerical model can be used for analysis only if model validation verifies it is reliable. The most commonly used method for model validation is to compare simulated data with observed data. If the calculated results agree well with the flow observations, the model can be used to simulate flow regulation of a cascade of reservoirs. The model used to simulate the flow modification by the cascade of reservoirs on the Lancang River was validated for the period of 2009–2011 (Shi 2013). During this period, Xiaowan, Manwan, Dachaoshan, and Jinghong dams were commissioning; thus the operation of these four reservoirs was simulated with the numerical model, thereby coupling reservoir operation and water movement.

After model validation, the impacts of hydropower operation were calculated with the validated model. The observed flow data at three gauging stations (Jiuzhou, Gajiu, and Yongjinghong) on the mainstem of the Lancang River from 1958 to 1985 were used as input data for the model. The regulating schemes of five hydropower projects on the middle and lower reaches of the Lancang River were programmed into the model (Gongguoqiao was not included because of its small regulating storage), and the discharge downstream of each dam was calculated with the validated model.

Hydropower operation changes not only the flow regime but also the hydraulic conditions of a river. Fast-flowing rivers, or portions thereof, are changed to stagnant reservoirs during hydropower development and operation. Hydraulic conditions are critical for many aquatic species, including fish, and changes to hydraulic conditions will affect habitat suitability for different species. The change in hydraulic conditions of a river was categorized using the hydraulic index, which is the portion of river reach with unchanged hydraulic conditions. The hydraulic index (HC) is expressed as:

$$\mathrm{HC} = \frac{L_{un}}{L}$$

where L_{un} is the length of river with unchanged hydraulic conditions and L is the total length of the river reach. For this project we defined five categories of change in hydraulic conditions based on HC (Table 2.2). Categories of good and critical were defined as 80% and 30% of a river reach with natural hydraulic conditions, respectively.

Once HC was determined, the hydrological index was calculated from HC and the index for flow change (HF). The hydrological index (HI) was expressed as:

Table 2.3 Assessment criteria for river connectivity

Indicator	Very good	Good	Fair	Poor	Critical
DCI	0.86–1.00	0.66–0.85	0.46–0.65	0.26–0.45	≤0.25

$$\mathrm{HI} = 0.5\mathrm{HF} + 0.5\mathrm{HC}$$

where HI, HF, and HC are the hydrological index, flow change index, and hydraulic index, respectively. The flow change index and hydraulic index are the corresponding grades (1–5) assigned against the criteria in Tables 2.1 and 2.2.

Connectivity

The Dendritic Connectivity Index (DCI) developed by Cote et al. (2009) quantifies river fragmentation by barriers, including dams. The number, location, and passability of the barriers are considered in the calculation of the DCI for potadromous species, such that:

$$\mathrm{DCI} = \sum_{i=1}^{n}\sum_{j=1}^{n} c_{ij} \frac{L_i}{L}\frac{L_j}{L}$$

where L is the length of all stream sections in the drainage network; L_i and L_j are the lengths of section i and j, respectively; and c_{ij} is the passability between sections i and j, respectively. If there are M barriers between sections i and j, then c_{ij} is defined as:

$$c_{ij} = \prod_{m=1}^{M} p_m^u p_m^d$$

where p_m^u and p_m^d are the upstream and downstream passabilities of the barrier, respectively.

If there are no barriers on the river section, DCI equals 1.0 which can be set as the target or reference value of river connectivity. The connectivity index was divided into five categories from very good to critical based on the percentage connectivity decrease relative to the reference value (Table 2.3).

Sediment and River Morphology

Over long time frames, rivers tend to be in a natural balance of erosion and deposition. However, after the impoundment of water within a hydropower project's reservoir, sediment begins to deposit in the reservoir due to the decrease of sediment carrying capacity upstream of the dam. As a result, the discharged flow has a reduced sediment concentration and tends to scour the downstream river channel because of

Table 2.4 Assessment criteria for sediment release rate and sediment deposition rate

Indicator	Very good	Good	Fair	Poor	Critical
Sediment release rate (S_r)	≥80%	60–79%	40–59%	20–39%	<20%
Reservoir deposition rate (S_d)	<20%	20–39%	40–59%	60–79%	≥80%

sediment starvation. However, the sediment concentration of the discharged flow will increase over time; thus, provided that the hydropower development has been in operation for some time, sediment outflow volume will once again become approximately equal to the inflow volume, and sediment transport will return to natural conditions. The sediment release rate (S_r) is therefore a physical river characteristic that changes over time and is calculated as:

$$S_r = \frac{S_{\text{out}}}{S_{\text{in}}}$$

where S_{out} and S_{in} are annual sediment outflow and inflow volume, respectively.

Prior to the operation of a hydropower project, the sediment release rate at the project location is 100%, which indicates sediment transport is in equilibrium condition. Observational data show that sediment release rates at some reservoirs may be less than 10%. Consequently, the assessment criteria for S_r are equally divided between 100% and 0% (Table 2.4). Because sediment inflow and outflow equalize over time, the sediment release rate cannot sufficiently quantify the geomorphic changes caused by a hydropower project. For example, a reservoir that has reached sediment transport equilibrium having been operational for a long time still affects river morphology. Thus, a second indicator (deposition rate) was used to further assess the impacts of hydropower projects on sediment transport and river channel geomorphology. Sediment deposition rate (S_d) was defined as:

$$S_d = \frac{V_d}{V_r}$$

where V_d and V_r are sediment deposition volume and reservoir volume, respectively.

Sediment deposition rates in the upstream reach are 0% before the operation of a hydropower project but during infilling of the reservoir deposition rates may be 100%. The assessment criteria for S_d is therefore equally divided between 0% and 100% (Table 2.4).

The lower grade of the two indicators was attributed to the component of sediment and river morphology. For example, sediment and river morphology was characterized as poor if sediment release rate and deposition rate were poor and fair, respectively.

Table 2.5 Assessment standard for relevant water quality indicators (mg/L)

Grade	I	II	III	IV	V
COD_{Mn}	2	4	6	10	15
BOD_5	3	3	4	6	10
NH_3-N	0.15	0.5	1	1.5	2
DO	Saturation rate 90% (or 7.5)	6	5	3	2

Taken from the GB 3838-2002 Environmental Quality Standards for Surface Water

Water Quality

The organic pollution index (OPI) has been used to assess river water quality status. The basic formula quantifies the organic pollution index as:

$$\mathrm{OPI} = \frac{1}{n}\sum_{i=1}^{n}\frac{C_i}{C_{i0}}$$

where n is the number of water pollutant indicators, C_i is the concentration of the water pollutant indicator, and C_{i0} is the critical value of the water pollutant indicator which is defined in the Chinese national standard GB 3838-2002 Environmental Quality Standards for Surface Water. Four water pollutant indicators including COD_{Mn}, BOD_5, NH_3-N, and DO are typically included in the calculation of the OPI. Hence, the OPI can be calculated as:

$$\mathrm{OPI} = \frac{1}{4}\left(\frac{C_{\mathrm{COD}}}{C_{\mathrm{COD0}}} + \frac{C_{\mathrm{BOD}}}{C_{\mathrm{BOD0}}} + \frac{C_{\mathrm{NH3N}}}{C_{\mathrm{NH3N0}}} - \frac{C_{\mathrm{DO}}}{C_{\mathrm{DO0}}}\right)$$

In China, the water quality of rivers and lakes is assessed according to GB 3838-2002. The standard defines five grades that are suitable for certain uses (I–V) and a sixth grade that is not suitable for any purpose (worse than V) (Table 2.5). In these six grades, grade III is critical because it represents the threshold requirement for water used for drinking and direct human contact. Using grade III values as critical values, Wang et al. (2000) and Yu et al. (2011) proposed five criteria to categorize water quality in the upstream reach of the Manwan dam (Table 2.6). We adopted these criteria to characterize water quality in the Lancang River.

Eutrophication is a prominent water quality issue in reservoirs due to reduced flows compared to the natural riverine condition. The trophic level index (TLI), which quantifies eutrophication (Jin et al. 1995), was calculated using the equations below:

$$\mathrm{TLI}(\Sigma) = \sum_{j=1}^{m} W_j \cdot \mathrm{TII}(j)$$

$$\mathrm{TLI}(\mathrm{chla}) = 10(2.5 + 1.086 \ln \mathrm{chla})$$

Table 2.6 Assessment criteria for organic pollutant index

Indicator	Very good	Good	Fair	Poor	Critical
Organic pollution index	<0	0.0–1.0	1.1–2.0	2.1–3.0	3.1–4.0

Taken from Wang et al. (2000) and Yu et al. (2011)

$$\mathrm{TLI}(\mathrm{TP}) = 10(9.436 + 1.624 \ln \mathrm{TP})$$

$$\mathrm{TLI}(\mathrm{TN}) = 10(5.453 + 1.694 \ln \mathrm{TN})$$

$$\mathrm{TLI}(\mathrm{SD}) = 10(5.118 - 1.94 \ln \mathrm{SD})$$

$$\mathrm{TLI}(\mathrm{COD}) = 10(0.109 + 2.661 \ln \mathrm{COD})$$

where TLI(Σ) is the integrated TLI in which chla, total phosphorus (TP), total nitrogen (TN), transparency (SD), and chemical oxygen demand (COD) are considered; W_j is the weighted coefficient of specific water quality parameters; TLI(chla), TLI(TP), TLI(TN), TLI(SD), and TLI(COD) are TLI of chla, TP, TN, SD, and COD, respectively; and chla, TP, TN, SD, and COD are observed values of specific parameters. According to the Guidelines of Eutrophication Assessment and Grading of Lake and Reservoir in China, the water body can be clarified as oligotrophic ($0 \leq \mathrm{TLI} \leq 20$), mesotrophic ($20 < \mathrm{TLI} \leq 50$), eutrophic ($50 < \mathrm{TLI} \leq 60$), supertrophic ($60 < \mathrm{TLI} \leq 80$), and hypertrophic ($80 < \mathrm{TLI} \leq 100$).

2.2.4.2 Biological Indicators

Fish Community

A variety of environmental components have been used to evaluate the status of aquatic species including phytoplankton, zooplankton, and fish community. Fishes have valuable properties as an indicator because they tend to integrate the effects of lower trophic levels and they are relatively easy to sample and identify in the field.

The construction and operation of a hydropower project have the potential to directly or indirectly impact various metrics of fish community health, including abundance, density, diversity, biomass, size at age, distribution, timing of migration, and survival. Among these, species diversity is a valuable indicator and is usually incorporated into river health assessment. The change in the number of fish species present is based on a comparison with reference values (Geng et al. 2006; Yu et al. 2011), such that change in fish species before and after dam construction (C_f) is calculated as:

$$C_f = (n_{fb} - n_{fa}) / n_{fb}$$

where n_{fb} and n_{fa} are numbers of fish species before and after dam operation, respectively.

Table 2.7 Assessment criteria for fish species change

Indicator	Very good	Good	Fair	Poor	Critical
Change of fish species	≤5%	5–10%	11–15%	16–20%	>20%

Taken from Geng et al. (2006)

Geng et al. (2006) proposed five categories of fish species change (Table 2.7) and applied them to the river health assessment of the Lancang River. Yu et al. (2011) also applied these categories to assess the impacts of the Manwan dam on the river ecosystem. The decline of fish species numbers was classified from very good to critical depending on the percentage of species decline from 0% to 20%.

Riparian Vegetation

The impacts of hydropower projects include those that occur during construction activities and those that occur through project operations, and human activity changes after the project becomes operational. During the construction of the dam and ancillary facilities, activities may disturb the land and remove riparian vegetation. During project operation, the land may be flooded by the reservoir, and the fluctuation of water level of the reservoir may adversely impact the growth of riparian vegetation. Moreover, human activities may change owing to the development of the hydropower project that may also impact the land use and vegetation types nearby, the effects of which may propagate downstream (e.g., through changes in fine sediment inputs).

The Normalized Difference Vegetation Index (NDVI) is an index of plant "greenness" or photosynthetic activity and is one of the most commonly used vegetation indices. Because of its ease of use and correlation with many ecosystem parameters, NDVI has seen widespread use in rangeland ecosystems. The uses of this index include assessing or monitoring vegetation dynamics, plant phenological changes over time, biomass production, changes in rangeland condition, and vegetation or land cover classification (Fuller 1998; Geerken and Ilaiwi 2004; Wang et al. 2008; Fan et al. 2012).

The change of NDVI can be used to evaluate the trend of vegetation cover over time by analyzing historical NDVI data. The rate of change of NDVI over time (θ) is calculated from the following equation:

$$\theta = \frac{n \times \sum_{j=1}^{n} j \times \mathrm{NDVI}_j - \left(\sum_{j=1}^{n} j\right)\left(\sum_{j=1}^{n} \mathrm{NDVI}_j\right)}{n \times \sum_{j=1}^{n} j^2 - \left(\sum_{j=1}^{n} j\right)^2}$$

where n is the number of years and NDVI_j is the maximum value of NDVI in year j. If $\theta > 0$, it indicates NDVI increased in n years, whereas if $\theta < 0$, it indicates NDVI

Table 2.8 Assessment criteria for vegetation cover

Indicator	Very good	Good	Fair	Poor	Critical
Change in NDVI	>0.020	0.001–0.020	−0.001–0.001	−0.001–−0.020	<−0.020

Take from Liu et al. (2015)

decreased in n years. θ was used as indicator of riparian vegetation for this project by calculating the NDVI change relative to the reference condition, with the reference condition defined as the condition of vegetation 1 year before significant impacts of human activities occurred (Gippel et al. 2012). Five categories were defined by Liu et al. (2015) in the assessment of changes in riparian vegetation in the Lancang River Basin (Table 2.8).

2.2.4.3 Social Indicators

Flood Protection

Flooding can have both positive and negative impacts on river ecosystems, and the development of hydropower projects along large rivers that flow through populated areas decreases the risks of flooding. Flooding is a natural ecological process that plays an integral role in ensuring biological productivity and diversity in the floodplain. However, floods can also have devastating consequences on the environment as well as on people and the economy. Many people have died or been injured in flash floods, and flooding can be extremely damaging and costly in urban areas as it can negatively impact infrastructure, homes, and businesses. Flooding can also increase disease and infection rates including military fever, pneumonic plague, dermatopathia, and dysentery. In addition to the adverse effects on people and the economy, chemicals and other hazardous substances may be carried away by floods and eventually contaminate downstream water bodies.

Jinghong City is a large city (with a population of 528,600 in 2013) located alongside the mainstem Lancang River, so flood protection is a critical task for hydropower projects on the Lancang River. According to the National Standard for Flood Control of China (GB 50201-2014), the flood protection standard is the 50-year to 100-year recurrence interval (i.e., the city's flood protection measures should resist floods with recurrence periods of 50–100 years). Zhao and Yang (2005) classified the criteria for flood control of urban rivers into five categories based on the standard for flood protection (Table 2.9).

Table 2.9 Assessment criteria for the national flood control standard

Indicator	Very good	Good	Fair	Poor	Critical
Standard for flood protection (recurrence interval)	>100 years	50–100 years	30–49 years	5–29 years	<5 years

Taken from Zhao and Yang (2005)

Water Supply

Water supply capacity is usually a critical indicator for assessing water resources management within a river basin or region. It can be used to evaluate the water supply capacity of a water facility (e.g., reservoir) or integrated facilities within a river basin or region. The indicator of water supply was based on the guarantee rate, which refers to the percentage of time that key water uses (e.g., agriculture, industrial production, and domestic water use) can be met in the Lancang River Basin. Under normal circumstances, the river's social value is stronger if the river can supply more water for urban and rural development.

In China, domestic and industrial water supply has higher priority over other water uses such as agriculture, electricity, navigation, and recreation. Relevant guidelines (e.g., Code for design of outdoor water supply engineering GB50013-2006) require that the guarantee rates for domestic and industrial water use are 95% and 90–95%, respectively. By comparison, the guarantee rate for agricultural water use is 75–85% for xerophilous crops and 80–95% for rice in moist regions (Code for design of irrigation and drainage engineering GB50288-99). According to the Water Resources Bulletin of Changjiang and Southwest Rivers 2014 (Changjiang Water Resources Commission 2015), the total water use in 2014 was 2.837 billion m^3 in which 78.3% of water was used by agriculture, with domestic and industrial water use accounting for 10.2% and 10.5%, respectively. Agricultural water use in the Lancang River Basin is therefore approximately four times the total amount of domestic and industrial water uses. Based on the amount of water use between agricultural and urban water demands, this project used water supply status for agriculture to define the water supply indicator. The ratio of effective irrigated area is usually used as the indicator of irrigation status in agricultural and social statistics and planning in China, so it was used as the water supply index (SI) in this project.

The thresholds for the ratio of effective irrigation area were determined based on relevant planning and critical values for agricultural water supply in China and Yunnan Province. In the planning of water resources allocation of key regions in Yunnan Province (2012), the ratio of effective irrigation area was set to 75–80% by 2030. The value of the ratio of effective irrigation area in the national water resources planning was set to 58% by 2020, and it was 41.4% in Yunnan Province by 2015. Five categories were defined to categorize the water supply indicator for this project by ratio of effective irrigated area (Table 2.10). A ratio of 80% was used as the threshold for the very good category because it reflects the long-term objective in the region. The value of 60% was used as the threshold for the good category because it reflects the nationally set short-term objective. The current average level of agricultural irrigation in Yunnan Province (40–60%) was defined as fair based on

Table 2.10 Assessment criteria for water supply indicator

Indicator	Very good	Good	Fair	Poor	Critical
Water supply index (%)	80–100	60–79	40–59	20–39	0–19

the general state of agricultural water supply. Finally, the ratio of 20% was used as the threshold for the poor category.

Navigation

River navigation is an important component of modern transportation due to its relatively low cost and large traffic volume. Thus, improved potential for navigation increases the social value of the river. The development of river navigation is affected by a variety of components including the channel and hydrological conditions, port facilities, and security. With the improvement of channel conditions, the tonnage of ships that can navigate on a river will increase. For this project, we used the navigation index developed by Gippel et al. (2012) to measure the indicator of navigation based on navigable length of the river and the maximum weight of boat that can travel on the river. The navigation index (NI) is calculated as:

$$NI = 0.5NL + 0.5NW$$

$$NL = \frac{NL_{act}}{NL_{max}}$$

$$NW = \frac{NW_{act}}{NW_{max}}$$

where NL and NW are the navigation length index and navigation weight index, respectively; NL_{act} and NL_{max} are the maximum observed and expected maximum navigable length of the river, respectively, and NW_{act} and NW_{max} are the maximum observed and expected maximum boat weight on the river.

For this project, the assessment criteria for the navigation index were categorized into five grades from 0 to 1.0, and the values for each grade were evenly divided between 0 and 1.0 (Table 2.11).

Electricity Generation

Electricity generation is an important social value that a river can provide to improve societal and economic development. The Lancang River has great potential for hydropower development due to its steep channel and plentiful water; thus electricity generation was identified as a primary target of the water resources development of the Lancang River Basin. A hydroelectricity generation index was developed for

Table 2.11 Assessment criteria for the navigation index

Indicator	Very good	Good	Fair	Poor	Critical
Navigation index	0.80–1.00	0.60–0.79	0.40–0.59	0.20–0.39	0–0.19

Table 2.12 Assessment criteria for the electricity generation index

Indicator	Very good	Good	Fair	Poor	Critical
Hydroelectricity index	0.80–1.00	0.60–0.79	0.40–0.59	0.20–0.39	0.00–0.19

this project based on hydropower potential and actual hydroelectric power generation. The electricity index (EI) of a specific river basin or river reach was calculated as:

$$EI = \frac{HE_{act}}{HE_{max}}$$

where HE_{act} is actual electricity generation and HE_{max} is the theoretical hydropower potential, which can be regarded as the maximum electricity generation of the river basin or river reach.

The assessment criteria for the electricity generation index were categorized into five grades from 0 to 1.0 based on generation rate, and the values for each grade were evenly divided between 0 and 1.0 (Table 2.12).

Recreation

The Millennium Ecosystem Assessment (Sarukhán and Whyte 2005) defined cultural ecosystem services as "the nonmaterial benefits people obtain from ecosystems through spiritual enrichment, cognitive development, reflection, recreation, and aesthetic experiences." Aylward et al. (2005) categorized the cultural services of freshwater as recreation (river rafting, kayaking, hiking, and fishing as a sport), tourism (river viewing), and existence values (personal satisfaction from free-flowing rivers). The activities of recreation and tourism are similar, so for this project, the positive effects of recreation and tourism were integrated as recreation services. It is difficult to assess the personal satisfaction gained from free-flowing rivers due to different perspectives; therefore in this study we only evaluate the recreational value (including recreation and tourism values) of the Lancang River for cultural ecosystem services.

Cultural ecosystem services have been estimated in economic valuations of river ecosystems, often in relation to recreation and ecotourism, with market price methods having been by far the most frequently employed (Milcu et al. 2013; Ouyang et al. 2004; Wilson and Carpenter 1999). Although the significance of cultural values to river ecosystems has been recognized in river health studies, only a small portion of publications included the recreation indicator and undertook quantitative assessments. The lack of work in this area reflects the difficulty in determining reference or expected values that are required for river health assessment (Gippel et al. 2012). Recreation was included in the assessment indicators but was not evaluated in the assessment of river health in the Lower Yellow River by the International Water Centre (Gippel et al. 2012). Meanwhile, two indicators were developed and applied in the assessment of the health of the Haihe River: the development level of travel services and the degree of water for amenity (Gao et al. 2009). The thresholds were defined by the researchers, and responses to questionnaires were used to support the assessment. The degree of comfort when water viewing has also been used in river health assessments to assess recreational value, with thresholds developed using a qualitative method (Lin et al. 2006; Zhang et al. 2010; Peng et al. 2014). Any qualitative assessment is dependent on subjective judgment, and the results may thus vary significantly depending on who undertook the assessment or completed the questionnaire.

For this project, the percentage of "tourism value added" in gross domestic product (GDP) was selected as the indicator to evaluate the level of recreation in the river health assessment. Tourism value added, also referred to as gross domestic product (GDP) by the tourism industry, is calculated by subtracting the costs of supplies and services used to produce goods or services from total revenues. This highlights the specific contribution the tourism industry makes to the economy of a region. The percentage of tourism value added in GDP is therefore a parameter that quantifies the degree of tourism development in a specific region. It should be noted that the tourism value added is contributed by a variety of sectors such as, but not limited to, food, accommodation, transportation, viewing and recreation sites, and communication. However, we can assume that the development level of recreation value of a river is close to the development degree of the tourism industry because river viewing and recreation are fundamental to the tourism industry. The data necessary for the tourism value-added assessment was compiled from social and economic bulletins in China.

The thresholds for assessing recreation value were determined based on relevant national standards and planning goals for the tourism industry. A pillar industry is an industry that plays an important role in GDP and greatly advances economic and social development; in China, a pillar industry's value added is commonly recognized as making up 5% of GDP of a nation or region. A strategic pillar industry's value added accounts for 8% of the GDP (Zheng 1995; CNTA 2014). The national average of the percentage of tourism value added in GDP reached 4.9% in 2015 (Liu 2016), and the percentage of tourism value added in GDP of Yunnan Province was planned to reach 8% in 2020 (People's Government of Yunnan Province 2016). These percentages were used to derive the five categories for the recreation index in the Lancang River Basin (Table 2.13).

Table 2.13 Assessment criteria for the recreation index

Indicator	Very good	Good	Fair	Poor	Critical
Percentage of tourism value added in GDP (%)	>8.0%	5.0–7.9%	3.0–4.9%	1.0–2.9%	≤1.0%

2.2.4.4 Summary of Assessment Indicators

The indicators, thresholds, and methods of determining reference values for the river health assessment conducted for this project are summarized in Table 2.14. The indicators and thresholds of river health assessment in this study were generally based on (1) national standards in China, e.g., water quality; (2) international recognized standards or study results, e.g., hydrology, connectivity, and sediment; or (3) relevant studies in China, e.g., fish community, riparian vegetation, navigation, and electricity. Few studies quantitatively assessed the performance of flood, water supply, and recreation. This project defined the indicators, thresholds, and classification values for these three categories according to relevant technical requirements (e.g., flood and water supply) or economic data (recreation) in China.

To a large extent, the development of indicators was based on the data and/or standards available, and the ability to characterize river health accurately is therefore limited by the sensitivity of the indicators. For example, the evaluation of the fish community by examining the number of species captured during historical and recent surveys only incorporates one component of the fish community (species diversity), and it is possible that additional, more sensitive metrics could be added to the river health assessment as more data become available. Examples for fish community include the abundance or density of important endemic species, esti-

Table 2.14 Summary of indicators and thresholds used for the assessment of river health for this project

Category	Indicator	Value	Method for determining threshold value
Hydrology	Amended annual proportional flow deviation (AAPFD)	0–4.0	Index of Stream Condition (ISC)
	Rate of unchanged hydraulic characteristics	0–1	0.3 as critical
Connectivity	River fragmentation index (RFI)	0–1	75% decrease as critical
Sediment and river morphology	Sediment release rate (trap rate)	0–1	20% as critical
	Reservoir deposition rate	0–1	80% as critical
Water quality	Water pollution index in the reservoir	0–4	National surface water quality standard of China
Fish community	Change of number of fish species	0–1	20% as critical
Riparian vegetation	Change of the Normalized Difference Vegetation Index (NDVI)	0.02, −0.02	<−0.02 as critical
Flood	Flood protection frequency	5%, 30%, 50%, 100%	5 years as critical
Water supply	Water supply index	50–100%	20% as critical
Navigation	Navigation index	0–1	0.2 as critical
Electricity	Hydroelectricity index	0–1	0.2 as critical
Recreation	Percentage of tourism value added in GDP (%)	1–8%	1% as critical

mates of the biomass of fish produced within a given reach (i.e., productivity), or the yield of commercially valuable fish species.

2.3 Results and Discussion

2.3.1 Physical and Chemical Indicators

2.3.1.1 Hydrology

River Health Assessment

The calculated results of reservoir outflow produced by the numerical Muskingum model (Shi 2013) were compared to observed data at Manwan, Xiaowan, and Jinghong dams (data for Dachaoshan were not available). The time-series plot of daily average outflow data and model output are shown in Fig. 2.4. The correlation coefficients of reservoir outflow were 0.89, 0.89, and 0.94 at Manwan, Xiaowan, and Jinghong, respectively, which indicates that the simulation results were generally in good agreement with observed data and the reliability and accuracy of the model are acceptable.

The AAPFD, which was calculated based on monthly mean flow of each month in reference and operational values (Fig. 2.5 and Table 2.15), was 2.09. The index of flow change (HF) was therefore categorized as poor indicating that the flow regime of the Lancang River is substantially different from natural flow conditions.

The hydraulic index was examined according to the hydraulic conditions in the assessment area. In the middle and lower reaches of the Lancang River (712 km in length), the reach between two dams was impounded by the reservoir, and hydraulic conditions were changed from fast-flowing rivers to still reservoirs. Natural hydraulic conditions remain mostly unchanged only in the river reach downstream of the Jinghong dam (85 km in length). Thus, the hydraulic index (HC) was calculated to be 0.12 (HC = 85/712), which falls into the critical category.

The hydrological index (HI) is 1.5 based on the weighted average values of the flow change index (HF) and the hydraulic index (HC). Based on this hydrological index score, the state is categorized as poor indicating there are significant changes of flow and hydraulic conditions compared to reference conditions.

Hydropower Impacts

Two major hydraulic changes generally occur with the operation of a reservoir. First, the area upstream of the dam changes from lotic (i.e., running water) to lentic (i.e., standing water) conditions, which has implications for changes in hydrologic and ecological processes. Second, diurnal and seasonal variations in the demand for

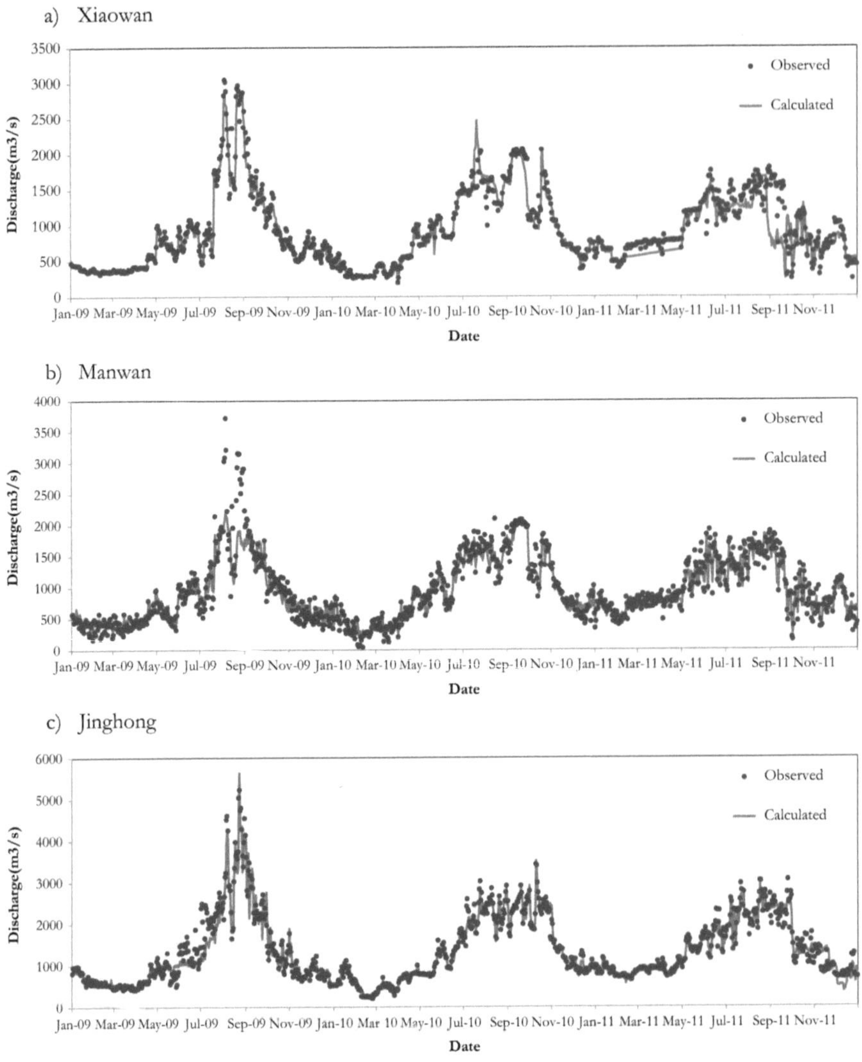

Fig. 2.4 Daily average outflow monitoring data (line) and model output (dots) at (**a**) Xiaowan, (**b**) Manwan, and (**c**) Jinghong dams

water or power will cause short- and long-term discharge variations in the downstream reach that are quite different from the river's natural flow regime (Prowse et al. 2004). Hydrological alteration, defined as changes in the magnitude and pattern of a flow regime caused by the storage, regulation, diversion, and/or extraction of water by dams and other infrastructure, is one of the primary contributors to the degradation and decline of freshwater habitats (Postel and Richter 2003). Hydropower projects with higher regulatory capacity tend to have greater impact on the natural flow regime than those with lower regulatory capacity. In addition to

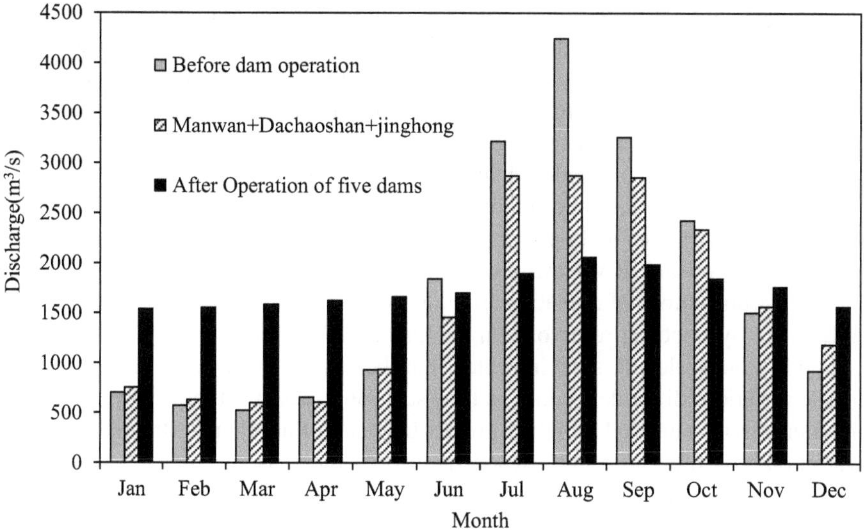

Fig. 2.5 Monthly average flow before and after cascade dams operation

Table 2.15 Monthly mean flow before and after the operation of dams on the Lancang River

	Flow (m^3/s)	
Month	Reference value	Operational value
Jan	705	1538
Feb	577	1554
Mar	525	1584
Apr	657.5	1624
May	933	1661
Jun	1843	1703
Jul	3220	1899
Aug	4250	2063
Sep	3260	1987
Oct	2430	1846
Nov	1505	1761
Dec	923	1565

reservoir operation, climate change also measurably alters some aspects of the flow regime. By comparison, however, reservoir operations will alter flow regimes much more than climate change (Rheinheimer and Viers 2014; Mittal et al. 2016).

Simulation results from six dams on the mainstem Lancang indicate that the mean monthly flows (MMFs) in the dry season (November to May of the following year) are enhanced and MMFs in rainy season (June to October) are decreased significantly relative to natural flow. Storage volumes of the Xiaowan and Nuozhadu reservoirs are 15.0 and 22.7 km^3, respectively, and regulating volumes are 10.0 and 11.3 km^3, respectively. These values indicate that the regulating capacities of Xiaowan and Nuozhadu are much higher than those of the other projects (see Table 2.1). Although enhanced regulating capacity of these reservoirs is beneficial to business purposes (e.g., electricity), this comes at the cost of substantial alterations of natural flow (as indicated by results presented in the section above).

The significance of hydropower projects and climate change in hydrologic alteration of the Lancang River was studied. Tang et al. (2014) quantitatively assessed relative contributions of climatic variations and human activities to flow changes in relation to hydropower reservoirs on the Lancang River at the yearly, seasonal, and monthly time scales. Results indicated that the relative impacts of human and climatic variations differ depending on time scale and season. At the yearly time scale, human activities exerted a slightly greater impact on flow changes than did climatic variations (54.6% and 45.4%, respectively). At the seasonal time scale, climatic variations made a greater contribution (65.8%) during the wet season, while the contribution of human activities became the dominant factor during the dry season (85.3%). At the monthly time scale, the contribution of climatic variations in January, June, August, and September was greater than that of human activities, while in the remaining 8 months, human activities exerted a greater contribution than did climatic variation. However, the effects of dams commissioned after 2008 were not considered in the study of Tang et al. (2014) due to the lack of hydrological data. The regulating capacity of cascade reservoirs on the Lancang River increased remarkably after Xiaowan and Nuozhadu were commissioned in 2010 and 2012. The increased regulating capacities were also associated with significant changes of flow regime. The discharge decrease in the wet season and flow increase in the dry season were greater after the operation of Xiaowan and Nuozhadu.

Tang's research, which was based on data before 2008, indicated that hydropower projects and climate changes have similar effects on the flow regime and hydropower's impacts may offset the effects of climate change. However, the effects of hydropower projects on flow regime are anticipated to be greater than that of climate change after Xiaowan and Nuozhadu become operational because the regulating volume of the reservoirs increased from 0.94 km^3 to 22.24 km^3.

Hydropower projects on the Lancang River have altered the natural flow regime but are also able to offset some of the effects of climate change. So we conclude hydropower projects impact the flow regime positively and negatively. The magnitude of change in river discharge due to climate change is anticipated to be smaller than that of hydropower projects on the Lancang River. Consequently, the impact of hydropower projects is assessed as significant negative, due to the negative effects on the natural flow regime, and moderate positive due to the ability of hydropower projects to offset the effects of climate change. The impact of hydropower projects on river hydrology is illustrated in the state-impact matrix (Fig. 2.6).

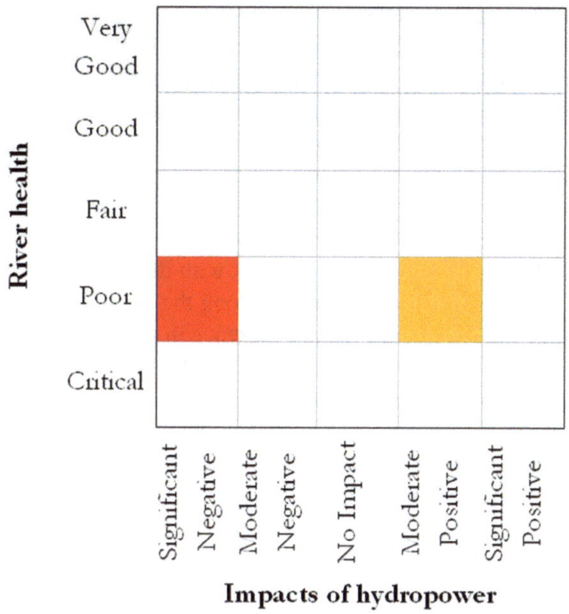

Fig. 2.6 Matrix of the state of hydrology and hydropower impacts on the Lancang River

2.3.1.2 Connectivity

River Health Assessment

High effectiveness in mitigating river fragmentation was not expected for the Lancang River because no fish passage facilities were constructed at the six dams in the study area (Gongguoqiao, Xiaowan, Manwan, Dachaoshan, Nuozhadu, and Jinghong). As a mitigation measure, fishes are collected with fishing nets and transported across the dam at Nuozhadu HEP, but the effectiveness is limited because it can only be implemented at several locations and times. Thus, DCI was assigned a value of 0, which indicates that the connectivity index was 0.20, and therefore categorized as critical.

Hydropower Impacts

Habitat fragmentation is an important causal agent in species decline (Allan et al. 1997). The development of water control barriers, such as dams, has altered freshwater habitats and has had a profound effect on aquatic organisms around the world (Schilt 2006). Dams may block or delay upriver fish migrations, and downriver passage through turbines or over spillways is often unsuccessful. Such effects contribute to the decline and even the extinction of species that depend on longitudinal movements along the stream continuum during certain phases of their life cycle. Consequences of reduced connectivity following dam construction on fish species include (1) reduction or elimination of the ability of fish to disperse to or reach upstream habitats, (2) eventual extirpation of mobile life history types from upstream populations, (3) fragmentation and isolation of upstream populations, (4) increased vulnerability to stochastic environmental and habitat disturbances, (5) restriction of upstream populations to habitats that may be marginal or degraded, (6) prevention of recolonization of disturbed upstream habitats, and (7) population-level genetic impacts (Hoffman and Dunham 2007). The influence of dams with different levels of river connectivity on fish community is variable. Santos et al. (2013) studied the influence of three types of dams with different degrees of river connectivity on the structure of fish communities along the Paraíba do Sul River in Brazil. Not surprisingly, the most significant difference in fish community structure between the reservoir and downriver was found at the dam that totally blocked fish passage. The other two dams represented partial blockages: one had a permanently open 4-m-wide gate allowing water passage at an average velocity of ca. 5 m/s, and the other had a fish ladder that was open during the wet season. The dam with the permanently open gate showed broader differences in fish fauna between the upstream and downstream reaches than the dam with the fish ladder. Santos et al. (2013) hypothesized that the velocities through the narrow gate prevented upstream fish movement and led to differences in fish assemblages. This study highlights that fish passage facilities can help to mitigate the effects of connectivity loss but that the effectiveness of the mitigation may be variable depending on the design and its implementation.

The six dams on the mainstem of the Lancang River are the only barriers on this river; thus the impact of hydropower projects on connectivity was assigned the category of significant negative. The state impact of connectivity is illustrated in the matrix of river health and hydropower impacts in Fig. 2.7.

2.3.1.3 Sediment and River Morphology

River Health Assessment

By trapping sediment in reservoirs, dams interrupt the continuity of sediment transport through rivers, resulting in a loss of reservoir storage and reduced usable life, and depriving downstream reaches of sediments essential for channel form and aquatic and riparian habitats (Kummu and Varis 2007; Kondolf et al. 2014). To reduce the impacts on fine sediment transport and river morphology, the "storing clear and releasing muddy" mode of reservoir operation can be used. This consists of storing water with relatively low concentrations of fine sediment in non-flood seasons and releasing muddy water in flood seasons. Although this method was applied to HEPs on the mainstem Lancang and reduced sediment deposition in the reservoirs to a certain degree, a considerable portion of the river's sediment load was accumulated in the reservoirs. An evaluation of sediment deposition volumes at the Manwan Reservoir by Mei et al. (2006) calculated changes of reservoir volume under normal operating level, dead storage, and active storage. Results indicated that 54.2% of the total volume of the Manwan Reservoir, which became operational in 1993, was filled by sediment after 12 years. However, the loss of storage was

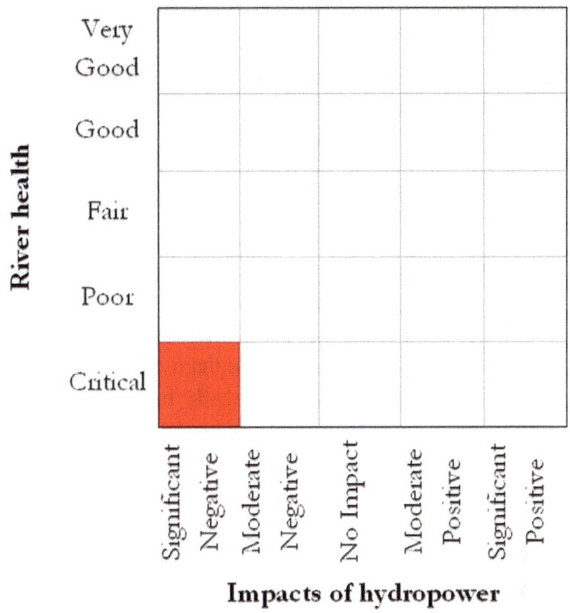

Fig. 2.7 Matrix of the state of connectivity and hydropower impacts on the Lancang River

dominated by a loss of dead storage (70.7%), with only 11.6% of the active storage zone lost. The operational life of the Manwan dam with respect to reservoir storage has also been extended by construction of the Xiaowan dam upstream.

Sediment trapping efficiency (TE) was estimated for a number of reservoirs on the mainstem Lancang River. Fu and He (2007) estimated that the multi-year average TE at Manwan dam was 60.5%, and the trapping efficiencies of Gongguoqiao, Dachaoshan, and Jinghong dams were estimated at 30.2%, 66.1%, and 63.5%, respectively. Two dams with yearly regulating capacity, i.e., Xiaowan and Nuozhadu, have trapping efficiencies as high as 92%, and the cascade of six dams on the Lancang River has a total theoretical trapping efficiency of 94% (Kummu and Varis 2007). Lu and Siew (2006) examined the sediment trapping impacts of Lancang dams on the Mekong River and found that statistically significant decreases in sediment were only evident at the nearest gauging station below the China dams (i.e., Chiang Saen, ~325 km downstream of the Jinghong dam). Areas located further downstream (e.g., Nong Khai, ~1140 km downstream of the Jinghong dam) showed less sensitivity to the operation of the Manwan dam, as sediment fluxes have remained stable or even increased in the post-dam period. Research results by Fu et al. (2007) and Liu et al. (2013) verified the above conclusions on transboundary sediment effects of dam construction on the Lancang River.

Data are not available to calculate the overall sediment deposition rate (S_d) for the cascade of reservoirs on the Lancang River, but the sediment release rate (S_r) for the cascade of reservoirs was expected to be below 10% (Kummu and Varis 2007). Given the value of S_r and the principle that the lower grade of the two indicators will be attributed to the component of sediment and river morphology, sediment and river morphology was categorized as critical based on the impacts of the six existing reservoirs on the mainstem Lancang River.

Hydropower Impacts

Increased erosion and sediment loads from poor land use and the expansion of human development are important concerns for soil conservation; however, most river systems around the world actually show decreased sediment loads because of sediment trapping by upstream dams (Walling and Fang 2003). Vorosmarty et al. (2003) extrapolated estimates from 633 large reservoirs to over 44,000 smaller reservoirs and estimated that more than 53% of the global sediment flux in regulated basins is trapped in reservoirs, or 28% of the global sediment flux in all river basins, with an estimate of 4–5 billion tons of sediment trapped per year. Sediment management techniques for reservoirs, including reservoir flushing, upstream check structures, protecting dam outlets, mechanical removal, and increasing the dam's height, may be physically and economically feasible measures for maintaining reservoirs' storage and long-term operation.

As a result of sediment accumulation in hydropower reservoirs, significant drops in sediment load downstream of the dams are evident from sediment survey data on the Lancang River. The annual sediment load at Gajiu gauge station, which is located downstream of the Manwan dam, decreased by over 60% after construction of the Manwan dam (Liu et al. 2013). Such a sudden and dramatic change illustrates the magnitude of impact that hydropower reservoirs can have on sediment transport and river morphology. Accordingly, the impact of hydropower projects on sediment and river morphology is graded as significant negative. The state impact of sediment/river morphology is illustrated in the matrix of river health and hydropower impacts in Fig. 2.8.

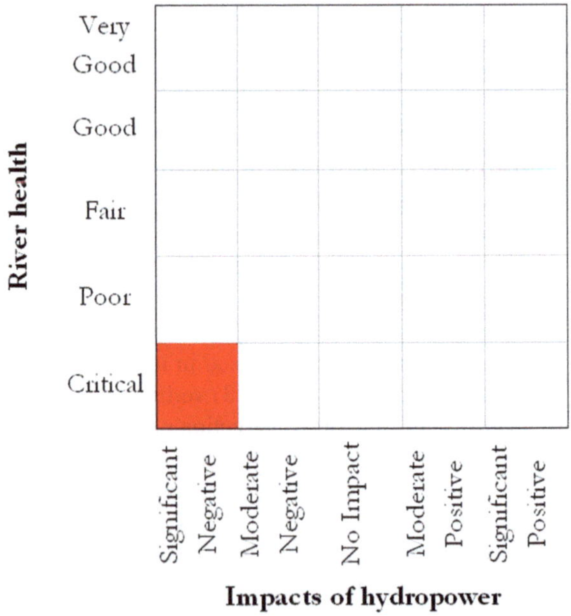

Fig. 2.8 Matrix of the state of sediment and river morphology and hydropower impacts on the Lancang River

2.3.1.4 Water Quality

River Health Assessment

The damming of rivers has the potential to result in a deterioration of water quality due to reduced oxygenation and dilution of pollutants by relatively stagnant reservoirs, flooding of biomass, and reservoir stratification. However, the change of water quality in river reaches affected by hydropower is determined by specific hydrological, hydraulic, and pollutant conditions.

Water quality of the Lancang River was evaluated at two spatial scales: the entire mainstem and specific gauge stations. According to the Environmental Quality Communique of Yunnan Province in 2014 (YEPD 2014), 21, 10, 4, 1, and 2 gauge stations in the Yunnan reach of the Lancang River were categorized as grades II, III, IV, V, and worse than V, respectively. Hence, water quality at 31 of 38 gauge stations met grade III or higher. Seven of the gauge stations with grades lower than III are located on the tributaries of the Lancang River. The most downstream gauge station (Guanlei), which is close to the China-Myanmar border, was categorized as grade II. These results indicate that water quality on the mainstem Lancang and at the outflow to the Mekong River is good.

Gajiu station is located 2 km downstream of the Manwan dam, and there is no tributary between the dam and the gauge station. Data from the Gajiu station are therefore used to evaluate the status of water quality of the river reach where the Manwan dam is located. The observed water quality data and grade before and after construction of the Manwan dam are shown in Table 2.16 (Wang et al. 2000). These results indicate that although there have been changes in water quality since dam construction and operation, the changes have not been significant enough to result in changes to the water quality grade. The organic pollution index of the river reach of Manwan was calculated as −0.074 and was therefore classified as very good according to the standard described in Sect. 2.2.4.1.

The eutrophication level of Manwan Reservoir was assessed based on data collected in the reservoir in 2003 (Wang et al. 2004). COD_{Mn}, TN, TP, chla, and transparency (SD) were recorded at three sites (Luodihe, Sanjiacun, and Goutoupo) in the Manwan Reservoir, and averaged values are shown in Table 2.17(Wang et al. 2000). The trophic level index (TLI) of the Manwan Reservoir is 29 and therefore can be classified as mesotrophic (see Sect. 2.2.4.1). This indicates the reservoir was

Table 2.16 Water quality data at Gajiu station in the Manwan reach

Time frame	Value	DO	COD_{Mn}	BOD_5	NH_3-N
Baseline	Concentration	8.70	3.42	0.73	0.29
	Grade	I	II	I	II
Post-project	Concentration	7.90	2.07	1.27	0.29
	Grade	I	II	I	II

Taken from Wang et al. (2004)

Table 2.17 Water quality data in the Manwan Reservoir

	Concentration (mg/L)				
Value	chla	COD_{Mn}	TN	TP	SD (cm)
Concentration	0.252	6.47	0.183	0.197	79

Taken from Wang et al. (2000)

not eutrophic, despite this being a common water quality issue for reservoirs in China.

Hydropower Impacts

The assessment of hydropower's impact on river water quality was based on theoretical analysis and a review of past studies. We first used a continuously stirred tank reactor model to examine water quality in the reservoir. Flow that enters the reactor is assumed to instantaneously mix throughout the full reactor volume. The governing equation can be derived by considering mass conservation in the tank:

$$V\frac{\mathrm{d}C}{\mathrm{d}t} = W - QC - f(C)V$$

where V is the volume of the tank, W is the pollutant flux into the control volume, Q is water inflow to the tank, and $f(C)$ is a sink reaction term that can be expressed as a first-order die-off reaction function.

The solution for the above equation can be expressed as an exponential function, and the steady state concentration in the tank is:

$$C = \frac{W}{Q + kV}$$

where k is the coefficient of the first-order die-off reaction function.

Water quality change in the reservoir can generally be analyzed based on the equation above. In the backwatered area upstream of the dam, the reduced flow may result in a reduction in the capacity of the river to assimilate pollutants, which is referred as water self-purification capacity. It means the die-off reaction coefficient k will be decreased and pollutant concentration will increase accordingly. The chemical, thermal, and physical changes due to the dam have the potential to contaminate the reservoir and the river downstream. The retention time of the reservoir is a critical parameter for measuring the potential impact of a hydropower reservoir on water quality. The impact will be higher if the retention time is longer. However, the reduced flow does not necessarily result in deterioration of water quality in the reservoir because the volume of the reservoir will be substantially increased. Additionally, water quality change is determined by pollutant loads into the reservoir. The creation of hydropower reservoirs therefore has the potential to have both positive and negative impacts on water quality.

For the river reach downstream of the dam, water quality will be improved if the huge water volume in the reservoir can dilute and assimilate pollutants effectively. Additionally, positive impacts may result from a dilution effect downstream of the reservoir because of the greater pollutant assimilation capacity associated with increased minimum river flows (Hearnshaw et al. 2010).

According to the above theoretical analysis, hydropower can have both positive and negative impacts on water quality in the reservoir and downstream reach. The actual change of water quality in the reservoir is dependent on a number of factors including pollutant load, volume of the reservoir, and inflow discharge. Water quality in the reservoir and the manner in which water is released from the reservoir are critical in determining water quality downstream of the dam.

Wei et al. (2009) studied the impact of dam construction and operation on water quality in the middle and lower reaches of the Lancang River by analyzing water quality and river flow data over 20 years (1985–2005). Three periods were compared: period (1) from 1985 to 1992, during which there were no dams on the river reach; period (2) from 1993 to 2000, during which only the Manwan dam was in operation; and period (3) from 2001 to 2010, during which both Manwan and Dachaoshan were in operation. From the pre-dam period (period 1) to the first 7 years following Manwan dam construction, water quality in the Manwan Reservoir became worse due to the accumulation of pollutants. In the next 5 years, water quality improved due to the dilution of pollutants in the large reservoirs. The joint operation of the Manwan and Dachaoshan dams had cumulative positive impacts on water quality in the river reach downstream of Dachaoshan (Dachaoshan-Jinghong) but no impacts on the reach further downstream (Jinghong-Ganlanba).

Based on the theoretical analysis and review of relevant studies on the Lancang River, it was concluded that hydropower projects have positive and negative impacts on water quality. Moreover, the contributions from hydropower projects are considered to have a similar level of influence as water pollutant loads. The impacts of hydropower on water quality are therefore assessed as moderate negative and moderate positive. The state impact of water quality is illustrated in the matrix of river health and hydropower impacts in Fig. 2.9.

2.3.2 *Biological Indicators*

2.3.2.1 Fish Community

River Health Assessment

The fish community assessment is based on surveys of the mainstem Lancang River as a whole, as well as surveys focused on specific reaches before and after dam construction. Based on published papers, the baseline condition of fish species

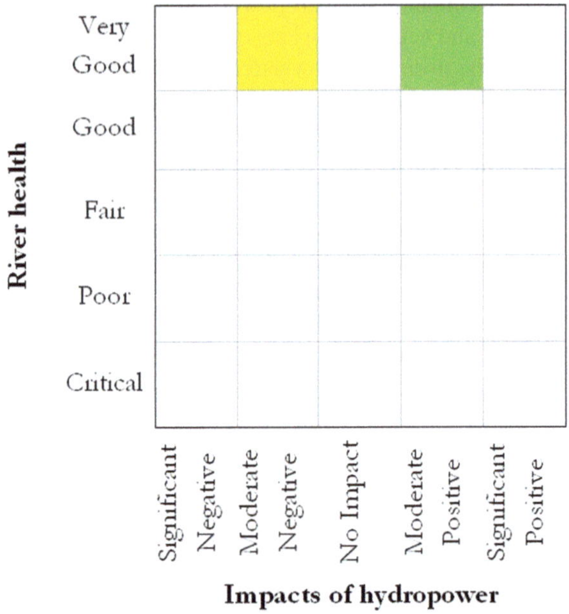

Fig. 2.9 Matrix of the state of water quality and hydropower impacts on the Lancang River

present in the Lancang River was analyzed by Kang and He (2007). There were 6 orders present including 21 families, 86 genera, and 162 species in the Lancang River, consisting of:

- *Cypriniformes*: 117 species belonging to 60 genera in 4 families
- *Siluriformes*: 27 species belonging to 13 genera in 7 families
- *Perciformes*: 13 species belonging to 9 genera in 7 families
- *Cyprinodontiformes*: 3 species belonging to 2 genera in 1 family
- *Synbranchiformes*: 1 species
- *Tetraodontiformes*: 1 species

In different parts of the Lancang River, 11 species were in the headwaters (upstream of Changdu), 22 were in the upper reaches (Changdu to Gongguoqiao), 44 were in the middle reaches (Gongguoqiao to Jinglinqiao), 142 were in the lower reaches (Jinglinqiao to Nanla confluence), and 9 were in the affiliated lake named Erhai. In total, 13 species in the Lancang River were recorded in the "China Red Data Book of Endangered Animals: Pisces" in 2004, accounting for 14.1% of the total endangered fishes in China. According to the habitat, fishes in this area could be divided into four groups: pelagic fishes, demersal species, bottom dwellers, and shallow water species. The feeding habits of the fishes contained six types: algivore, planktivore, aquatic insectivore, piscivore, detritivore, and invertebrate feeder. The number of species greatly increased from the river's headwaters to its lower reaches, and the species present shifted based on habitat preferences; for example, *Schizothoracinae* and *Triplophysa,* which are adapted to higher altitudes, were substituted by *Barbinae*, *Labeoninae*, and *Sisoridae,* which are adapted to warm water and slow current.

Fish biodiversity post-project was evaluated based on field surveys and historical records. Zheng et al. (2013) found that there were a total of 165 species of fish in the middle and lower Lancang River reaches. However, in recent years, only 71 of these species were captured, which shows a decline in the diversity of the fish resources in the middle and lower Lancang River, especially the medium to large species and those that are rare or endemic. In the 4787 fish samples collected on the Lancang River in 2009 and 2010 by Liu et al. (2011), 5 orders, 14 families, and 80 species of native fishes were identified. In past studies in the same reach, a further 59 species were recorded, indicating that historically there were a total of 139 species of fish in the Yunnan section of the Lancang River (Liu et al. 2011).

The change of fish species present at specific project sites was also examined. Yu et al. (2011) surveyed the fish species in the Manwan reach in 2008 by sampling in four locations: downstream of the dam, in front of the dam, in the middle of the reservoir, and at the end of the reservoir. A total of 34 fish species were identified in the 1535 fish samples collected, and 21 were endemic species. According to the fish survey in the same area in 1995, the total fish species and endemic fish species were 61 and 47, respectively (Wang et al. 2000). The number of fish species and endemic fish species therefore decreased significantly after 13 years of dam operation. After reservoir impoundment, the flow environment in the reservoir changed from lotic to lentic conditions. Consequently, fish species adapted to lentic conditions increased significantly in the reservoir. Alien species of fish including *Ctenogobius giurinus*, *Pseudorasbora parva*, *Abbottina rivularis*, *Cyprinus carpio carpio*, *Carassius auratus auratus*, *Silurus asotus*, *Misgurnus anguillicaudatus*, *Hypophthalmichthys molitrix*, *Aristichthys nobilis*, *and Neosalanx taihuensis* were dominant both in abundance and biomass. Endemic fish species adapted to lotic flow, including *Tor douronensis*, *Scaphiodonichthys acanthopterus*, *Schizothorax griseus*, *Acrossocheilus krempfi*, *Cosmochilus cardinalis*, *Bagarius yarrelli*, *Platytropius sinensis*, *Mastacembelus armatus,* were more difficult to find because of the change in habitat conditions. Other endemic fish species, including *Tor sinensis*, *Barbodes huangchuchieni*, *Percocypris pingi retrodorslis*, and *Sikukia flavicaudata*, are more readily captured, but the number of individuals captured in recent surveys was very small. The changes to fish species numbers and abundance varied between the different study areas. Only three species of fish were native (*Tor sinensis*, *Onychostoma elongates*, and *Botia longiventralis*) among the 12 species captured in the lentic area, but there were 6 species of endemic fish among the 11 species captured at the end of the reservoir where flow remained lotic. At the survey site downstream of the dam, which is not subject to the dramatic shift from lotic to lentic conditions, there was only 1 alien fish species among the 17 species captured.

The trends observed at the Manwan dam were verified by surveys before and after the operation of Xiaowan project. Survey data indicated that 15 endemic species of fish disappeared after the impoundment of the Xiaowan Reservoir (Li et al. 2013). The inspection and acceptance of environmental protection measures at Xiaowan and Gongguoqiao hydropower projects also showed that the fish assemblages were dominated by lentic species instead of lotic species after the impoundment of the reservoirs (MEP 2015).

The results on the decline of fish species in the Lancang River differ by study, so the assessment of fish community was based on data from the river reach assessment. The decline of fish species numbers in the Lancang River is 57.0% based on the survey of Zheng et al. (2013), 42.4% in the Yunnan section of the Lancang River based on the survey of Liu et al. (2011), and 45.9% in the Manwan dam reach based on the survey of Yu et al. (2011). The river health assessment in this study is focused on the middle and lower reaches of the Lancang River; based on the aforementioned surveys, the decline rates of species of fish in these reaches were determined to be 40–45%. According to the thresholds for fish community defined in Sect. 2.2.4.2, the indicator of fish community was assigned the grade critical.

Hydropower Impacts

The construction and operation of dams have a major impact on the local fish community via a number of pathways of effect, including barriers to fish migration, reservoir impoundment and natural flow alteration, fish habitat degradation, and water temperature changes. Globally, dams eliminate 10–60% of fish species in their vicinity (Baran et al. 2009). A dam can block or delay upstream fish migration and thus contribute to the decline and even the extinction of species that depend on longitudinal movements along the stream continuum during certain phases of their life cycle (Larinier 2011). Flow availability and variability can affect fish populations at different life stages. For example, high flows can serve as a cue for fish migration and reproduction (Welcomme 1979). As noted above, it is common that fishes adapted to lotic flow conditions disappear from reservoirs after impoundment. Water releases from dams often have low sediment concentrations, which may result in the removal of smaller sediments from the riverbed downstream (Sect. 2.3.1.3) and a degradation of fish habitat. A reduction in water temperatures below dams that draw water from depth may delay the natural migration and spawning of specific fishes and reduce the scale of spawning (Guo et al. 2011). Reservoir fisheries can compensate for the decline of fish from the above impact pathways to some extent; however, they cannot compensate for the loss of diversity. Compensation for a loss in yield from river fisheries can also be difficult to achieve through development of reservoir fisheries (Jackson and Marmulla 2011).

The obvious decline in diversity of fish species from the middle and lower reaches of the Lancang River can be attributed to reservoir impoundment and flow alteration resulting from the development of a cascade of dams. Almost no lotic habitat was maintained in the middle and lower reaches of the Lancang River because the backwatered area of the lower dam in the cascade reaches the toe of the dam upstream. This has resulted in a dramatic loss of habitat for lotic fish species. In addition, the blockage of migration also contributes to the decline in fish diversity. The fish species that migrate long distances from the Mekong River only migrate to the Jinghong dam (Kang et al. 2009); hence hydropower projects on the Lancang River have limited impacts on long-distance migrating fishes from the Mekong River. Nevertheless, the cascade of dams still blocks short-distance migra-

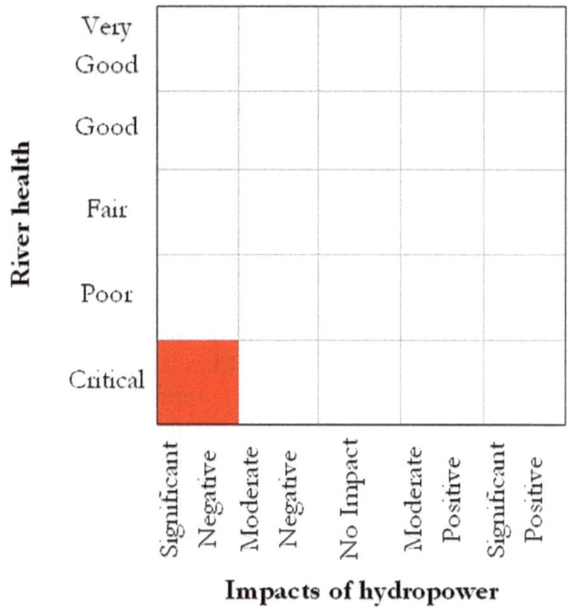

Fig. 2.10 Matrix of the state of fish community and hydropower impacts on the Lancang River

tions of fish species native to the Lancang River. No fish passage facilities were constructed at hydropower projects on the middle and lower reach of the Lancang River, and data are not available to verify the mitigation measures for the loss of fish diversity (e.g., hatchery and release, netting and release).

Based on the declines in fish diversity following hydropower project development, the impacts of hydropower projects on fish community are assessed as significant negative. The state impact of the fish community is illustrated in the matrix of river health and hydropower impacts in Fig. 2.10.

2.3.2.2 Riparian Vegetation

River Health Assessment

Fan et al. (2012) analyzed the temporal and spatial variation of vegetation cover change by using remote sensing data and GIS techniques. The study results indicated that overall there was generally a significant increasing trend of riparian and terrestrial vegetation cover in the Lancang River Basin ($R^2 = 0.52$, $P = 0.02$) from 2001 to 2010. The trends of the upper basin, middle basin, and lower basin, when considered separately, were similar to the trend for the entire basin; however, the degree of change was not the same. In general, the rate of change in the upper basin was higher than other areas, and vegetation in the lower basin generally remained stable during the 10 years studied.

Liu et al. (2015) analyzed changes to vegetation cover from 1998 to 2012 in the Yunnan reach of the Lancang River Basin. Manwan was in operation and Dachaoshan was in construction in 1998, but these two reservoirs are relatively small compared with other dams in the middle and lower reaches of the Lancang River. Thus, although vegetation changes may exist, they are not expected to be substantial. As such, 1998 was used as a reference year. The vegetation cover within a 20 km strip along the Yunnan reach of the Lancang River was analyzed based on the SPOT data from 1998 to 2012. Study results showed that the NDVI decreased during construction of hydropower projects and then increased once project operation began, indicating that vegetation cover was disturbed by clearing reservoir during construction but recovered gradually afterward. The values of NDVI varied in the segment from 4 km upstream of the dam to 4 km downstream of the dam, but NDVI remained stable outside of this 8 km reach, further supporting the conclusion that dam construction activities were responsible for the changes to vegetation cover.

The results indicate that vegetation cover in the majority of the area assessed improved over the past 15 years (overall percentage is 62.8%). According to the overall values of six dam locations, the overall rate of change in NDVI in the Yunnan section of the Lancang River was 0.004 (Liu et al. 2015), which is categorized as good based on the definitions in Sect. 2.2.4.2.

Hydropower Impacts

The impacts of hydropower projects on riparian vegetation are different in the construction and operational phases. The most significant impact of dam construction on riparian vegetation is the biomass lost upstream of the dam due to reservoir impoundment and the submergence of riparian vegetation. Construction activities also disturb the vegetation in areas where dam and ancillary facilities are constructed, but vegetation in these areas may recover slowly upon the completion of construction activities and implementation of vegetation restoration measures. This trend was observed at the Three Gorges Project in China (Wang et al. 2008).

During operations, variation in the water levels of reservoirs can have a negative impact on plants in the immediate vicinity of the reservoir such that riparian vegetation cover is extremely sparse adjacent to the reservoir, which often results in a barren strip across the landscape in the drawdown zone. Downstream of dams, the altered hydrological characteristics (Paetzold et al. 2008), habitat fragmentation (Newman et al. 2013), and dispersal of organisms by water (Nilsson et al. 2010) can negatively impact riparian vegetation. New and Xie (2008) studied the impacts of the Three Gorges Reservoir on local plant communities by evaluating the response to increased water levels, altered hydrological characteristics, and other effects associated with the construction of the dam. This study found that vegetation responses were diverse and changed over time but ultimately resulted in a markedly different landscape and riparian zone. Kellogg and Zhou (2014) further examined the riparian vegetation cover changes at different elevation zones in response to the construction of the Three Gorges Dam. Results showed that the non-vegetated area increased in the inundated zone (below 175 m in elevation, which is the maximum water level), that the area of densely vegetated land cover increased within the elevation zone from 175 to 775 m, and that there was no change in vegetation cover above 775 m in elevation. The observed increase in vegetation from 175 to 775 m in elevation may have been due to an increase in moisture in the atmosphere and subsequent precipitation within these elevation zones caused by the creation of a large reservoir.

The overall improvement of vegetation cover in the Yunnan reach of the Lancang River over time may have been due to the conversion of low productivity farmland into forest and grassland, conservation of natural forest, and changes in precipitation related to the creation of a hydropower reservoir (Liu et al. 2015). Among these factors, the increase of regional precipitation in the vicinity of the reservoir was beneficial to the vegetation growth. However, water fluctuations can affect the physical, chemical, and biological components of aquatic environments. Although vegetation cover increased over a broad spatial scale, fine-scale loss of vegetation within the seasonally submerged riparian zone of a reservoir that experiences substantial water fluctuation is still a critical problem.

Based on the above, we assumed that hydropower projects on the Lancang River have both positive and negative impacts on riparian vegetation. In addition to hydropower impacts, the effects of land use initiatives and natural forest conservation are direct and significant such that hydropower projects are only partly responsible for the improvements in vegetation cover observed. Consequently, the positive impacts of hydropower projects on vegetation are classified as moderate positive. With respect to adverse effects, the impacts of hydropower projects are limited in spatial extent because the submerged and disturbed areas represent a small portion of the river basin. The adverse effects of hydropower are therefore classified as moderate negative. The state impact of riparian vegetation is illustrated in the matrix of river health and hydropower impacts in Fig. 2.11.

2.3.3 Social Indicators

2.3.3.1 Flood Protection

River Health Assessment

The Lancang River is a river prone to flooding. The precipitation in the Lancang River Basin is unevenly distributed in time, and the precipitation in flood season (May to October) accounts for 85–90% of the annual rainfall. Severe flooding occurred in 1905, 1924, and 1966. In 1905, the flood peak discharge at Yunjinghong reached 17,100 m^3/s (based on a historical flood survey), and the regions along the downstream Lancang River were severely damaged. In 1924, the flood peak discharge at Yunjinghong reached 15,000 m^3/s (based on a historical flood survey), and the flooding caused significant loss of life and property. In 1966, the flood peak

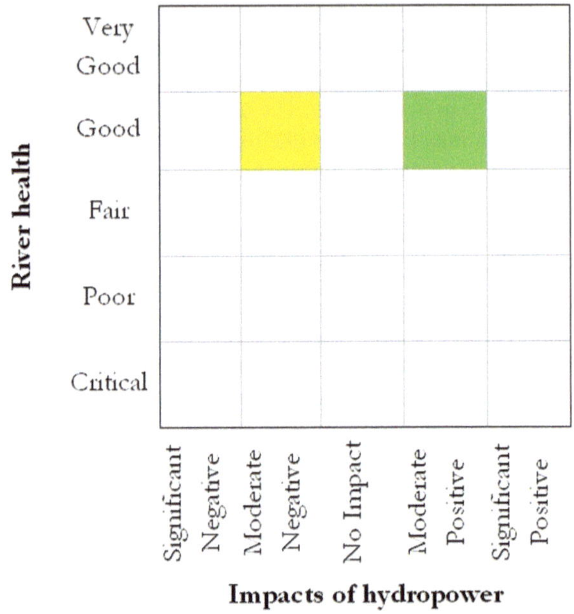

Fig. 2.11 Matrix of the state of riparian vegetation and hydropower impacts on the Lancang River

discharge at Yunjinghong reached 12,800 m^3/s (20-year recurrence interval), and 3000 ha of cropland were damaged by the flooding. The most recent flooding in the Lancang River Basin occurred in October 2006, and discharge at Jinghong reached 8030 m^3/s (Jinghong is located at the downstream end of Yunjinghong; thus the flood peak at Jinghong is larger than that for Yunjinghong). The flooding affected 96,463 people; 6 people are known to have died and an additional 5 people were lost, and 41 houses collapsed with a further 140 houses damaged. The economic loss was estimated to be 217.37 million RMB.

Hydropower projects play an important role in reducing flood risk in the Lancang River Basin. Prior to dam development, flood control in Jinghong City mainly depended on levees along the Lancang River and its tributaries. The standard of the Jinghong City levees for flood control was the 50-year recurrence interval. The Nuozhadu hydropower project has a reservoir with a regulating volume of 113.35×10^8 m^3. Calculations of hydrologic regulation indicate that the average discharge released from the Nuozhadu reservoir has decreased by 46.8% compared to the natural discharge. The maximum released flows have decreased by 23.3% and 15.1% for floods with 10-year and 100-year recurrence intervals, respectively. Given the regulation capacity of the Nuozhadu hydropower project, the flood control capacity of Jinghong City has improved from the 50-year to the 100-year recurrence interval.

Flood control for both the Lancang River and the lower Mekong is accomplished by hydropower projects. It is estimated the joint operation of cascade reservoirs on the Lancang River could retain and store about 13 billion m^3 of flood waters in the wet season, which substantially reduces the potential that the river will flood its banks downstream. It was estimated that the water level from the Chiang Saen to Vientiane section decreases approximately 1.0 m on average during July to August (Ma 2014). Based on a simulation conducted by Ma (2014), a flood with a recurrence interval of 50 years could be managed: the flow and water level at the Chiang Saen station could be decreased by 3300 m^3/s and 1.2 m, respectively, with joint operation of upstream reservoirs. Given these statistics on flood control for the Lancang River, flood protection was categorized as good.

Hydropower Impacts

Dams are important in the prevention of loss of life and property due to flooding. Flood control dams capture floodwaters and can then release them under controlled circumstances to the river below, or the water can be stored and utilized to meet other needs. Flood control is a key objective for many existing dams and for some of the major dams currently under construction around the world. Most flood control reservoirs are of the multipurpose type, which means that they are used to store water for irrigation, power generation, navigation, water supply, and recreation. The flood control pool of a reservoir, or flood storage capacity, is always above the multipurpose pool level. In China, the flood control capacity of a reservoir is usually expressed as the frequency of flood it can regulate during flooding.

In addition to reservoir regulation, other engineering works, including strengthening of dikes, river bank reinforcement, improving waterways, installing pumping stations, and establishing flood management systems, can also help improve flood control capacity. In recent years, strengthening of dikes and bank stabilization, river channel modification, and dredging were implemented in major tributaries and within the urban reach of the Lancang River. Hydropower reservoirs, however, still contribute the main form of flood control for the Lancang River. Based on the importance of hydropower projects to flood control for the Lancang River, flood protection was assigned the category of significant positive. The state impact of flood protection is illustrated in the matrix of river health and hydropower impacts in Fig. 2.12.

2.3.3.2 Water Supply

River Health Assessment

Water shortage caused by a lack of water storage projects is a common issue in the Lancang River Basin (Liu et al. 2014). This problem has been particularly prominent in recent years due to the threat of severe drought. According to the Water Resources Bulletin of the Changjiang (Yangtze) River, almost all of the water in the Lancang River Basin comes from surface water (98.7%). Of this, 35.8% comes from water storage facilities (e.g., reservoirs and lakes) and 58.3% from water diversion facilities (e.g., channels and aqueducts). However, under normal

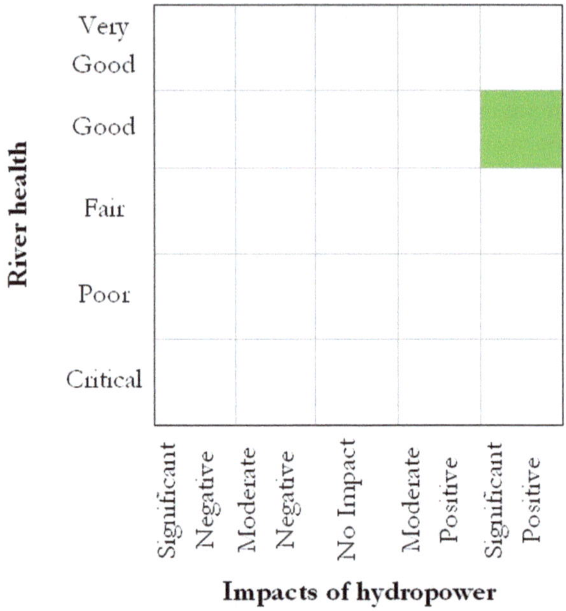

Fig. 2.12 Matrix of the state of flood and hydropower impacts on the Lancang River

circumstances, the water supplied by storage facilities is greater than that supplied by diversion facilities. For example, in the Changjiang River Basin, the contributions of storage and diversion facilities were 34.6% and 19.0%, respectively. The lower proportion of water supplied by storage facilities in the Lancang River Basin indicates that water supply is relatively low because of an uneven temporal distribution of water resources. Rainfall in the wet season accounts for over 80% of the annual total rainfall in the Lancang River Basin, and rainfall in the dry season accounts for less than 20% (Liu et al. 1998). Water storage facilities are particularly critical in this kind of region. If water storage was higher, sufficient water could be stored in the dry season, and a higher proportion of water supply needs could be met through the release of stored water.

Statistical data for agricultural irrigation in the Lancang River Basin were used to evaluate water supply. It was not available for the scale of the Lancang River Basin; thus the ratio of effective irrigated areas was derived from data for cities along the Lancang River. The irrigation data of Baoshan, Lincang, Jinghong, and Xishuangbanna were collected from statistical yearbooks, government work reports, and research reports. The percentage of effective irrigated areas of these four regions in 2015 was 43.6%, 45.0%, 25.4%, and 38.7%, respectively. The average value (38.2%) was taken as the guaranteed ratio for the Lancang River Basin. Based on this guaranteed rate and the categories defined in Sect. 2.2.4.3, the water supply indicator was categorized as poor.

Hydropower Impacts

Large reservoirs on the Lancang River can store water in the flood season and release it in the dry season to alleviate drought in downstream regions. This storage capacity improves water supply capacity in the Lancang River and lower Mekong River regions. In March 2016, Vietnam proposed that China increase its outflow from the Jinghong dam to the Mekong River to cope with drought and salt intrusion in a number of Vietnam's Mekong Delta provinces. In response to Vietnam's request, China agreed to release more water from the Jinghong reservoir into the lower Mekong River. The water released from the Jinghong dam was increased from 1000 m^3/s to 2190 m^3/s on March 15, 2016, and was maintained around 2000 m^3/s from March 15 to April 10, 2016. The downstream Mekong countries welcomed the increased water releases that alleviated the severe drought.

Although the large reservoirs on the mainstem Lancang River assist in meeting water supply needs in China, it is not realistic to expect them to meet all water supply requirements. A robust water supply system should be comprised of water storage and distribution facilities at various scales that have the ability to supply water to water users widely distributed in space. The water storage capacity of hydropower reservoirs on the Lancang River is most useful in providing water supply for urban and rural regions close to the mainstem Lancang (e.g., domestic and industrial water demands of Jinghong City), whereas water supply for regions that are far away from the mainstem Lancang needs to be supported by local water supply systems. However, Yunnan Province does not have enough small-scale infrastructure (e.g., ponds, small reservoirs, and canals) to store and distribute clean water to distant areas (Qiu 2010). Furthermore, most medium and small reservoirs in Yunnan Province were built more than 50 years ago, and half are either disused or do not function properly. This problem is also common in the cities along the Lancang River including Baoshan, Lincang, and Xishuangbanna. For example, the severe drought from October 2009 to June 2010 affected 1.36 million people and 0.3 million hectares of cropland and resulted in direct economic losses of 1.75 billion yuan in Lincang City (Ye 2012).

Based on the above considerations, the impact of hydropower projects on water supply was assessed as moderate positive. Hydropower projects on the mainstem Lancang aid in improving water supply to some extent; however, they are not able to solve the existing problems of water supply at the watershed scale. The state impact of water supply is illustrated in the matrix of river health and hydropower impacts in Fig. 2.13.

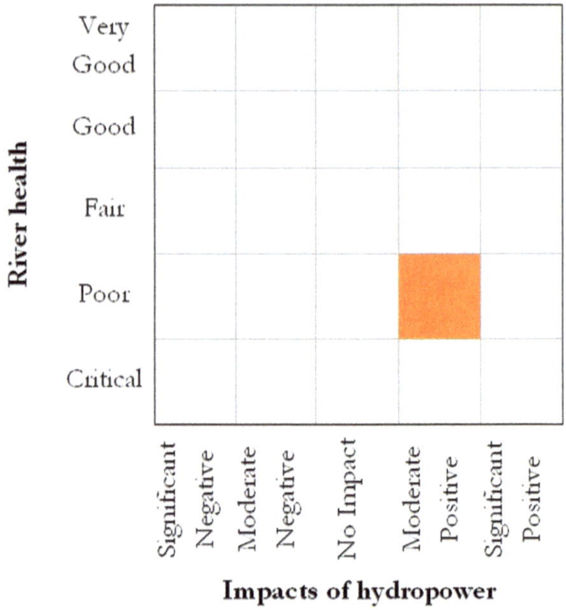

Fig. 2.13 Matrix of the state of water supply and hydropower impacts on the Lancang River

2.3.3.3 Navigation

River Health Assessment

Due to the importance of the Lancang River for navigation in Yunnan Province, multiple upgrades to the channel have occurred since the 1970s to improve the ability of large ships to travel within various sections. Before the 1970s, only small wooden boats travelled on the Lancang River due to rapid, turbulent flow, numerous rock obstructions, shoals, and sharp bends in the river. After the 1970s, the Chinese government invested several million RMB to improve channel conditions from Simao Port to the China-Myanmar border, a river section 158 km in length. By 1987, the channel of this section had been improved to the extent that it reached grade VI of the Chinese channel navigation standard and 100 ton ships could navigate this section of the river. In 2002–2004, the Chinese government invested 5 million US dollars to improve the channel conditions of the upper Mekong section, which has a length of 301 km. Following the improvements, this portion of the river channel also reached grade VI of the Chinese standard. In recent years, the Chinese government has continued to invest in the improvement of navigation conditions on the Lancang River. For example, the channel from Jinghong Port to the China-Myanmar border (71 km) was further improved to grade V, and 300 ton ships can now navigate this section. In 2015, China, Myanmar, Laos, and Thailand reached a consensus on "International Navigation Planning of the Lancang-Mekong River." According to this planning exercise, 500 ton ships will be able to travel from Simao Port in China to Luang Prabang in Laos, a distance of 890 km.

There are almost no cargo or passenger transport boats upstream of Jinghong dam due to the dam blocking navigation. Therefore, the maximum observed and expected maximum navigable lengths of the Lancang River are 71 km and 158 km, respectively. With regard to the navigation weight index, we used 500 tons as the expected maximum ship weight on the Lancang River, and the current maximum boat weight of 300 tons. Synthesizing navigation length and weight, the navigation index is 0.52 and can be classified as fair according to the definitions in Sect. 2.2.4.3.

Hydropower Impacts

Hydropower projects have both positive and negative impacts on navigation. The negative impacts are mainly the blockage of the navigable channel, as well as possible flow fluctuations resulting from dam operation. The positive impacts result from flow and water-level regulation that improve navigation conditions especially in the dry season.

The Jinghong dam blocked the navigable channel from Simao Port to the China-Myanmar border, so ships can only navigate from Jinghong Port to the China-Myanmar border. A ship lift was completed at the Jinghong hydropower project; however, use of the facility has been limited. A critical problem for the operation of the ship lift at Jinghong is the operational costs. The operational costs include the operation and maintenance costs of the facility together with the water required for its operation. Water use by the ship lift therefore reduces the electricity generation of the hydropower facility.

In addition to the blockage of the dam, water-level fluctuations downstream the Jinghong dam can adversely affect navigation. Water released from Jinghong dam varies significantly to meet electrical demands of the power grid, and the water-level fluctuations downstream of the dam may significantly affect navigation (Wang et al. 2006; Liu and Wang 2012). According to navigation requirements, the change in rate of the flow release at Jinghong dam should be controlled below 300 m^3/h so that water-level fluctuations in the downstream channel will not exceed 1.0 m/h. To reduce flow fluctuations, Ganlanba dam is proposed to be constructed downstream of Jinghong dam and act as a re-regulation hydropower project that adjusts the unsteady flow releases from Jinghong dam to meet navigation requirements (KECL 2014). When Jinghong hydropower project is operated at maximum capacity, it is predicted that water-level alterations downstream of Ganlanba dam can be decreased from 2.57 m/h to 0.95 m/h (NHRI 2014). It is also expected that water-level fluctuations between Jinghong and Ganlanba dams can be controlled below 1.0 m/h if water releases from Jinghong dam can be optimized (NHRI 2014).

Flow regulation by hydropower projects on the Lancang River may also improve channel conditions for navigation through reservoir impoundment and increased water levels in the dry season. The impoundment of Jinghong and Nuozhadu reservoirs had a significant benefit to navigation through increased water depths in the reservoirs. Moreover, the regulation of the reservoirs may increase the flow downstream of the dam during the dry season, improving water depths in the channel. In the first quarter of 2011, the watershed runoff yield in the Lancang River Basin was 1.15% lower than average. However, discharge from the Jinghong dam increased by 37.6% compared with average outflow under the regulation of the Xiaowan project, meaning that the water level at Jinghong Port was 1.0 m higher than the previous year. Flow regulation of reservoirs on the Lancang River can also improve navigation in the Mekong regions. Large cargo ships couldn't pass through Chiang Saen port in Chiang Rai Province, Thailand, due to low water levels in drought seasons. Recently, the situation has been ameliorated as more water has been released from the Lancang River dams to increase water levels in the Mekong River so that large cargo ships can navigate through Chiang Saen (National News Bureau of Thailand 2015).

The positive impact of hydropower projects on navigation is graded as moderate positive because the navigation facilities (e.g., channel construction, navigational aids, and communication) and management are also vital for waterway navigation. The negative impact of hydropower projects on navigation is graded as moderate negative because the deficiency of navigation conditions mentioned above also limits the sufficient development of waterway navigation. The state impact of navigation is illustrated in the matrix of river health and hydropower impacts in Fig. 2.14.

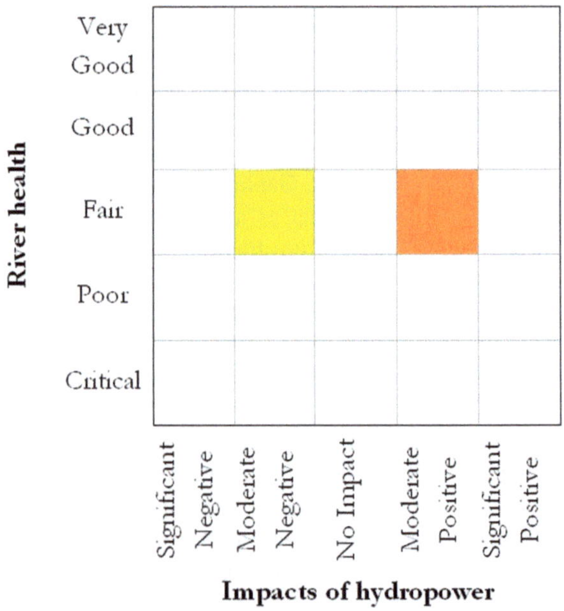

Fig. 2.14 Matrix of the state of navigation and hydropower impacts on the Lancang River

Table 2.18 Electricity generation of HEPs on the mainstem Lancang River (TWh)

Year	HydroLancang facilities	Dachaoshan	Total
2011	31.88		n/a
2012	40.31		n/a
2013	50.31		n/a
2014	57.71	6.50	64.21
2015	51.26	5.58	56.84

2.3.3.4 Electricity Generation

River Health Assessment

The technical potential for hydropower on the mainstem Lancang River is approximately 32 GW for installed capacity and 146 TWh for annual electricity generation (Wang 2015). To determine actual rates of hydroelectric power generation in the middle and lower reaches of the Lancang River, we compiled reported generation from 2011 to 2015 from hydropower projects operated by HydroLancang and SDIC Power (Dachaoshan dam) from relevant reports (e.g., annual audit report, credit rating report, and annual report) (Table 2.18). Hydroelectric power generation of Dachaoshan project in 2011–2013 was unavailable, and the hydroelectric generation fluctuated due to water availability. Total hydroelectric power generation of 2014 and 2015 on the mainstem Lancang River was 64.21 TWh and 56.84 TWh, respectively. The average electricity generation of 2014 and 2015 was 60.53 TWh, and this was used as the actual hydroelectric power generation for hydropower projects on the mainstem Lancang River.

Relative to the technical potential for hydropower in the middle and lower reaches of the Lancang River, the electricity generation index of the mainstem Lancang River was 0.94 and was categorized as very good. However, electricity loss is an emerging problem on the Lancang River due to the economic slowdown, grid connectivity, and other industry problems in China. According to the estimates of HydroLancang, hydropower facilities on the Lancang River were forced to release 14.6 km^3 of water that could have generated an additional 8.1 TWh in 2014 (Fu 2015).

Hydropower Impacts

The beneficial uses of hydropower projects, including electricity generation, can be maximized by joint operation of multi-reservoir systems on the mainstem and key tributaries. Hydropower plays an important role in today's electricity mix, contributing to more than 16% of electricity generation worldwide and about 85% of global renewable electricity (IEA 2015). Hydropower can produce electricity with negligible greenhouse gas emissions compared to the fossil energy sources currently in widespread use. For this reason, hydropower has an important future role to play in mitigating climate change (Kumar et al. 2011). However, climate-induced changes in precipitation and river flow may have adverse impacts on the technical and economic viability of some hydropower projects (Kumar et al. 2011). China is the world's largest hydroelectricity producer, with 320 GW of total installed capacity (including 23 GW of pumped storage) and 1126 TWh of electricity generation in 2015 (IHA 2016).

After the reservoirs of Xiaowan and Nuozhadu reached normal operating levels in Oct 2012 and Oct 2013, respectively, the reservoir system on the Lancang River can be jointly operated. The Xiaowan project has total storage of 15 km^3, regulating storage of 10 km^3, and installed capacity of 4200 MW. The Nuozhadu project has total storage of 23.7 km^3, regulating storage of 11.3 km^3, and installed capacity of 5850 MW. The total regulating storage of these two large projects is over 20 km^3, which can improve the regulating capacity of multi-reservoirs. Water can be stored in flood season and then released in low season and can also be adjusted among wet and dry years. In this way, hydroelectric power generation of the multi-reservoirs can be improved in the long term. Hydropower projects produce the electricity by using water in the Lancang River, and the contribution of hydropower projects on electricity is vital. Therefore, the impact of hydropower projects on the factor of electricity generation is significant positive. The state impact of electricity generation is illustrated in the matrix of river health and hydropower impacts in Fig. 2.15.

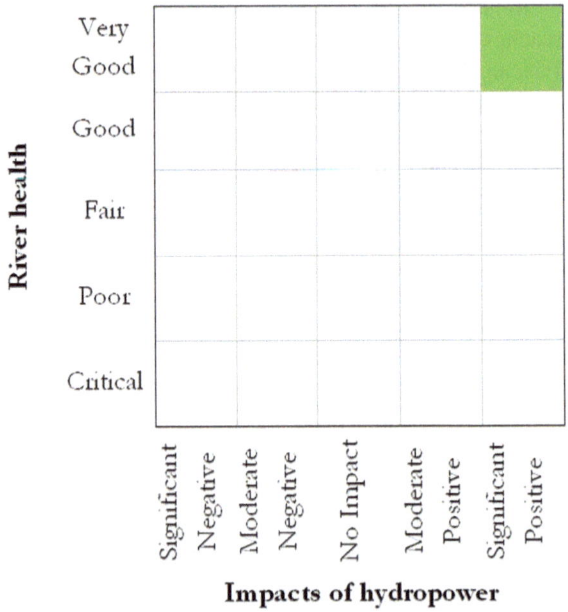

Fig. 2.15 Matrix on the state of electricity generation and hydropower impacts on the Lancang River

2.3.3.5 Recreation

River Health Assessment

The Lancang River Basin is rich in tourism resources that mainly consist of geographical landscape, water viewing, bio-landscape, cultural heritage, and humanistic activities (Wang et al. 2012). Major tourism resources are distributed along the mainstem of the Lancang River as well as its tributaries. Researchers in China and Thailand have suggested promoting subregional tourism cooperation along the Lancang-Mekong River based on a waterway tourism corridor (Li 2003; Amnuay-ngerntra and Sonoda 2013). The tourism value added of Yunnan Province was 90.7 billion RMB and accounted for 6.6% of GDP in 2015 (People's Government of Yunnan Province 2016). We used the percentage of tourism value added in GDP in 2015 to assess the development degree of recreation in the Lancang River. According to the categories developed for the recreation indicator (Sect. 2.2.4.3), the recreation value of the Lancang River is categorized as good.

Hydropower Impacts

Recreational and aesthetic values of hydropower reservoirs are important ecosystem services that have remarkable economic significance. Many reservoirs enable recreational uses such as fishing, sailing (canoe, kayak, water-skiing, swimming, or even use small sailboats), and other activities. These activities are becoming increasingly important for the people living close to the reservoirs and attract tourists (Branche 2015; Bonnet et al. 2015). Hydropower reservoirs generally improve the recreational values of the river (Hearnshaw et al. 2010), and the values can be estimated with methods including the travel cost method (TCM), the contingent valuation method (CVM), and unit day values (Bonnet et al. 2015). Wei et al. (2008) evaluated the recreational value of Manwan hydropower project using the TCM. The recreational value of Manwan was 50 million RMB/year, which accounts for 4.4% of the total value of its positive benefits.

The operation of the cascade of hydropower projects on the Lancang River has created new tourism opportunities and resources that have benefited Yunnan Province's tourism sector. Three big hydropower projects (Manwan, Dachaoshan, and Xiaowan) were built on the river reach within 200 km of Lincang City. The reservoirs of these three hydropower projects have led to the creation of a new scenic zone called the "One Hundred Miles High Gorge Landscape" which consists of the Lancang River canyon, river and lake viewing, hydropower dams, and historical sites. These tourist attractions allow Lincang to be a popular tourist destination for a variety of purposes, for example, industrial tours, historical and cultural research, exploration, leisure and holidays, sightseeing, ecological tours, aquatic recreation, and the discovery of folk customs. Similarly, the ecological mitigation measures employed at the Nuozhadu project, such as the fish hatchery, wildlife rescue, and the transplant of valuable and rare plants, led to the establishment of a Demonstration

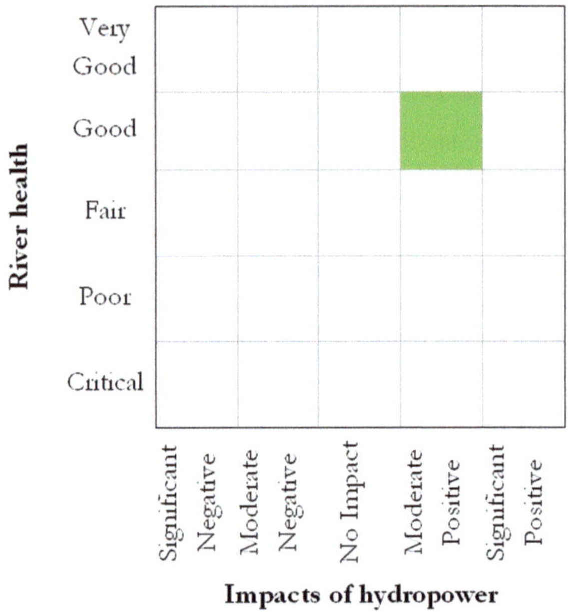

Fig. 2.16 Matrix on the state of recreation and hydropower impacts on the Lancang River

and Education Base of Biodiversity Conservation of Yunnan Province. This base acts as a place for educating people on biodiversity conservation and practices relevant to hydropower projects. However, the recreational values of the Lancang River are contributed by a variety of components such as recreation resources, transportation, accommodation, and service facilities. Hydropower projects positively impact the recreation values by creating or improving recreational resources, so the contribution is assessed as moderate positive. The state impact of recreation is illustrated in the matrix of river health and hydropower impacts in Fig. 2.16.

2.3.4 Combining Indicator Scores

The spider diagram (Fig. 2.17) illustrates the results of the river health assessment conducted for this Project. The numbers and indicators for each grade are presented in Table 2.19. The assessment results for the three categories of indicators (physical, biological, and social) are summarized below.

The assessment results for physical and chemical indicators illustrate these indicators were degraded, with the exception of water quality. Two indicators (connectivity, sediment and river geomorphology) in the physical and chemical group were evaluated as critical, and the status of hydrology on the Lancang River was classi-

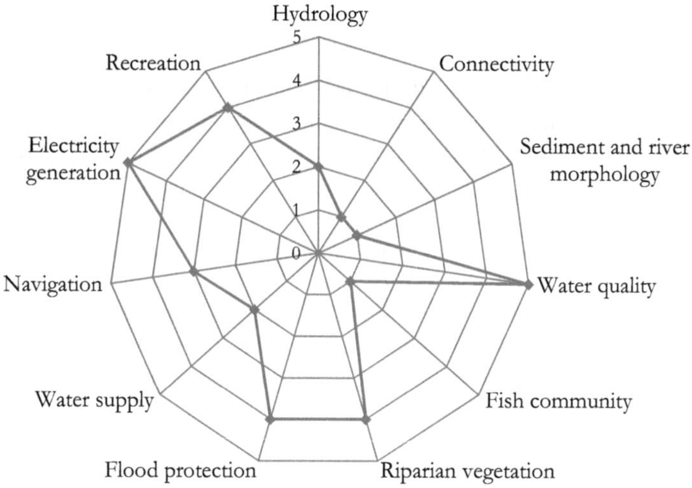

Fig. 2.17 Profile of the river health assessment for the Lancang River

Table 2.19 Summary of grades assigned to the river health indicators

Grade	Number of indicators	Indicators
Critical	3	Connectivity, sediment and river geomorphology, fish community
Poor	2	Hydrology, water supply
Fair	1	Navigation
Good	3	Riparian vegetation, flood protection, recreation
Very good	2	Water quality, electricity generation

fied as poor. Only water quality of the mainstem Lancang River was categorized as very good.

Of the biological indicators, fish community on the Lancang River was characterized as critical, and riparian vegetation was good. The assessment results of biological indicators illustrate significant degradation of fish biodiversity, but riparian habitat remains stable, with the degradation of fish species likely the result of dramatic changes in hydrology and connectivity.

The assessment results for social indicators were generally better than those of the ecological indicators. Only one indicator was assessed as poor (water supply) and fair (navigation), respectively.

2.4 Conclusions

The impacts of HEPs to river health for the Lancang River were analyzed through the use of the matrix that combines river health and hydropower impacts (Fig. 2.18). The number of times that the indicators fall into each combination of river health state and hydropower is marked in the matrix. The impacts of hydropower for a number of indicators have two values (i.e., both positive and negative); thus, although there are a total of 11 indicators, the sum of the numbers in the matrix is 15.

Analysis of the four quadrants in the matrix can help to identify different strategies for managing hydropower's impacts on river health. There are five indicators in "green zone": water quality, riparian vegetation, flood protection, electricity, and recreation. Thus, for these indicators, river health is good, and hydropower has contributed positively to achieving this status. Nevertheless, efforts are still needed to maintain and enhance the healthy state of these indicators. Three indicators are in the "yellow zone," water quality, riparian vegetation, and navigation, indicating that while river health is good for these indicators, they have been negatively impacted by hydropower dams. For these indicators, therefore, efforts need to be made to manage hydropower in order to potentially improve river health. Four indicators are located in the red zone: hydrology, connectivity, sediment and river morphology, and fish communities, indicating river health is poor and that hydropower has caused or exacerbated these problems. We recommend that relevant government regulators and hydropower operators should pay more attention to the impacts of HEPs on these indicators for improving health condition of the Lancang River. Three indicators are in the orange zone, hydrology, water supply, and navigation, indicating that river health is generally poor but that hydropower has contributed to improving it. Further alleviation of the impacts on these indicators may help improve river health.

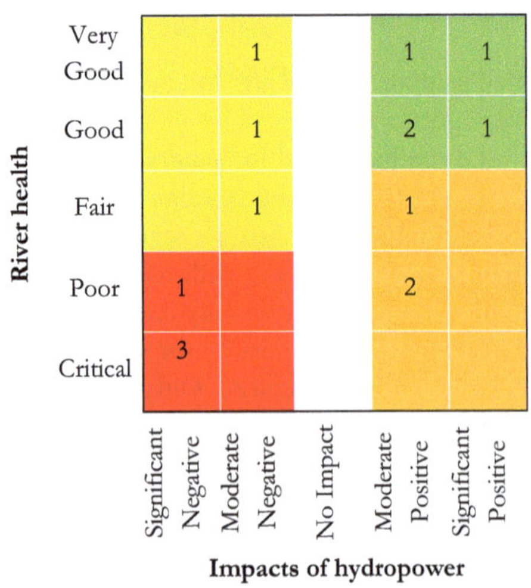

Fig. 2.18 Profile of the river health assessment and hydropower impacts for the Lancang River

Note that the assessment of river health and hydropower impacts is dynamic and may vary with time and the information available. For example, the baseline data of fish community was usually based on historical surveys over several decades, but the post-project data was generally obtained from field surveys over several years. The assessment result for fish community may be different with additional monitoring data collected over a longer period of time, or with the integration of higher resolution indicators. Additionally, improvements in mitigation measures also have the potential to improve the assessment results of river health and hydropower impacts.

There are both positive and negative impacts to the Lancang River as a result of hydropower development. The most significant concern is the dramatic decrease in the number of fish species compared with the pre-project conditions, mainly due to the river fragmentation and changes to hydrology. In addition, hydropower traps large quantities of sediment in the reservoirs, such that the sediment and river morphology is assessed as being of critical concern. Conversely, water quality, riparian vegetation, flood protection, electricity, and recreation are in "good" or "very good" states. The conclusion of our assessment is that the middle and lower reaches of the Lancang River cannot be considered "healthy" because a number of indicators fall into the "critical" category; but, at the same time, the river cannot be considered "dead" because the river has not lost most of its functions and values. Consequently, our study also indicates that the aquatic ecosystem of the Lancang River is, to some extent, resilient to hydropower's disturbance with the implementation of mitigation or compensation measures.

There are those who would argue that the inclusion of social indicators artificially improves the results of the river health assessment. In a highly developed catchment that supports essential economic and social activity, it is important to recognize the importance of these social and economic values of the river, and it is unrealistic to expect that rivers can be returned to a near natural state. Strategies and programs for river health should aim to balance environmental, economic, and social values.

In general, hydropower enhances socially and economically desirable values while decreasing ecological values. Post-project monitoring and mitigation measures should be improved to mitigate or compensate "poor" or "critical" impacts. Our project has examined the implementation and effectiveness of existing mitigation measures and then proposed improvements to river health monitoring and environmental mitigations (Chap. 3). The objectives of the river health monitoring were to evaluate the effectiveness of mitigation measures undertaken during the construction and operation of hydropower projects and to evaluate the project's effects on aquatic ecosystem. In this way, the monitoring program may encourage hydropower developers and government agencies to protect and enhance river health. Greater consideration of environmental flow requirements and fish habitat restoration are two major approaches for further mitigating adverse impacts. These efforts require hydropower operators to develop a fish mitigation program to modify hydropower operation and habitat restoration.

References

Allan, J., D. Erickson, and J. Fay. 1997. The influence of catchment land use on stream integrity across multiple spatial scales. *Freshwater Biology* 37 (1): 149–161.

Amnuay-ngerntra, S., and H. Sonoda. 2013. Develping river tourism on the Upper Mekong: Challenges and opportunities. *European Journal of Business and Social Sciences* 1 (10): 36–51.

Aylward, B., J. Bandyopadhyay, J.-C. Belausteguigotia, P. Börkey, A. Cassar, L. Meadors, L. Saade, M. Siebentritt, R. Stein, S. Tognetti, C. Tortajada, T. Allan, C. Bauer, C. Bruch, A. Guimaraes-Pereira, M. Kendall, B. Kiersch, C. Landry, E. Mestre Rodriguez, R. Meinzen-Dick, S. Moellendorf, S. Pagiola, I. Porras, B. Ratner, A. Shea, B. Swallow, T. Thomich, and N. Voutchkov. 2005. Freshwater ecosystem services. In *Ecosystems and human well-being: a framework for assessment human well-being: policy responses*, Volume 3, findings of the Responses Working Group of the Millennium Ecosystem Assessment, ed. K. Chopra and R. Leemans, 213–255. Washington, DC: Island Press.

Baran, E., P.J. Meynell, Y. Kura, A.A. Agostinho, G. Cada, O. Hamerlynck, T. Nao, and K. Winemiller 2009. *Dams and fish: Impacts and mitigation*, Unpublished draft report from WorldFish Center to the MRC.

Barbour, M.T., J. Gerritsen, B.D. Snyder, and J.B. Stribling. 1999. *Rapid bio-assessment protocols for use in streams and Wadeable rivers: Periphyton, benthic macro-invertebrates, and fish*. 2nd ed. EPA 841-B-99-002.

Beanlands, G.E., and P.N. Duinker 1983. *An ecological framework for environmental impact assessment in Canada*. Institute for Resource and Environmental Studies, Dalhousie University, Halifax, and Federal Environmental Assessment Review Office, Hull.

Biemans, H., I. Haddeland, P. Kabat, F. Ludwig, R.W.A. Hutjes, J. Heinke, W. Von Bloh, and D. Gerten 2011. Impact of reservoirs on river discharge and irrigation water supply during the 20th century. *Water Resources Research* 47 (3). https://doi.org/10.1029/2009WR008929.

Bonnet, M., A. Witt, K. Steware, B. Hadjerioua, and M. Mobley. 2015. *The economic benefits of multipurpose reservoirs in the United States-Federal hydropower fleet*. Oak Ridge: Oak Ridge National Laboratory.

Boulton, A.J. 1999. An overview of river health assessment: Philosophies, practice, problems and prognosis. *Freshwater Biology* 41: 469–479.

Branche, E. 2015. *Sharing the water uses of multipurpose hydropower reservoirs: The SHARE concept*, Multipurpose water uses of hydropower reservoirs framework. Le Bourget du Lac Cedex: Electricité de France & World Water Council.

Bunn, S., E. Abal, M. Smith, S. Choy, C. Fellows, B. Harch, M. Kennard, and F. Sheldon. 2010. Integration of science and monitoring of river ecosystem health to guide investments in catchment protection and rehabilitation. *Freshwater Biology* 55 (Suppl 1): 223–240.

Chang B., L. Geng, and C. Liu. 2006. Index system for ecological effect evaluation of hydropower projects. *Advances in Science and Technology of Water Resources* 26 (6): 11–15 (in Chinese).

Chinese Research Academy of Environmental Sciences (CRAES). 2012. Technical regulation for assessment of River Health (in Chinese).

CNTA (China National Tourism Administration). 2014. *Tourism as strategic pillar industry positioning has become increasingly clear*. http://www.cnta.gov.cn/zwgk/fgwj/zcjd/201506/t20150625_462371.shtml. Accessed 23 Sept 2016 (in Chinese).

Costanza, R., I. Kubiszewski, P. Paquet, J. King, S. Halimi, H. Sanguanngoi, N.L. Bach, R. Frankel, J. Ganaseni, A. Intralawan, and D. Morell. 2011. *Planning approaches for hydropower development in the Lower Mekong Basin*. Portland: Institute for Sustainable Solutions, Portland State University.

Cote, D., D.G. Kehler, C. Bourne, and Y.F. Wiersma. 2009. A new measure of longitudinal connectivity for stream networks. *Landscape Ecology* 24: 101–113.

Dong, Z. 2007. Exploring eco-hydraulic engineering. *Engineering Science* 9 (1): 1–7 (in Chinese).

Fan, N., G. Xie, C. Zhang, L. Chen, W. Li, and S. Cheng. 2012. Spatial-temporal dynamic changes of vegetation cover in Lancang River Basin during 2001–2010. *Resources Science* 34 (7): 1222–1231 (in Chinese).

Feng, Y., B. Kang, and L. Yang. 2012. Feasibility analysis of widely accepted indicators as key ones in river health assessment. *Journal of Geographical Sciences* 22 (1): 46–56.

Fu, Y. 2015. Lost power of hydropower in Yunnan. *China Energy News*. http://paper.people.com.cn/zgnyb/html/2015-08/10/content_1597850.htm. Accessed 23 Aug 2016 (in Chinese).

Fu, K., and D. He. 2007. Analysis and prediction of sediment trapping efficiencies of the reservoirs in the mainstream of the Lancang River. *Chinese Science Bulletin* 52 (Suppl 2): 134–140 (in Chinese).

Fu, K., D. He, Y. Chen, C. Ye, and Y. Li. 2007. Impacts of dam constructions on the annual distribution of sediment in Lancang- Mekong River Basin. *Acta Geographica Sinica* 62 (1): 14–21 (in Chinese).

Fu, B., Y. Wang, P. Xu, K. Yan, and M. Li. 2014. Value of ecosystem hydropower service and its impact on the payment for ecosystem services. *Science of the Total Environment* 472: 338–346.

Fuller, D. 1998. Trends in NDVI time series and their relation to rangeland and crop production in Senegal, 1987–1993. *International Journal of Remote Sensing* 19 (10): 2013–2018.

Gao, X., S. Zhao, C. Zhang, and X. Tu. 2009. Index system and method for assessing the health status of river. *Journal of Hydraulic Engineering* 40 (8): 962–968 (in Chinese).

Geerken, R., and M. Ilaiwi. 2004. Assessment of rangeland degradation and development of a strategy for rehabilitation. *Remote Sensing of the Environment* 90 (4): 490–504.

Gehrke, P.C., P. Brown, C.B. Schiller, D.B. Moffatt, and A.M. Bruce. 1995. River regulation and fish communities in the Murray-Darling river system, Australia. *Regulated Rivers Research and Management* 15: 181–198.

Geng, L., H. Liu, H. Zhong, and C. Liu. 2006. Indicators and criteria for evaluation of healthy rivers. *Journal of Hydraulic Engineering* 37 (3): 253–258 (in Chinese).

Gippel, C., X. Jiang, X. Fu, X. Liu, L. Chen, H. Wang, S. Jin, L. Li, L. Wang, Y. Sun, H. Pang, R. Song, F. Sun, H. Shang, and R. Speed. 2012. *Assessment of river health in the Lower Yellow River*. Brisbane: International WaterCentre.

Gore, J.A. (ed.). 1985. *The restoration of rivers and streams*, 280 pp. Stoneham: Butterworth.

Gu, Y., S. Lei, and J. Liu 2008. Effects of cascade hydropower development on hydrological regime of Lower Lancang River. *Water Resources and Hydropower Engineering* 39 (4): 20–23 (in Chinese).

Guo, J., and W. Huang. 2008. Exploration on assessment methods for healthy Yangtze River. *Journal of Yangtze River Scientific Research Institute* 25 (4): 1–4 (in Chinese).

Guo, W., H. Wang, J. Xu, Z. Xia. 2011. Effects of Three Gorges reservoir on the downstream eco-hydrological regime during the spawning of important fishes. *Journal of Hydrelectric Engineering* 30 (3): 22–26 (in Chinese).

He, D., Y. Feng, S. Gan, and H. You. 2006. The cross-border hydrological effect of hydropower development along main stream of the Lancang River. *Chinese Science Bulletin* 51 (S1): 14–20 (in Chinese).

Hearnshaw, E., R. Cullen, and K. Hughey. 2010. Ecosystem services review of water projects. In *Australian Agricultural and Resource Economics Society Annual Conference (54th)*, Adelaide, Australia, February 10–12.

Hegmann, G., C. Cocklin, R. Creasey, S. Dupuis, A. Kennedy, L. Kingsley, W. Ross, H. Spaling, D. Stalker, and Axys Environmental Consulting. 1999. *Cumulative effects assessment practitioners guide*. Hull: Canadian Environmental Assessment Agency.

Hoffman, R., and J. Dunham. 2007. *Fish-movement ecology in high gradient headwater streams: Its relevance to fish passage restoration through stream culvert barriers*. Reston: U.S. Geological Survey.

Huang, G. 2013. Practices in environmental protection of hydropower development in the Lancang River Basin. *Proceeding of the Hydropower 2013 – CHINCOLD 2013 Annual Meeting and the 3rd International Symposium on Rockfill Dams*, Kunming, China (in Chinese).

IEA (International Energy Agency). 2000. *Hydropower and the environment: Effectiveness of mitigation measures*, IEA Hydropower Agreement, Annex III – Subtask 6. Paris: International Energy Agency.

———. 2015. *Medium-term renewable energy market report 2015*. Paris: OECD/IEA.

IHA (International Hydropower Association). 2016. *Hydropower status report 2016*. London.

International Finance Corporation (IFC). 2012. *Good practice handbook cumulative impact assessment and management: Guidance for the private sector in emerging markets*. Washington, DC.

Jackson, D.C., and G. Marmulla. 2011. *The influence of dams on river fisheries. Chapter in Dams, fish and fisheries: Opportunities, challenges and conflict resolution*, FAO Fisheries Technical Paper, Rome.

Jin, X., et al. 1995. *Environment of Chinese Lakes*. Beijing: The Ocean Press (in Chinese).

Jones, I., C. Abrahams, L. Brown, K. Dale, F. Edwards, M. Jeffries, M. Klaar, Mark Ledger, L. May, A. Milner, J. Murphy, A. Robertson, and G. Woodward. 2013. *The impact of extreme events on freshwater ecosystems*, 67pp. London: British Ecological Society.

Kang, B., and D. He. 2007. Research progress of biodiversity of fish species in the Lancangjiang River. *Resources Science* 29 (5): 195–200 (in Chinese).

Kang, B., L. Perreit, Y. Li, and D. He. 2009. Are the fish of the upper and lower Mekong interconnected? *Chinese Journal of Oceanology and Limnology* 27 (2): 400–407.

Karr, J.R. 1981. Assessment of biotic integrity using fish communities. *Fisheries* 6 (6): 21–27.

———. 1991. Biological integrity: A long-neglected aspect of water resource management. *Ecological Applications* 1: 66–84.

———. 1996. Ecological integrity and ecological health are not the same. In *Engineering within ecological constraints*, ed. P.C. Schulze, 97–109. Washington, DC: National Academy Press.

———. 1999. Defining and measuring river health. *Freshwater Biology* 41: 221–234.

KECL (Kunming Engineering Corporation Ltd). 2014. *Environmental impact assessment of the Ganlanba hydroelectric project on the Lancang River*. Kunming: KECL (in Chinese).

Kellogg, C.H., and X. Zhou. 2014. Impact of the construction of a large dam on riparian vegetation coverat different elevation zones as observed from remotely sensed data. *International Journal of Applied Earth Observation and Geoinformation* 32: 19–34.

Kleynhans, C.J. 1996. A qualitative procedure for the assessment of the habitat integrity status of the Luvuvhu River. *Journal of Aquatic Ecosystem Health* 5: 41–54.

Kleynhans, C.J, M.D. Louw, and M. Graham. 2008. Module G: EcoClassification and EcoStatus determination in River EcoClassification: Index of habitat integrity (Section 1, Technical manual) Joint Water Research Commission and Department of Water Affairs and Forestry report, WRC Report No. TT 377-08.

Kondolf, G.M., et al. 2014. Sustainable sediment management in reservoirs and regulated rivers: Experiences from five continents. *Earth's Future*. https://doi.org/10.1002/eft22013EF000184.

Kumar, A., T. Schei, A. Ahenkorah, Rodriguez R. Caceres, J.M. Devernay, M. Freitas, D. Hall, Å. Killingtveit, and Z. Liu. 2011. Hydropower. In *IPCC special report on renewable energy sources and climate change mitigation*, ed. O. Edenhofer, R. Pichs-Madruga, Y. Sokona, K. Seyboth, P. Matschoss, S. Kadner, T. Zwickel, P. Eickemeier, G. Hansen, S. Schlömer, and C. von Stechow. Cambridge/New York: Cambridge University Press.

Kummu, M., and O. Varis. 2007. Sediment-related impacts due to upstream reservoir trapping, the Lower Mekong River. *Geomorphology* 85 (3–4): 275–293.

Ladson, A., and L. White. 1999. *Index of stream condition: Reference manual*. Department of Natural Resources and Environment, Melbourne, Australia.

Ladson, A.R., L.J. White, J.A. Doolan, B.L. Finlayson, B.T. Hart, P.S. Lake, and J.W. Tilleard. 1999. Development and testing of an Index of Stream Condition for waterway management in Australia. *Freshwater Biology* 41: 453–468.

Larinier, M. 2011. *Environmental issues, dams and fish migration. Chapter in dams, fish and fisheries: Opportunities, challenges and conflict resolution*, FAO Fisheries Technical Paper. Rome.

Li, B. 2003. *Theory and practice of tourism integration in international region: Case study in Lancang-Mekong region*, Master's dissertation of Yunnan University (in Chinese).

Li, G. 2004. Keeping the Yellow River healthy- ultimate target of the Yellow River management. *China Water Resources* 1: 5–7 (in Chinese).

Li, J., S. Dong, M. Peng, Z. Yang, S. Liu, X. Li, and C. Zhao. 2013. Effects of damming on the biological integrity of fish assemblages in the middle Lancang-Mekong River basin. *Ecological Indicators* 34: 94–102.

Lin, M., X. Li, and M. Yang. 2006. Proble into the index system for evaluating the health of the Rivers in the Pearl River basin. *Pearl River* 4: 1–3 (in Chinese).

Liu, S. 2016. Tourism-emerging energy for economic and social development of China. *China Tourism News*, August 16. http://www.cnta.gov.cn/xxfb/hydt/201608/t20160816_780922.shtml. Accessed 2 Sept 2016 (in Chinese).

Liu, Y., and Y. Wang. 2012. Influence of daily regulation schemes of Jinghong hydropower station on downstream navigation flow conditions. *Hydro-science and Engineering* (4): 103–107 (in Chinese).

Liu, H., J. Liu, and H. Tang. 1998. Present and future of water resources development in Lancang River Basin in Yunnan Province. *Advances in Water Science* 9 (1): 70–76 (in Chinese).

Liu, X., J. Zhang, and Y. Zhang. 2006. Indicators of the healthy Yellow River. *Acta Geographica Sinica*, 61 (5): 451–460 (in Chinese).

Liu, M., D. Chen, X. Duan, K. Wang, and S. Liu. 2011. Ichthyofauna composition and distribution of fishes in Yunnan section of Lancang River. *Journal of Fishery Sciences of China* 18 (1): 156–170 (in Chinese).

Liu, C., Y. He, D.E. Walling, and J. Wang. 2013. Changes in the sediment load of the Lancang-Mekong River over the period 1965–2003. Science *China Technological Sciences* 56 (4): 843–853 (in Chinese).

Liu, B., H. Tao, B. Kong, and HE B. 2014. Research on the water resource security and its temporalspatial distributions of Yunnan Province, China. *Journal of Natural Resources* 29 (3): 454–465 (in Chinese).

Liu, S., N. An, S. Dong, H. Zhao, L. Deng, and C. Zhao. 2015. The effects of hydropower stations construction on vegetation dynamics based on NDVI: A case study of cascade hydropower stations of Lancang River. *Mountain Research* 33 (1): 48–57 (in Chinese).

Lu, X., and R. Siew. 2006. Water discharge and sediment flux changes over the past decades in the lower Mekong River: possible impacts of the Chinese dams. *Hydrology and Earth System Sciences* 10 (2): 181–195.

Ma, J. 2014. *The role of cascade reservoirs on the Lancang River for downstream flood control and water resources utilization*. http://www.mrcsummit.org/presentations/track1/2.1-c-Impacts-of-cascade-reservoirs-in-the-Lancang-River.pdf. Presentation at the Second MRC Summit and International Conference. Accessed 25 Aug 2016.

Mei, Z., C. He, and B. Liu. 2006. Analysis of sedimentation of Manwan Reservoir. *Proceedings of the International Symposium on Hydropower 2006*, Kunming (in Chinese).

MEP (Ministry of Environmental Protection of China). 2013. *Guideline for ecological health assessment of river basin*. Beijing (in Chinese).

———. 2015. Inspection and acceptance of environmental protection of Gongguoqiao HEP. http://www.mep.gov.cn/gkml/hbb/spwj1/201507/t20150721_306951.htm. Accessed 16 Feb 2016 (in Chinese).

Meyer, J.L. 1997. Stream health: Incorporating the human dimension to advance stream ecology. *Journal of the North American Benthological Society* 16: 439–447.

Meynell, P.J., and Z. Nazia. 2014. *EIA guidelines for large-scale hydropower in Pakistan*. Islamabad: IUCN Pakistan.

Milcu, A.I., J. Hanspach, D. Abson, and J. Fischer. 2013. Cultural ecosystem services: A literature review and prospects for future research. *Ecology and Society* 18 (3): 565–598.

Mittal, N., A.G. Bhave, A. Mishra, and R. Singh. 2016. Impact of human intervention and climate change on natural flow regime. *Water Resources Management* 30 (2): 685–699. https://doi.org/10.1007/s11269-015-1185-6.

National News Bureau of Thailand. 2015. China released more water from the dams, hope to increase commercial barge traffic. http://thainews.prd.go.th/website_th/news/news_detail/WNSOC5901210010013. Accessed 16 Feb 2016.

New, T., and Z. Xie. 2008. Impacts of large dams on riparian vegetation: Applying global experience to the case of China's Three Gorges Dam. *Biodiversity and Conservation* 17: 3149–3163.

Newman, B.J., P. Ladd, M. Brundrett, and W.D. Kingsley. 2013. Effects of habitat fragmentation on plant reproductive success and population viability at the landscape and habitat scale. *Biological Conservation* 159: 16–23.

NHRI (Nanjing Hydraulic Research Institute). 2014. *Research on key technologies of the Ganlanba Navigation and Electricity Project*. http://www.transst.cn/swf/9/20142201805661776 90.swf. Accessed 25 Apr 2016 (in Chinese).

Nilsson, C., R.L. Brown, R. Jansson, and D.M. Merritt. 2010. The role of hydrochory in structuring riparian and wetland vegetation. *Biological Reviews* 85 (4): 837–858.

Norris, R.H., and M.C. Thoms. 1999. What is river health? *Freshwater Biology* 41: 197–209.

Ouyang, Z., T. Zhao, X. Wang, and H. Miao. 2004. Ecosystem services analyses and valuation of china terrestrial surface water system. *Acta Ecologica Sinica* 24 (10): 2091–2099 (in Chinese).

Paetzold, A., C. Yoshimura, and K. Tockner. 2008. Riparian arthropod responses to flow regulation and river channelization. *Journal of Applied Ecology* 45 (3): 894–903.

Peng, B., H. Wang, R. Wang, Y. Han, W. Huang, and Y. Zhu. 2014. Development of a river health assessment system for the Lower Yellow River. *Journal of Hydroecology* 35 (6): 81–87 (in Chinese).

People's Government of Yunan Province. 2016. *2nd Press Conference on Economic Development of the Yunan Province*. http://www.yn.gov.cn/yn_fbh/201606/t20160630_25851.html. Accessed 2 Sept 2016 (in Chinese).

Petts, G.E. 1984. *Impounded rivers: Perspectives for ecological management*. New York: Wiley.

Poff, N.L., J.D. Allan, M.B. Bain, J.R. Karr, K.L. Prestegaard, B.D. Richter, R.E. Sparks, and J.C. Stromberg. 1997. The natural flow regime. *BioScience* 47: 769–784.

Postel, S., and B. Richter. 2003. *Rivers for life: Managing water for people and nature*. Washington, DC: Island Press.

Prowse T.D., F.J. Wrona, and G. Power. 2004. *Chapter 2: Dams, reservoirs and flow regulation. Threats to water availability in Canada. Environment and climate change Canada*. https://www.ec.gc.ca/inre-nwri/default.asp?lang=En&n=0CD66675-1&offset=7&toc=show. Accessed 8 Aug 2016.

Qiu, J. 2010. China drought highlights future climate threats. *Nature* 465: 142–143.

Rankin, E.T. 1989. The Qualitative Habitat Evaluation Index (QHEI): Rationale, methods, and application. Division of Water Quality Planning and Assessment, Ecological Assessment Section, Columbus, OH.

———. 1995. The use of habitat assessments in water resource management programs. In *Biological assessment and criteria: Tools for water resource planning and decision making*, ed. W. Davis and T. Simon, 181–208. Boca Raton: Lewis Publishers.

Rapport, D.J. 1991. Myths in the foundations of economics and ecology. *Biological Journal of the Linnaean Society* 44: 185–202.

Rheinheimer, D.E., and J.H. Viers. 2014. Combined effects of reservoir operations and climate warming on the flow regime of hydropower bypass reaches of California's Sierra Nevada. *River Research and Applications* 31 (3): 269–279.

Santos, A.B.I., R.J. Albieri, and F.G. Araujo. 2013. Influences of dams with different levels of river connectivity on the fish community structure along a tropical river in Southeastern Brazil. *Journal of Applied Ichthyology* 29 (1): 163–171.

Sarukhán, J., and A. Whyte, eds. 2005. *Ecosystems and human well-being: Synthesis (Millennium Ecosystem Assessment)*. Washington, DC: Island Press, World Resources Institute.

Schilt, C. 2006. Developing fish passage and protection at hydropower dams. *Applied Animal Behaviour Science* 104: 295–325.

Segar, D.A. 1998. Coastal Oceans and Estuaries. In *Introduction to ocean sciences*, 359–387. Belmont: Wadsworth Publishing Company.

Shi, W. 2013. *Study on hydrological cumulative effects of cascade reservoirs on lower Lancang River*, PhD dissertation, Beijing (in Chinese).

Speed, R., C. Gippel, N. Bond, S. Bunn, X. Qu, Y. Zhang, W. Liu, and X. Jiang. 2012. *Assessing river health and environmental flow requirements in Chinese rivers*. Brisbane: International WaterCentre.

Tang, J., X. Yin, Y. Pan, and Z. Yang. 2014. Assessment of contributions of climatic variation and human activities to streamflow changes in the Lancang River, China. *Water Resources Management* 28 (10): 422–432.

The World Bank. 2012. *Sample guidelines: Cumulative environmental impact assessment for hydropower projects in Turkey*. Ankara.

UN DESA (United Nations Department of Economic and Social Affairs). 2018. *Sustainable development knowledge platform*. https://sustainabledevelopment.un.org/sdg6. Accessed 23 Feb 2018.

UNEP (United Nations Environment Programme). 2007. *A compendium of relevant practices for improved decision-making, planning and management of dams and their alternatives*. United Nations environment programme dams and development project, Nairobi, Kenya.

Vorosmarty, C.J., M. Meybeck, B. Fekete, K. Sharma, P. Green, and J.P.M. Syvitski. 2003. Anthropogenic sediment retention: Major global impact from registered river impoundments. *Global and Planetary Change* 39: 169–190.

Walling, D.E., and D. Fang. 2003. Recent trends in the suspended sediment loads of the world's rivers. *Global and Planet Change* 39: 111–126.

Wang, S. 2004. On dams and ecology. *Water Power* 30 (4): 1–4 (in Chinese).

Wang, Y. 2015. *Sustainable Hydropower development on the Lancang River*. http://www.hydropower.org.cn/showNewsDetail.asp?nsId=15588. Accessed 16 Feb 2016 (in Chinese).

Wang, Z., X. Zhang, and Z. Jin. 2000. *The ecological environment and biological resources of Manwan hydropower station reservoir along Lancang River in Yunnan, China*. Kunming: Yunnan Science and Technology Press (in Chinese).

Wang, H., Y. Liu, Y. Shen, B. Xiao, and Y. Liu. 2004. Preliminary research on water bloom of Dinophyceae in Yunnan Manwan hydropower station reservoir. *ACTA Hydrobiologica Sinaca* 2: 213–215 (in Chinese).

Wang, Y., Y. Liu, and Y. Shi. 2006. Study of the project of the daily regulation of Jinghong Hydroelectricity Station on Lancang River. *Journal of Chongqing JianZhu University* 28 (5): 43–46 (in Chinese).

Wang J., X. Bai, X. Deng, and M. Wang. 2008. Research on spatial and temporal changes of riparian vegetation cover in the Three Gorges Dam area based on NDVI. *Geo-Information Science* 10 (6): 808–815 (in Chinese).

Wang, L., L. He, S. Cheng, Q. Min, Z. Xu, and Z. Yuan. 2012. Characteristics of spatial differentiation for tourism resources and the development modes in the Lancang River Basin. *Resources Science* 34 (7): 1266–1276 (in Chinese).

Wei, G., B. Cui, S. Dong, and Z. Yang 2008. Impact of hydropower development on river ecosystem service: A case study from the Manwan Hydropower Project. *Acta Scientiae Circumstantiae* 28 (2): 235–242 (in Chinese).

Wei, G., Z. Yang, B. Cui, B. Li, H. Chen, J. Bai, and S. Dong. 2009. Impact of dam construction on water quality and water self-purification capacity of the Lancang River, China. *Water Resources Management*. 23: 1763–1980.

Welcomme, R.L. 1979. *Fisheries ecology of Floodplain Rivers*. London: Longman Group LTD.

Wilson, M.A., and S.R. Carpenter. 1999. Economic valuation of freshwater ecosystem services in the United States: 1971–1997. *Ecological Applications* 9 (3): 772–783.

WWF (World Wide Fund for Nature). 2016. *Living Planet Report 2016. Risk and resilience in a new era*. Gland: WWW International.

Xiao, J., G. Shi, C. Mao, M. Wang, and Z. Xing. 2007. Evaluation of effects of dams on river ecosystem service functions in China. *ACTA Ecologica Sinica* 27 (2): 526–537 (in Chinese).
Ye, J. 2012. Improving water infrastructure development in Lincang City for adapting severe drought. *China Flood & Drought Management* 22 (1): 65–67 (in Chinese).
YEPD (Yunnan Environmental Protection Department). 2014. *Environmental quality communique of Yunnan Province*. Kunming: YEPD (in Chinese).
Yu, X., J. Xia, J. Yang, and W. Ma. 2011. Preliminary study on the index system and assessment method of green hydropower. *Journal of Hydroelectric Engineering* 30 (3): 71–77 (in Chinese).
Zhang, J., Z. Dong, D. Sun, and J. Wang. 2010. Complete river health assessment index system based on eco-regional method according to dominant ecological functions. *Journal of Hydraulic Engineering* 41 (8): 883–892 (in Chinese).
Zhao, Y., and Z. Yang. 2005. Preliminary study on assessment of urban river ecosystem health. *Advances in Water Science* 16 (3): 349–355 (in Chinese).
Zheng, X. 1995. *Development strategy for China's pillar industry*. Beijing: China Planning Press (in Chinese).
Zheng, L., X. Chen, and J. Yang. 2013. Status and conservation of fishes in the middle and lower Lancangjiang River. *Zoological Research* 34 (6): 680–686 (in Chinese).

Chapter 3
Improving River Health Through Mitigation and Monitoring

3.1 Background

Chapter 2 identified components of river health that are in poor or critical condition and have been adversely affected by the development of hydropower. In this chapter, we review the existing mitigation, management, and monitoring activities that are employed on the Lancang River and make recommendations for improvements that, if implemented, will further mitigate the adverse effects of hydropower development.

The adverse impacts of hydropower on river ecosystems should be identified and assessed through an environmental impact assessment (EIA), and in certain cases strategic environmental assessment (SEA), prior to project development. The adverse impacts should be addressed by employing measures to avoid, mitigate, or compensate for the impacts. Avoidance measures aim to avoid the adverse impacts through site selection or project design modifications. Mitigation measures are expected to partially or fully address adverse impacts by reducing the spatial scale, duration, or magnitude of impacts by employing procedures to protect the environment (e.g., erosion and sediment control measures, clearing vegetation prior to reservoir filling). Compensation measures are employed to replace or provide substitute resources or environments to compensate for residual environmental effects that cannot be mitigated.

Ideally, a combination of avoidance, mitigation, and compensation measures will be designed prior to project development, and adverse environmental impacts will be effectively minimized. However, monitoring of the avoidance, mitigation, and compensation measures is critical to ensure that the measures were implemented and are effective and that the hydropower project has not had adverse impacts on the physical and biological environment that were not predicted. If avoidance, mitigation, and compensation measures are not effective, or unexpected impacts arise, monitoring data can be used to guide adaptive management and improve environmental performance.

X. Yu et al., *Balancing River Health and Hydropower Requirements in the Lancang River Basin*, https://doi.org/10.1007/978-981-13-1565-7_3

In addition to reviewing the effectiveness of the mitigation measures implemented at hydropower projects on the Lancang River, in this chapter we also develop a framework for a river health monitoring program that draws upon other monitoring programs from around the world.

3.2 Methods

To characterize the current mitigation, management, and monitoring activities in the Lancang River Basin, we compiled, reviewed, and analyzed relevant peer-reviewed publications, industry and news reports, and publications by government and non-government organizations. These documents, along with our professional experience, were used to evaluate the effectiveness of the existing activities and provide recommendations for improvements (Sects. 3.4.1 and 3.4.2).

Our review examined the mitigation and compensation measures employed in the Lancang River Basin, along with the existing environmental monitoring program. We also analyzed the strengths and weaknesses of the existing monitoring program in light of long-term monitoring programs established on other large river systems, or those developed specifically to monitor the effects of hydropower development (Sect. 3.3.3.2). The principles of these monitoring programs and protocols are used to develop the framework of a watershed scale monitoring program for the Lancang River with the objective of improving the existing monitoring program (Sect. 3.4.3).

Our review of the existing monitoring program considered three types of monitoring: surveillance, compliance, and effectiveness monitoring according to the following definitions. Surveillance monitoring is undertaken to monitor and evaluate the status and trend of river health components. Compliance monitoring is undertaken to verify the compliance of the projects with the applicable laws, regulations, and technical requirements. Effectiveness monitoring is undertaken to evaluate the effectiveness of any measures taken to mitigate the adverse environmental effects of the projects. These three types of monitoring can collectively establish: (1) a platform for monitoring and assessing current river health status and long-term changes, (2) key indicators by which regulatory agencies can measure compliance, and (3) tools that can be utilized to evaluate the relative success of mitigation and compensation measures designed to minimize or offset environmental impacts (Gippel and Speed 2010; Lewis et al. 2013).

In designing an appropriate monitoring program, a number of factors need to be considered in addition to the objectives of the monitoring program and the corresponding types of monitoring required:

- *Indicators and parameters*: monitoring data are collected on indicators that reflect values to be protected (Suter 1990). In this study, the indicator categories of the river health monitoring program are the same as the river health assessment (Chap. 2): physical and chemical, biological, and social indicators. For a given indicator, multiple parameters may be monitored. For example, for the water flow indicator, the parameters measured are typically stage and discharge. For the fish community indicator, the parameters measured may include the number of species and their density and/or biomass.

- *Monitoring scale*: the scale at which different parameters and indicators are monitored depends on the objectives of the monitoring program and the extent to which they are influenced by individual projects.
- *Frequency*: the frequency with which monitoring occurs is determined by the variable features of the parameters monitored and the complexity/cost of the monitoring work.
- *Duration*: the length of time that specific parameters are monitored depends on the objective of the monitoring. For example, if assessing whether a particular parameter is influenced by development of a specific project, then the parameter should be monitored until the parameter stabilizes or the impact of the project is clear. On the other hand, if assessing whether the parameter changes over long periods of time and is influenced by watershed or landscape level changes, then the parameter should be monitored over the long term to track trends.

3.3 Results and Discussion

3.3.1 Existing Avoidance, Mitigation, and Compensation Measures

A variety of environmental impact avoidance, mitigation, and compensation options were designed to minimize the adverse impacts on VECs as a result of hydropower project development on the Lancang River (Table 3.1). The avoidance, mitigation, and compensation measures were proposed in the SEA and EIA documents for the hydropower projects and have been implemented on constructed projects. In addition to these measures, environmental monitoring and management plans were developed and implemented.

Environmental measures at the Gongguoqiao project cost RMB 137 million (approximately US$21.1 million); at the Xiaowan project, they cost RMB 406 million (approximately US$62.5 million), accounting for 1.8% and 1.2% of total investment costs, respectively (MEP 2015a, b). The investment on environmental measures at the Nuozhadu project was estimated to be 1.0 billion RMB (approximately US$153.9 million), with an estimated 240 million RMB (approximately US$36.9 million) to be spent on a multilevel water intake structure to reduce downstream water temperature changes (Yunnan News 2012).

It is commonly agreed that the most effective environmental management measure to avoid adverse impacts is good site selection and design (Ledec and Quintero 2013; Harrison et al. 2007). This principle has been applied to hydropower planning on the Lancang River. For example, the Guonian hydropower project was conceived

Table 3.1 Environmental impacts and mitigation options for hydropower projects on the Lancang River

Environmental impacts	Measures	Project or region
Downstream hydrological changes	Dam site selection and operational modifications Management of water releases	All HEPs (environmental flow release); Xiaowan, Nuozhadu (flood peak reduction and low flow augmentation in dry season)
Low temperature water release	Selective water intake	Huangdeng, Nuozhadu
Deterioration of water quality	Water pollution control Forest clearing in the reservoir	All project reservoirs
Reservoir sedimentation	Operation of storing clear and releasing muddy water Watershed soil erosion control	Xiaowan, Manwan, Dachaoshan, Nuozhadu (storing clear and releasing muddy); Lancang River Basin (erosion control)
Flooding of natural habitats	Dam site selection and operational modifications Fish habitat compensation Fish nature reserve	Jidu River at Miaowei (dam removal and fish habitat restoration); Luosuo (Buyuan) River, Nanla River downstream of Jinghong (rare and endemic fish nature reserve); eight tributaries identified for fish habitat preservation
Fish and other aquatic life	Dam site selection and operational modifications Management of water releases Fish passage facilities Fish hatcheries and release Fishing regulations	Ganlanba (fish ladder); Wunonglong, Huangdeng, Dahuaqiao (fish lift); Wunonglong, Lidi, Tuoba, Huangdeng, Dahuaqiao, Miaowei (fish attraction barge); Huangdeng, Gongguoqiao, Nuozhadu (fish hatcheries and release); Gongguoqiao, Nuozhadu (fish netting and transport)
Greenhouse gases (GHG)	Forest clearing in the reservoir	All project reservoirs
Loss of riparian vegetation	Transplant of valuable and rare plants	Lidi, Xiaowan, Nuozhadu, Jinghong
Loss of terrestrial wildlife	Wildlife rescue	Nuozhadu, Xiaowan

in the original hydropower plans for the upstream Lancang River (prior to the 2005 SEA). The proposed site is excellent in terms of electricity generation and construction cost. However, the proposed dam was located in close proximity to the Mingyong Glacier, which is unique in terms of its low altitude, low latitude, its monsoon oceanic climate, and the fact that it is China's southernmost glacier. The Guonian project was canceled due to concerns that the reservoir may contribute to increased melting of the glacier. The Mengsong project (Fig. 1.2) was also canceled to protect the migration routes of four long-distance migratory fish species. The capture records of these four species that migrate between the Lancang and the Mekong indicate that the populations have been declining since 1970 due to a number of natural changes and human activities (Kang 2013; Yang et al. 2007). However, dams on the Lancang River are unlikely to be blocking fish migration routes from the lower Mekong River because studies have shown the species only migrate as far as Jinghong. Nevertheless, the Mengsong dam would have prevented fish from reaching the Buyuan (Luosuo) and Nanla rivers, two tributaries of the lower Lancang River that provide critical habitat for migratory fish from the Mekong River (Yang et al. 2007; Kang et al. 2009b). Fish conservation measures, including establishing a fish reserve area and removing fishing facilities from the river, have also been implemented on these two important tributaries.

Other examples of the avoidance of adverse effects through design include the reduction of the dam height of Wunonglong, which was lowered to avoid impacts on the Three Parallel Rivers of Yunnan Protected Area, a UNESCO World Heritage site (Wang 2015). The dam height of the Gushui project was also reduced to avoid inundation of the salt fields in Naxi ethnic town, which are cultural relics with 1000-year-old history (Wang 2015). The function of the Ganlanba project was changed from electricity generation to that of a re-regulating reservoir to reduce water-level fluctuations caused by unsteady flow releases from the Jinghong project and thus improve navigation conditions on the Mekong River (Xinhua News 2011; Huang 2013). The installed capacity of the Ganlanba project was also reduced from 600 MW to 155 MW, which increased investment costs per kW to over four times the average cost per kW.

The dams on the middle and lower reaches of the Lancang River were constructed end to end (i.e., dams were constructed at the tail end of the reservoir of the dam immediately downstream). This resulted in a change in habitat conditions from the natural riverine condition (i.e., lotic flow) to a slow-flowing reservoir condition (i.e., lentic flow) (Sect. 2.3.1.1) and a subsequent dramatic decline in native fish species abundance (Sect. 2.3.2.1). In order to maintain more natural hydraulic conditions for fish habitat in the upper reaches of the Lancang River, the locations of the proposed dams were adjusted to increase the distance between dams so that some reaches with natural hydraulic conditions are retained. Under current plans, the lengths of river reaches with unaltered hydraulic conditions in Xizang (Tibet) and Yunnan provinces in the upper reaches of the Lancang River are 32.5 km and 77.9 km, respectively. The effects of these efforts on fish habitat and biodiversity will need to be verified with aquatic monitoring data. With the inclusion of the reach between the Ganlanba project and the China-Myanmar border (81 km), the total length of mainstem Lancang River reaches with unaltered hydraulic conditions will be approximately 190 km (Zhao and Li 2013).

3.3.1.1 Effectiveness of Existing Mitigation and Compensation Measures

Physical and Chemical Indicators

Hydrology

Hydropower project operations on the Lancang River change the magnitude, timing, duration, frequency, and rate of flow change (Chen et al. 2014; Lu et al. 2014). In accordance with minimum flow requirements set forth in the EIA, the Gongguoqiao and Xiaowan projects have released 168 m^3/s and 269 m^3/s, respectively, every dry season since commissioning to ensure portions of the mainstem channel downstream are not dewatered (MEP 2015a, b). These discharge releases in the dry season can be classified as fair and good, respectively, according to the Tennant method (Tennant 1976; Tharme 2003). The minimum flow releases for other projects were also set, and these can be considered generally acceptable following the implementation of mitigation. To improve navigation for large vessels downstream of Jinghong, variation in flow releases from the Jinghong project is also controlled at below 300 m^3/h so that water-level fluctuations at the Lancang-Mekong border are maintained under 1.0 m/h. However, this mitigation is the only operational measure addressing the timing, duration, frequency, and rate of flow change, and modeling results of the joint operation of Xiaowan and Nuozhadu indicated the frequency and duration of low and high flow events were significantly altered, which may disturb the life cycle of aquatic species (Chen et al. 2014; Li and Zhao 2016). Addressing these aspects of the flow regime through appropriate flow changes would improve the environmental performance of the hydropower projects.

Flow regulation by reservoirs on the Lancang River impacts the flow regime in the lower Mekong River. Chinese dam operations caused moderate changes to mean monthly flow at Chiang Saen (−15% to 9%) which is the nearest gauging station to the Chinese dams (Lu et al. 2014; Fig. 1.1). Meanwhile, the initial filling of hydropower reservoirs on the Lancang River was associated with moderate reductions in extreme low flows at Chiang Saen (Lu ct al. 2014). A comparison of the pre- and post-dam hydrology at Chiang Saen demonstrated modest changes in magnitude, timing, and duration (the water level increased in the dry season and decreased in the wet season compared to historical conditions), but the frequency and rate of water-level change were significantly altered by hydropower operations on the Lancang River (Cochrane et al. 2014). Similarly, studies have shown that hydropower projects on the Lancang River commissioned before 2010 caused only limited effects on flow magnitude on the lower Mekong (He et al. 2006; Hecht and Lacombe 2014), but the frequency and rate of water-level change should be given further consideration. Moreover, the operational effects of two large reservoirs (Xiaowan and Nuozhadu) after 2010 should be further examined.

On the positive side, flow regulation between flood and dry seasons by the Lancang River reservoirs has alleviated drought and improved navigation in the lower Mekong River. For example, historically large cargo ships could not pass through the Chiang Saen port in Thailand in the dry season due to low water levels. However, recently the situation has improved as increased water releases in the dry season have allowed large cargo ships to navigate normally (National News Bureau of Thailand 2015). In the spring of 2016, hydropower projects on the Lancang River also released additional water to alleviate severe drought in the lower Mekong River.

The occurrence of abnormally low and wet season flows in the lower Mekong River in the 1990s and 2000s triggered international debates regarding the magnitude and extent of the hydrological impacts of the mainstem Lancang dams. Recent studies indicate that there is no evidence that upstream reservoir storage or releases in China significantly altered the flood (e.g., 2008) or low flow (e.g., 1996–1997 and 2003–2004 dry seasons) events that triggered the debates. Instead, the flood and drought events in the lower Mekong River were attributed primarily to extreme climatic conditions and accelerating deforestation (Adamson et al. 2009; Campbell 2007; Fan et al. 2015; Lu and Siew 2006; MRC 2008).

Sediment

The mitigation of "storing clear and releasing muddy water," a reservoir operation method to store clear water after the flood season and release muddy water by lowering the water level of the reservoir at the start of the flood season, is applied by hydropower projects on the Lancang mainstem. This technique may reduce fine sediment aggradation in the reservoirs to a degree, but a considerable portion of large sediment still settles in the reservoirs. Fu and He (2007) estimated the multi-year average trapping efficiency (TE) at the Manwan HEP was 60.48%, and the trapping efficiencies of Gongguoqiao, Dachaoshan, and Jinghong dams were estimated at 30.23%, 66.05%, and 63.50%, respectively. Based on field data, it was estimated that 54.2% of the total volume of the Manwan Reservoir was filled by sediment after 12 years of operation (Mei et al. 2006). However, the loss of storage was dominated by a loss of dead storage (70.7%), with only 11.6% of the active storage zone lost (Sect. 2.3.1.3). The bigger reservoirs of Xiaowan and Nuozhadu have trapping efficiencies as high as 92%, and the whole cascade of dams has a total theoretical trapping efficiency of 94% (Kummu and Varis 2007). As to whether sediment trapping within Lancang reservoirs has an impact along the Mekong River, Lu and Siew (2006) found that statistically significant sediment reductions occur only at the nearest gauging station below the dam (i.e., at Chiang Saen; Fig. 1.1). Areas located further downstream (e.g., Nong Khai) show less sensitivity to the operation of the Chinese dams, as sediment levels have remained stable or even increased in the post-dam period. Research results by Fu et al. (2007) and Liu et al. (2013) verified the above conclusions on transboundary sediment effects of dam construction on the Lancang River. Based on the sediment trapping efficiencies and the field data available, the mitigation employed to limit sediment trapping can only be viewed as partially effective, and greater attention should be placed on ensuring adverse effects of sediment trapping are not transmitted downstream.

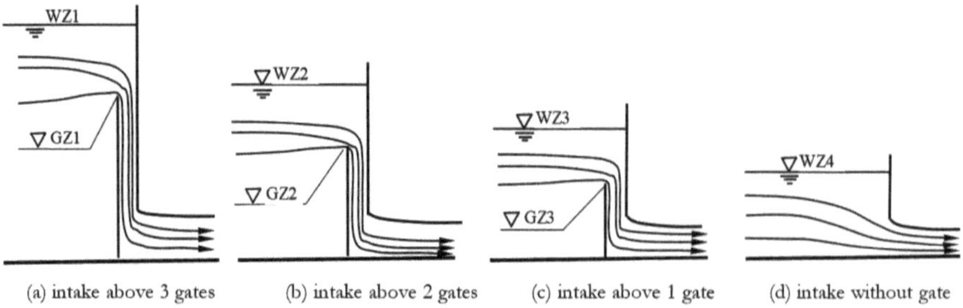

Fig. 3.1 Selective water intake structure installed at the Nuozhadu hydropower project. (**a**) Intake above three gates. (**b**) Intake above two gates. (**c**) Intake above one gate. (**d**) Intake without gate

Water Quality

The damming of rivers has the potential to reduce water quality as a result of reduced oxygenation and dilution of pollutants by relatively stagnant flow, flooding of biomass, and reservoir stratification. The hydropower developer and government improved pollutant control in the watershed of the reservoirs after the construction of the dams. So, despite the presence of reservoirs along the Lancang River, the Yunnan Province Environmental Quality Communique of 2014 showed that water quality on the mainstem (including the gauge at the Chinese border) was good (YEPD 2014), indicating water pollution control was effective and there was no water quality degradation on the mainstem of the Lancang River following hydropower development. Furthermore, the joint operation of the Manwan and Dachaoshan dams had cumulative positive impacts on water quality in the river reach downstream of Dachaoshan (Dachaoshan-Jinghong) but no impacts on the reach further downstream (Jinghong-Ganlanba) (Wei et al. 2009). These results demonstrate that low industry and domestic pollution loads in the Lancang River Basin mean that water quality can be maintained in good status following hydropower development.

Water Temperature

Stratification refers to the formation of layers of water with different temperatures in a lake or reservoir, with the bottom layer tending to be colder than the surface layer. In reservoirs, the temperature differences can be considerable, and the lack of mixing between water layers results in high oxygen consumption at the bottom of the reservoir (Elçi 2008; Thornton et al. 1996). Monitoring data at the Xiaowan project indicated that its reservoir is steadily stratifying and that the water temperature of released flow is 0–5.2 °C lower than normal in summer and 0.8–1.2 °C higher than normal in winter (MEP 2015b). To minimize the potential adverse effects of water releases with atypical temperatures, HydroLancang invested in a selective water intake structure (Fig. 3.1) at the Nuozhadu project, another deep-water reservoir, to minimize low water temperature releases in summer. According to numerical and physical modeling results, the multilevel withdraw structure will allow the temperature of water releases to be increased by up to 4.3 °C (Gao et al. 2013). However, monitoring data and reports are not currently available to assess the effectiveness of the structure.

Greenhouse Gases

Carbon dioxide and methane emissions are a water quality problem for hydropower reservoirs as a result of the decay of biomass at the reservoir bottom (Galy-Lacaux et al. 1997; Gile 2006). Due to the mitigation of clearing forests prior to reservoir filling, climatic characteristics, and topographical conditions in the Lancang River Basin, the greenhouse gas (GHG) emissions of the hydropower reservoirs on the Lancang River are quite low. Monitoring results indicate that Xiaowan, Manwan, and Nuozhadu emit 1.48 g, 8.40 g, and 3.20 gCO_{2eq} per kilowatt hour gross emissions, respectively. On a per kilowatt basis, the GHG emission values at these three reservoirs are considerably lower than that of many tropical reservoirs and are less than 1% of fossil fuel-fired generation facilities (CAE Project Team 2015).

Biological Indicators

Fish Community

The construction of dams on the mainstem Lancang River blocks fish migration and changes fish habitats from fast-flowing rivers to stagnant reservoirs. Biodiversity and populations of native fish in the middle and lower reaches of the Lancang River have undergone measurable declines, primarily due to hydropower development (Li et al. 2013; Kang et al. 2009a; Zheng et al. 2013, Sect. 2.3.2.1). Zheng et al. (2013) found that there were a total of 165 species of fish in the middle and lower Lancang River reaches. However, in recent years, only 71 of these species were captured, which shows that fish resources of the middle and lower Lancang River have been declining, especially the medium to large species and those that are rare or endemic. In the 4,787 fish samples collected on the Lancang River in 2009 and 2010 by Liu et al. (2011), 5 orders, 14 families, and 80 species of native fishes were identified. In past studies in the same reach, a further 59 species were recorded, indicating that historically there were a total of 139 species of fish in the Yunnan section of the Lancang River (Liu et al. 2011).

To compensate for these declines in fish populations, artificial fish hatchery and release programs are undertaken at the Nuozhadu and Gongguoqiao hydropower projects and will be implemented at the Huangdeng project (Huang 2013). Figure 3.2 shows the artificial hatchery and release program at the Nuozhadu hydropower project. Endangered and endemic fishes are artificially hatched at hatchery stations and then released to the Lancang River. The effects of the fish hatchery and release programs on fish populations in the Lancang River have not been monitored, so the effectiveness of these programs cannot be assessed.

Fig. 3.2 Artificial fish hatchery and release at Nuozhadu hydropower project (Left: photo by Xuezhong Yu; Right: photo by Agricture Bureau of Xishuangbanna)

Fig. 3.3 Dam removal and habitat restoration on the Jidu River. (Photo by Chengzhi Ding)

The Jidu River, a tributary of the upper Lancang, is an important spawning and rearing area for several endemic fish species. Four cascade hydropower projects were planned on the Jidu River, and the first one was completed in 2010. In order to reconnect the tributary to the mainstem Lancang River and compensate for fish habitat loss in the mainstem, HydroLancang spent 140 million RMB (approximately US$ 21.5 million) to purchase and remove the 12.6 MW hydropower project on the Jidu River in 2012 (Fig. 3.3). After dam removal, the number of fish species in the downstream reach (from the dam to 3 km downstream) increased significantly from 3 (2 endemic) in 2011 to 17 (10 endemic) in 2013, and the number of species 1 km upstream of the dam increased from 4 (1 endemic) in 2011 to 7 (2 endemic) in 2013 (Ding 2015).

Riparian Vegetation and Wildlife

Riparian vegetation loss and a decrease in the diversity of vegetation types have been observed in the middle and lower reaches of the Lancang River due to reservoir inundation and water-level variations (Li 2012). However, looking at changes

Fig. 3.4 Example of the transplanting of valuable and rare plants at the Nuozhadu hydropower project. (Photo by Xuezhong Yu)

over a larger spatial (20 km width on both sides of the river) and temporal scale indicates that vegetation cover decreased during dam construction periods but then increased during project operation (Liu et al. 2015). Decreases in the construction periods were due to construction activities, whereas increases during operation may be explained by regional changes in climatic conditions caused by the large reservoirs (Liu et al. 2015). To mitigate effects on riparian vegetation and associated wildlife, valuable and rare plant transplants and wildlife rescues were undertaken at hydropower projects on the Lancang River, with the species and number of individuals transplanted based on scientific investigation and assessment (Figs. 3.4 and 3.5). However, the effects of these measures on biodiversity and survival of transplanted species and rescued wildlife were not monitored, so the effectiveness of these measures cannot be assessed.

3.3.2 Existing Environmental Management Framework

Hydropower development and the associated environmental management processes in China can generally be divided into five phases:

1. Basin-wide hydropower planning, including a strategic environmental assessment (SEA)

Fig. 3.5 Wildlife rescue at the Nuozhadu hydropower project. (Photo by Xuezhong Yu)

2. Pre-feasibility studies, including a project-specific preliminary environment impact assessment (EIA)
3. Feasibility studies, including a project-specific EIA
4. Construction, including environmental mitigation design and implementation
5. Operation, including post-project environmental monitoring and impact assessment

The SEA conducted during the hydropower planning phase evaluates the potential environmental impacts of the river basin hydropower development plan and may propose modifications to the plan based on environmental constraints as noted in Sect. 3.3.1. The project EIA then evaluates the environmental impacts of a specific hydropower project and proposes measures to mitigate or compensate for adverse impacts.

In each phase, relevant technical documents are reviewed and approved before proceeding to the next phase. Environmental management agencies of the government will review and approve the SEA report, EIA report, and environmental mitigation designs. Environmental assessment and mitigation measures implemented at hydropower projects on the mainstem Lancang River have gradually improved over the last three decades. This is a result of increased environmental awareness and knowledge, increased concerns of government and the public, and the improvement of policy and standards for managing environmental impacts of hydropower projects.

The Environmental Protection Law of China requires the design, construction, and operation of appropriate mitigation measures to be synchronized with the design, construction, and operation of a new hydropower project. Once the construction of hydropower projects is completed, environmental inspection and approval by environmental administrations are required to ensure the prescribed mitigation measures were implemented. The environmental assessment of hydropower project on major rivers or with an installed capacity of over 250 MW are reviewed and approved by the Ministry of Environmental Protection (MEP). The Lancang River is one of the major rivers in China, so the SEA, EIA, and inspection and acceptance of HEPs on the mainstem Lancang River were all reviewed and approved by the MEP. It ensured that the projects on the mainstem Lancang met high standards and complied strictly with relevant requirements.

To improve environmental management of hydropower project impacts, the MEP issued the environmental impact assessment guidelines for instream flow, low temperature release, and fish passage of hydro projects in 2006 (EIA Department of the MEP 2006). In 2011, to enable ecological protection during hydropower development, the MEP proposed the principle of "ecological protection as the first priority with overall consideration, appropriate development and ensuring the bottom line" (Wu 2011). In 2012, the MEP issued the policy "Further Enhance the Environmental Protection of Hydropower Development." The initiatives of fish habitat compensation at tributaries, fish hatcheries and release, fish passages, and operational improvement for environmental purposes were highlighted in these specific requirements (MEP 2012).

The improvement in environmental protection measures is demonstrated by comparing the level of mitigation implemented at dams on the Lancang River mainstem over time. Mitigation measures at the Manwan project, the first dam on the Lancang commissioned in the 1980s, were relatively simple, whereas the environmental mitigation efforts at Jinghong in the 2000s were significantly improved. Since 2010, HydroLancang has further improved mitigation measures for the conservation of fish habitat and communities, and a number of measures including fish habitat restoration and fish passage were proposed or implemented after 2010 at the Huangdeng, Miaowei, and Ganlanba projects.

Since 2016, post-project environmental impact assessments will be required to examine the actual impacts of new projects and the effectiveness of mitigation measures and then further improve or modify the mitigation measures. The MEP required post-project environmental impact assessments to be conducted for Gongguoqiao, Xiaowan, and Jinghong 3–5 years after project commissioning. HydroLancang is also planning to improve the existing monitoring works and to integrate them into a watershed environmental monitoring system to monitor ecosystem changes caused by the cascade of dams on the mainstem Lancang. Post-project monitoring and assessment will provide a better basis for understanding project-ecosystem interactions on the Lancang River and hence enable better protection of its ecosystem.

Prior to 2006, the environmental assessment of small hydropower projects was managed by the municipal environmental department in Yunnan Province. Compared to the higher and stricter requirements for hydropower projects on the mainstem Lancang, the environmental management of small projects on the tributaries was relatively low. It resulted in a variety of environmental problems including construction without environmental assessment, low standards for setting instream flows, and harmful alteration of fish habitat and fish communities. Since 2006, the environmental assessments of small projects are reviewed and approved by the Environmental Protection Department of Yunnan Province. In 2016, the Yunnan provincial government halted new development and project expansion of medium and small projects (installed capacity ≤ 250 MW) in Yunnan Province for economic and environmental considerations. The inconsistency between environmental management and protection requirements for large and small hydropower has resulted in relatively serious environmental consequences arising as a result of small hydropower development. This is a vital lesson that China can share with other countries developing environmental management processes to mitigate the adverse effects of hydropower development.

3.3.3 Existing Monitoring Programs

3.3.3.1 Existing Monitoring on the Lancang River

Existing environmental monitoring on the Lancang River Basin can be categorized as regular hydrologic monitoring (surveillance monitoring) and hydropower-related monitoring (surveillance, compliance, and effectiveness monitoring). The monitoring components and locations of these two categories are summarized below.

Regular Hydrometric Monitoring

Regular hydrologic monitoring is conducted by the state-owned water resources agency. Monitoring parameters include precipitation, evaporation, water level, discharge, sediment concentration, water temperature, and water quality. The gauge stations are located on the mainstem Lancang River and key tributaries. The stations on the Lancang mainstem are Changdu, Liutongjiang, Jiuzhou, Gajiu, Yunjinghong, and Guanlei. Changdu gauge station is located in the Tibet Autonomous Region, and the other four stations are located in Yunnan Province. Liutongjiang, Jiuzhou, Gajiu, and Yunjinghong stations are located upstream of Gongguoqiao, Xiaowan, Dachaoshan, and Jinghong hydropower projects, respectively. Guanglei station is located 22 km upstream of the China-Myanmar border. Most of the tributary hydrologic gauge stations are located on the first-order tributaries of the Lancang River. Major stations include Tangshang station on Yongchun River, Jinding station on Pijiang River, Xincheng on Yinjiang River, Diannan, Yangzhuangping, and Tiankou on Heihuijiang River, Daluo on Nanlan River, Xiaoheijiang on Xiaoheijiang River, Manan on Buyuan River, and Mansala on Nanla River.

Hydropower Monitoring Programs

Hydropower-related monitoring is developed and conducted by the HydroLancang Corporation. HydroLancang operates a hydrologic monitoring and forecast system to serve the operation of the cascade of hydropower projects on the Lancang River. The system is used to collect precipitation, water level, and discharge data at gauge stations to monitor and forecast flows to the reservoirs for electricity and flood operation. The gauge stations span from Nuozika in the upstream reach to Guanglei station in the downstream reach. There are 19 hydrological stations measuring discharge, 37 water-level stations, 150 precipitation stations, and 1 meteorological station. In addition to hydrometric monitoring, a variety of parameters are monitored at each hydropower project to assess compliance and the effectiveness of some of the implemented mitigation measures. Table 3.2 provides an example of the monitoring program at one of the Lancang dams (Nuozhadu).

Table 3.2 Summary of the environmental monitoring at the Nuozhadu hydropower project

Category	Parameters	Location	Frequency	Duration
Water quality	20 parameters in the national surface water quality standard	3 in the reservoir (outflow of the upstream dam, middle of the reservoir, and dam site) and 1 downstream of the dam; 3 vertical layers at the middle of the reservoir and dam site	3 times a year (flood, average, and low season); twice per season	5 years during operations
Bottom sediment	As, Pb, Cd, Cu, Zn, and organic matter	2 in the reservoir (middle of the reservoir and at the dam site)	Once every 3–5 years	Not indicated
Water temperature	Water temperature	3 in the reservoir (outflow of the upstream dam, middle of the reservoir, and dam site) and 1 downstream of the dam; vertical temperature at the middle of the reservoir and dam site	Daily at outflow of the upstream dam and downstream of the dam; once a month at middle of the reservoir and dam site during first 5 years of operation, quarterly after 5 years operation	Until the change and distribution of water temperature are clear
Terrestrial plants	Phytophenology, endangered plants, and vegetation cover	Reservoir bank close to the dam and along the access road	Comprehensive survey after 3 years operation; annual for individual survey	5 years during operations

(continued)

Table 3.2 (continued)

Category	Parameters	Location	Frequency	Duration
Wild animals	Asian elephant, banteng, birds, amphibians, and reptiles	Reservoir bank close to the dam, access road, and nature reserves	Comprehensive survey after 3 years operation; annual for individual survey	5 years during operations
Zooplankton and benthic invertebrates	Number of individuals and biomass; family richness; family dominance; community structure	8 on the mainstem (4 in the reservoir and 4 downstream of the dam); 7 on key tributaries	Once a year	10 years during operations
Fish community	Number of species, density, and biomass	same as above	same as above	10 years during operations
Local climate	Temperature, humidity, precipitation, evaporation, air pressure, wind speed and direction, solar irradiance, cloud cover, frost days, fog days	3 meteorological stations (upstream of dam, middle of the reservoir, and dam site)	Daily and three times a day	Life of project
Bank stability and earthquake	Bank stability and reservoir-induced earthquake	Seismograph station close to the dam	not indicated	not indicated
Soil loss	Precipitation, sheet erosion, rill erosion, erosion modulus	Construction areas	not indicated	not indicated

Evaluation of Existing Monitoring Programs

The existing operational monitoring programs on the Lancang River were developed on a project by project basis. Given that there are now a number of large dams operating as a cascade of hydropower projects, the monitoring programs should be reviewed and updated to monitor effects not just at the project scale, but also at the watershed scale. The monitoring components, locations, and parameters should be updated to reflect this change of scope. Furthermore, existing monitoring programs were developed at different times and with different levels of technical requirements. The monitoring plan for each project should be analyzed and updated to the same level of technical requirements. The operational monitoring plan of the Nuozhadu project was developed in recent years, and the technical requirements are clearly higher than that of the Manwan project, which was commissioned 30 years ago. The components and parameters in Nuozhadu's monitoring program generally cover the environmental components and parameters that are impacted by project construction and operation. Moreover, the spatial range covers the mainstem and key tributaries impacted by the project. In contrast, the Manwan project was developed 30 years ago, and the environmental monitoring program is relatively simple with limited monitoring of the fish community. Manwan's monitoring program has not been updated to reflect a number of environmental quality standards that have been updated since its development.

Environmental components on the mainstem Lancang River are the focus of existing hydropower monitoring programs. However, as described in Sect. 2.3, the development of a cascade of dams on the mainstem Lancang River has drastically altered the fish habitat (i.e., from lotic to lentic environment) available and thus the diversity and distribution of fish species. A number of fish species have disappeared from the mainstem but remain in the tributaries. For example, two tributaries downstream of the Jinghong dam are critical habitat for migratory fishes from the Mekong River. A watershed level monitoring plan should be established and include monitoring of these two important tributaries, as well as others that are free-flowing and provide the type of fish habitat that is no longer available in the middle and lower reaches of the Lancang River mainstem.

Effectiveness monitoring should also be enhanced. The review and analysis of environmental mitigation measures in Sect. 3.3.1.1 demonstrated that data were not available to assess the effectiveness of a number of mitigation efforts. Hydropower developers have invested heavily in fish conservation measures, such as fish hatchery and release, and fish netting programs downstream of dams and the subsequent transport and release of captured fish in suitable tributary habitat. The effectiveness of these mitigation measures should be assessed by monitoring fish abundance, the distribution of fish catches, growth rates, and/or applying fish tagging techniques. Fish passage facilities are also under design or construction at hydropower projects on the Lancang River, and the effectiveness of these mitigation measures should be assessed through targeted monitoring programs.

Currently the national and provincial hydrologic gauges on the Lancang River are operated and managed by the provincial water resources agencies along the Lancang River, whereas the environmental monitoring programs associated with the hydropower projects are conducted and managed by each project. Since there is no integrated management of the environmental monitoring programs, it is not possible to monitor and manage environmental compliance and effectiveness at the watershed level. This is exacerbated by the generally weak analysis, reporting, and communication of the results of the environmental monitoring programs. A portion of the data and information collected by the environmental monitoring programs are available in the environmental inspection and approval reports, which are developed after 3 years of operation. However, after this there is no regular assessment and disclosure of the monitoring results. As the hydropower developer and operator, HydroLancang should establish a program to assess and report on the monitoring results to enable adaptive management of project operations and the environmental mitigation measures employed. The monitoring and assessment reports should also be disclosed to the public domestically and internationally.

3.3.3.2 Examples of International Monitoring Programs

For comparison with the existing monitoring of hydropower impacts conducted on the Lancang River, we reviewed several long-term monitoring programs established on other large river systems or those developed specifically to monitor the effects of hydropower development. The following three examples describe long-term monitoring programs or protocols in the USA, Australia, and Canada. The principles of these monitoring programs and protocols are used to develop the framework of a watershed scale monitoring program on the Lancang River (Sect. 3.4.3).

Monitoring Program on the Columbia River in the USA

In the USA, the National Oceanic and Atmospheric Administration (NOAA) Fisheries department is responsible for the stewardship of the nation's ocean resources and their habitat and the freshwater ecosystems that support productive and sustainable fisheries (e.g., Pacific salmon). To protect the endangered fisheries in the Columbia River Basin, NOAA Fisheries issued a biological opinion (BiOp) on the operation of the federally owned dams that make up the Federal Columbia River Power System (FCRPS) in May 2008. The FCRPS includes three US government agencies – the Bonneville Power Administration (BPA), the US Army Corps of Engineers (the Corps), and the Bureau of Reclamation (Reclamation). These agencies maximize the use of the Columbia River by generating power, protecting fish and wildlife, controlling floods, providing irrigation and navigation, and sustaining cultural resources. Based on the BiOp, the FCRPS developed and implemented the implementation plan for monitoring, analysis, and reporting with the objective to improve the survival of salmon and steelhead listed under the Endangered Species Act (ESA) (NOAA Fisheries 2008; Yu 2016).

As a component of the BiOp, research, monitoring, and evaluation (RM&E) provides information to support planning and adaptive management and demonstrate accountability related to the implementation of hydropower and off-site actions for all species. RM&E encompasses project implementation, compliance monitoring, fish status monitoring, action effectiveness research, and critical uncertainties research (FCRPS 2010). Based on the RM&E, the FCRPS regularly undertakes comprehensive and specific assessments to report on implementation progress, effectiveness, fish status, and environmental conditions.

South East Queensland Ecosystem Health Monitoring Program

The Ecosystem Health Monitoring Program (EHMP) delivers a regional assessment of the ambient ecosystem health for each of South East Queensland's 19 major catchments, 18 river estuaries, and Moreton Bay. The EHMP measures waterway health using a broad range of physical, chemical, and biological indicators of ecosystem health. These indicators provide essential information about the condition of South East Queensland's waterways. The monitoring program sits within a wider river health strategy known as the South East Queensland Healthy Waterways Partnership. The objective of monitoring is to annually observe and report on the spatial patterns of river health and to document any apparent trends in the data through time. These data are used within an adaptive management framework to maintain or improve river health (Gippel and Speed 2010).

The EHMP is managed by the South East Queensland Healthy Waterways Partnership on behalf of its various partners and is implemented by a large team of experts from the Queensland Government, universities, and the Commonwealth Scientific and Industrial Research Organisation (CSIRO). All major catchments have been monitored since 2003. Currently, 127 freshwater sites are monitored twice a year (in spring and autumn), and 248 estuarine and marine sites are monitored on a monthly basis. The results provide an assessment of the responses of aquatic ecosystems to human activities, such as catchment alterations and point source discharges (e.g., wastewater treatment plants), and also take into account natural processes such as rainfall. Monitoring values are compared against ecosystem health guideline values to derive standardized scores that are reported on annually.

Long-Term Aquatic Monitoring Protocols for Hydropower in Canada

In 2013, the Fisheries and Oceans Canada (DFO) released the long-term monitoring protocols for new and upgraded hydropower projects (Lewis et al. 2013). The protocols identify suitable methods to evaluate the effectiveness of mitigation and compensation activities undertaken during the development and operation of a project and to evaluate the project's effects on fish and fish habitat. Furthermore, this document is intended to promote standardized monitoring methodologies that will create consistency in the regulatory requirements of project proponents and allow for the comparison of data across multiple projects in order to evaluate environmental effects and generalize results across projects. Given the need for consistent monitoring over time, the document also details the requirements for baseline monitoring, which are necessary in order to complete an environmental impact assessment (EIA) to meet legislative and regulatory requirements under the Canadian Environmental Assessment Act and Fisheries Act. The geographic focus of this document is British Columbia and the Yukon territory, although it can be used elsewhere in Canada.

The protocol describes the parameters for hydropower projects situated within a stream channel and at the outlet of a lake or those intending to flood large areas to create new reservoirs. For each parameter, the protocol details the sampling location, number, methods, duration, frequency, timing, and comparison criteria.

3.4 Conclusions and Recommendations

3.4.1 Improving Environmental Mitigation and Compensation Measures

The effectiveness of environmental mitigation measures employed by hydropower projects on the Lancang River, as described in Sect. 3.3.1.1, can generally be classified into four types: effective, partially effective, not effective, and no data. The environmental mitigations for water quality and GHG emissions are effective. Water quality of the mainstem Lancang River remains good despite hydropower development and GHG emissions from the reservoirs are low relative to many tropical reservoirs. Mitigation measures played a positive role in controlling the negative impacts on water quality and GHG emissions, and other conditions (e.g., water pollutant loads, climatic conditions) also contributed to low environmental consequences. The minimum environmental flow releases partially reduce the adverse impacts of hydropower projects on natural flows, but the timing, duration, frequency, and rate of change in environmental flow releases have not been adequately addressed. Despite the implementation of mitigation measures, sediment trapping and changes to the mainstem fish community were dramatic. The mitigation and compensation measures employed to reduce the magnitude of these impacts, such as fish hatcheries and mainstem fish releases, and the conservation of tributary fish habitat, have not been evaluated and should be assessed as a matter of priority. Similarly, data are not available to understand the effectiveness of mitigation measures employed for water temperature, riparian vegetation, and wildlife.

The environmental mitigation measures implemented by hydropower projects in China lag behind international best practices and experience, capacity, and implementation should be improved, especially with respect to the conservation of fish habitat and communities (Yu 2016). The monitoring and assessment of a variety of fish conservation efforts on the Lancang River are poor. Environmental flow implementation should not only maintain minimum flows but should also seek to restore more features of the natural flow regime (e.g., the timing, frequency, and duration of flood flows). The standards and technical requirements for fish conservation measures should be specified, and the mitigation and compensation measures should be optimized at the watersed scale. The need for revised and/or additional mitigation and compensation measures to improve environmental performance should be evaluated and implemented for both new and existing projects. The assessment of river health and hydropower impacts (Sect. 2.3) identifies the environmental components that are in poor status and significantly impacted by hydropower projects. These components should be the focus of environmental mitigation or compensation. Based on the assessment results and analysis of existing environmental mitigation measures, mitigation improvements for hydrology, sediment transport and river geomorphology, connectivity, and fish community are recommended. Some recommendations are provided in the sections below, although these are not intended to be an exhaustive list of potential mitigation options.

The transboundary environmental effects of Chinese hydropower projects have become one of the focal points of public attention in the region and around the world, and Chap. 4 provides a summary of some transboundary effect studies completed to date. However, more in-depth studies on the transboundary environmental consequences of Chinese dams should be carried out, especially after the recent commissioning of Xiaowan and Nuozhadu dams. Measures for reducing flow fluctuations will be a focus for mitigating transboundary environmental impacts.

3.4.1.1 Physical and Chemical Indicators

Hydrology

Hydrological alteration caused by hydropower operation can be mitigated to a certain degree by modifying hydropower operations to implement appropriate environmental flow requirements. Such operational improvements have been implemented at a lot of dams in China, including the Three Gorges Dam. The regulatory agencies in China have also required hydropower owners to pay greater attention to environmental flow requirements in recent years.

Implementation of environmental flow on rivers affected by hydropower projects should consider both hydrological and hydraulics requirements. Dams on the middle and lower reaches of the Lancang River are located close together meaning the reservoir of the lower dam reaches the toe of the upstream dam. As a result, riverine hydraulic characteristics (depths and velocities) cannot be restored even if the upstream dam modifies its water releases to mimic natural flow discharges. The river reach where the implementation of environmental flows will improve flow conditions is therefore the reach downstream of the Jinghong project, and this could be achieved through integrated operation of the cascade of dams upstream of Jinghong.

Natural flow alteration by regulating discharge and short-term flow fluctuations caused by hydropeaking operations can result in significant hydrologic effects that need to be addressed. Large deviations from the natural flow regime can be addressed by releasing more water during the reservoir impoundment period and producing high flow pulses for aquatic communities. Short-term flow fluctuations downstream of Jinghong are expected to be reduced by regulation of the proposed Ganlanba dam. The Jinghong project is typically operated to meet peaking demand and thus produces rapid and large fluctuations of discharge and water level below the dam, despite the flow releases being controlled. The Ganlanba project has been designed as a re-regulation reservoir that will reduce the magnitude of flow fluctuations downstream of the Jinghong dam. When the Ganlanba project is operational, it will be possible to significantly reduce the downstream flow fluctuations while improving the electricity production capacity of the Jinghong dam.

Nevertheless, in general, the reoperation of hydropower projects to meet environmental requirements may be in conflict with electricity production. Electricity production will decrease if water releases from hydropower reservoirs mimic the natural process of river flow. Li and Zhao (2016) developed a cascade reservoir operation model to examine the trade-off between electricity generation and environmental flow releases and analyzed the cascade of Nuozhadu, Jinghong, and Ganlanba projects. The results show an obvious conflict between hydroelectricity generation and flow alteration – the more hydroelectric power generation, the more alteration of flow and vice versa. There is a tipping point in the plot showing the relationship between flow alteration and electricity generation (when flow alteration = 0.4; Li and Zhao (2016)) where the gain of electricity production can be made with only minor sacrifices in flow alteration. This makes this option attractive as a balance between the interests of electricity generation and maintenance of the natural flow regime. Li and Zhao's preliminary study should be expanded upon for engineering applications associated with the Lancang River cascade of dams.

Sediment and Geomorphology

Following commissioning of the Xiaowan and Nuozhadu projects, the sediment trapping and loss of storage in the Manwan, Dachaoshan, and Jinghong reservoirs were alleviated. River channel erosion in the middle and lower reaches of the Lancang River is not a big problem because of the coarse riverbed material. The focus of mitigation measures for sediment transport and river geomorphology should therefore be on reducing the adverse impacts of sediment trapping and the resulting decrease in sediment transport to the Mekong River. Sediment loads to the Mekong River are vital to transport nutrients that are are essential to ecosystem health and biodiversity along the river and in the delta.

We recommend hydropower project operators on the Lancang River continue to mitigate sediment trapping by flushing fine and large sediment from reservoirs through an operational drawdown program and by controlling sediment yield in the watershed and at dam construction sites through soil conservation measures.

3.4.1.2 Biological Indicators

Fish Community and Connectivity

The proximity of the dams in the middle and lower reaches of the Lancang River means that improving connectivity for fish in these reaches is not practical. The hydraulic conditions between the dams have been altered from lotic flow to lentic flow, and there is no remaining suitable habitat for endemic fish adapted to lotic flow on the mainstem. Nevertheless, trap and haul programs should be considered to maintain populations of migratory native and endemic fish species by transplanting these to suitable tributary habitat. Implementation of a trap and haul program, whereby migratory fish are attracted to a collection point at the base of the dam and transferred to suitable tributary habitat upstream, should help maintain viable native fish populations.

Considering the difficulties in restoring connectivity and habitat suitability in the mainstem Lancang River for native species adapted to lotic flow, habitat protection and restoration in the tributaries of the middle and lower reaches are critical for the conservation of the native fish community. Buyuan (Luosuo) River and Nanla River are two important tributaries for long-distance migratory fishes from the Mekong River, and they were protected by establishing fish reserves. In addition to these two tributaries, we identify another three tributaries in the middle and lower reaches for potential protection: Luozha River, Xiaohei River, and Weiyuan River. As a first step, the habitat conditions and fish community in these tributaries should be investigated, and then any existing issues for fish community conservation can be identified and the conservation plan developed accordingly.

3.4.2 Improving Environmental Management

The environmental management of hydropower development in China has evolved as environmental agencies and hydropower developers have realized that it isn't sustainable to ignore the environmental losses incurred over the last decade. The development concepts such as "protection giving way to development" and "protection lagging behind development" were dominant in the past and resulted in significant damage to the river ecosystem. From these environmental losses have emerged a number of valuable lessons. First, sound and tough laws relevant to hydropower development and environmental protection are fundamental for effective environmental management. Second, technical standards and guidelines relating to environmental assessment, design, implementation, and monitoring are vital. Last but not least, documenting complete procedures for environmental management of hydropower development may guarantee that the laws and standards set are complied with and implemented.

The technical and management experiences and lessons learned in China are beneficial to those Mekong countries with an interest in supporting social and economic development through hydropower development. In terms of technical practices, the SEA is critical for avoiding adverse impacts of hydropower projects before the construction of dams. Environmental administrations should deal carefully with the trade-offs between hydropower requirements and environmental and socioeconomic effects identified in the SEAs and EIAs for hydropower development. Environmental mitigation measures may reduce or compensate for negative impacts if designed well, but the costs of these measures may be expensive. Moreover, not all effects can be effectively mitigated. The degradation of fish biodiversity on the Lancang River is the most prominent environmental consequence of the cascade of hydropower dams due to habitat and flow regime changes and river fragmentation. Fish species are particularly important for Mekong countries given their significance to human livelihoods and biodiversity conservation. The impacts on fish communities should be carefully considered in hydropower planning at the river basin scale and in the design of environmental mitigation measures at the project scale. Post-project monitoring and assessment are critical for assessing the compliance and effectiveness of environmental mitigation measures and determining whether improvements to the mitigation measures are required based on actual responses of the river ecosystem.

3.4.3 Framework for a River Health Monitoring Program

The overall goal of a river health monitoring program is to provide information on the state and function of a river ecosystem. The development of this type of monitoring program is typically done at the watershed scale. However, as the Lancang River is a river system that has been significantly affected by hydropower projects, the framework we propose incorporates the collection of detailed monitoring data at a finer spatial scale to evaluate the effectiveness of mitigation measures undertaken during the development and operation of hydropower projects and to evaluate the projects' effects on the aquatic ecosystem. In this manner, the monitoring program will provide information to hydropower developers and government agencies to assist them in identifying suitable measures to protect and enhance river health despite the impacts of hydropower projects, while also providing information that can be used to understand river health at the watershed scale.

The development of a river health monitoring program for a large river such as the Lancang River is complicated and expensive. Detailed information on existing monitoring efforts should be collected and analyzed. Extensive monitoring data should be compiled and studied such that the removal, modification, or addition to existing monitoring components is based on existing data and the effectiveness of the existing monitoring programs. Field surveys are necessary to determine appropriate sampling locations with consideration of geographic conditions, monitoring requirements, and transport of samples. For these reasons, this study proposes a framework for a river health monitoring program (Fig. 3.6), rather than a comprehensive program. The framework outlines the objectives, monitoring types (as defined in Sect. 3.2), and components of the monitoring program. The reporting and assessment and public communication of the results are also discussed. The framework will serve as the basis for developing a complete monitoring program.

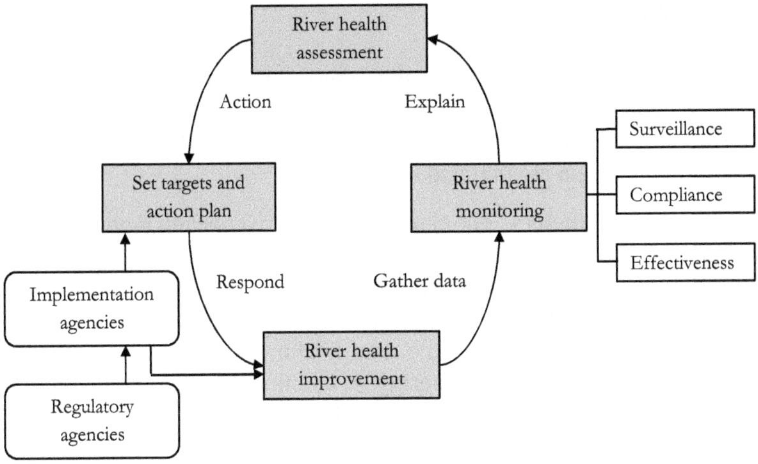

Fig. 3.6 Generic framework for river health monitoring in the Lancang River Basin

3.4.3.1 Objectives of the Monitoring Program

The river health monitoring program is intended to aid the government authorities and hydropower developers to evaluate, protect, and enhance river health in the Lancang River Basin. Specific objectives of the monitoring program include:

- Characterize current river health and trends in river health through time.
- Support tracking of responses, status, and trends of river health components as well as factors affecting them.
- Support evaluation and reporting on projects' compliance with statutory requirements and the effectiveness of actions in protecting, mitigating, and enhancing the basin's fish and wildlife resources.
- Facilitate adaptive management of factors affecting river health and provide sufficient information for making well-informed decisions.
- Facilitate reporting and sharing of monitoring and evaluation information with the public in an easily accessible and understandable manner.

Environmental monitoring of hydropower projects at the watershed scale is a requirement in the 13th Five-Year Plan for Hydropower Development (2016–2020) proposed by the National Energy Administration in November 2016. The plan requires the establishment of integrated ecological monitoring at a watershed scale that includes components of hydrology, sediment, environment, earthquake, dam safety, resettlement, and project benefits. The monitoring platform should be established as a real-time information sharing system to regulate and manage all processes in the river basin.

3.4.3.2 Monitoring Components

Based on the framework of river health assessment in this study, the indicators, parameters, monitoring scale, monitoring type, frequency, duration, and reporting for the river health monitoring program are listed in Table 3.3. The components of the monitoring program are outlined below.

Indicators and Parameters The categories of the river health monitoring program are the same as the river health assessment conducted in this study: physical and chemical, biological, and social indicators. The indicators included in the monitoring program are generally consistent with the assessment indicators. However, a number of indicators absent from the river health assessment are included in the monitoring program because monitoring of additional, complementary indicators aids in the interpretation of results and the development of appropriate mitigation measures, if required. For example, water quality is a direct indicator of the quality of flow, but we recommend riverbed sediment pollution monitoring as well because this influences water quality.

Monitoring Scale The monitoring program monitors river health at the watershed scale, and each indicator is monitored at three spatial scales: project, mainstem river reach, and watershed. As shown in Table 3.3, most of the indicators are monitored at the project and watershed scales. These indicators should be monitored at each project on the mainstem and in major tributaries. For instance, water flow should be monitored upstream and downstream of each dam to measure the variation in the flow regime caused by the dam's operations. Meanwhile, river flow should also be monitored in major tributaries to evaluate changes in the flow regime together with the monitoring data on the mainstem and to determine suitable areas to conserve fish habitat and/or transport and release migratory fish. Water temperature and sediment transport/river geomorphology are monitored at the scale of the project and the mainstem because the variation in these two indicators occur on the mainstem and are caused by hydropower project operation. Water supply should be monitored at the watershed scale, whereas effects on navigation should be monitored in the mainstem. Monitoring at different spatial scales allows the monitoring program to serve dual purposes: (1) to support an evaluation of projects' compliance with statutory requirements and the effectiveness of actions in mitigating adverse effects and (2) to characterize river health and trends in river health through time.

Monitoring Type Surveillance monitoring is applied to all of the indicators to monitor and assess the status of river health and trends over time. The monitoring of the indicators that are directly or indirectly affected by hydropower projects, including water flow, water temperature, water quality, sediment and river geomorphology, zooplankton and benthic invertebrates, fish community, riparian habitat, flood, and navigation, is also categorized as compliance and effectiveness monitoring and thus monitored at the project scale. Monitoring of sediment pollution and water supply is only classified as surveillance monitoring because the impacts of hydropower

Table 3.3 River health monitoring parameters and requirements in the Lancang River Basin

Category	Indicator	Parameters	Monitoring Scale	Monitoring type	Frequency	Duration[a]	Reporting[b]
Physical and chemical indicators	Water flow	Stage, discharge	Projects and watershed	Surveillance/ compliance/ effectiveness	Continuous	Life of project	Annually
	Water temperature	Water temperature	Projects and mainstem	Surveillance/ compliance/ effectiveness	Continuous	Year 1 to 5	Annually
	Water quality	20 parameters in GB3838–2002[c]	Projects and watershed	Surveillance	3 times a year (flood, average, and low season); twice per season	Year 1 to 5	Annually
	Sediment pollution	As, Pb, Cd, Cu, Zn, and organic matter	Projects and mainstem	Surveillance	Annually	Year 1, 3, 5, and 10	Annually
	Sediment and river geomorphology	Sediment load, transect topography	Projects and mainstem	Surveillance/ compliance/ effectiveness	Continuous for sediment load; year 3 (or 5), and 10 for geomorphology	Life of project for sediment load; year 3 (or 5), and 10 for geomorphology	Annually
Biological indicators	Zooplankton and benthic invertebrates	Number of individuals and biomass; family richness and dominance; Simpson's diversity; community structure	Projects and watershed	Surveillance/ compliance/ effectiveness	Annually	Year 1, 3, 5, and 10	Annually
	Fish community	Number of species, density, biomass, and migration behavior	Projects and watershed	Surveillance/ compliance/ effectiveness	Annually	Year 1, 3, 5, and 10	Annually

(continued)

Table 3.3 (continued)

Category	Indicator	Parameters	Monitoring Scale	Monitoring type	Frequency	Duration[a]	Reporting[b]
	Riparian habitat	Phytophenology, endangered plants, vegetation cover, and species at risk	Projects and watershed	Surveillance/ compliance/ effectiveness	Once for footprint impact; annually in operation	Immediately post-construction for footprint impact; year 3 (or 5), and 10	Annually
Social indicators	Flood	Flood control capacity, flood loss, flood operation of the reservoirs	Projects and watershed	Surveillance/ compliance/ effectiveness	Annually	Life of project	Annually
	Water supply	Water supply index, drought area and duration	Watershed	Surveillance	Annually	Life of project	Annually
	Navigation	Navigable length, navigation conditions	Mainstem	Surveillance/ compliance/ effectiveness	Annually	Life of project	Annually
	Electricity	Electricity generation, installed capacity, hydroelectricity wastage	Projects and watershed	Surveillance/ effectiveness	Annually	Life of project	Annually
	Recreation	Percentage of tourism value added in GDP, tourism relating to river or hydropower	Projects and watershed	Surveillance/ effectiveness	Annually	Life of project	Annually

[a]Monitoring may be extended past the prerequisite duration if additional changes (in trend or magnitude) are found
[b]Non-compliance, emergency, or unusual occurrences must be reported as required by the regulatory agencies
[c]GB3838-2002 is the national environmental quality standards for surface water in China

projects on the indicators are not significant, and no important mitigation measures are implemented. In addition to compliance monitoring, monitoring of electricity generation and recreation are also categorized as effectiveness monitoring to illustrate the effectiveness of the project's operational and management regimes.

Frequency The frequency with which monitoring occurs is determined by the variable features of the parameters monitored and the complexity/cost of the monitoring work. Continuous monitoring is applied to those indicators that change frequently with time, and monitoring is easy and relatively inexpensive. Water level, discharge, and water temperature change frequently with time and can be monitored with inline devices, so continuous monitoring (e.g., hourly) is applied to the monitoring of water flow and water temperature. A longer time interval is appropriate for indicators that remain generally stable within the time interval, and sampling/testing is costly or time-consuming. Water quality monitoring in flood, average, and low flow seasons can reveal the water quality status and variation of the river. The biological indicators and sediment pollution remain relatively stable over the course of a year, so annual surveys are sufficient to document the status of these indicators. For the same reason, the social indicators are monitored annually. River geomorphology changes slowly, and the surveys are particularly expensive, so river geomorphology is surveyed in years 3 (or 5) and 10 after project commissioning.

Duration For the monitoring of project effects, or assessing the effectiveness of mitigation measures, the monitoring duration is determined by the length of time that the indicator takes to stabilize or the time it takes for the project impact or the effectiveness of the mitigation measure to be clear. For the monitoring of river health over time, monitoring should be extended over the long term to track changes in trend or magnitude. For example, water flow and sediment load are regular monitoring parameters of the gauging stations, so the monitoring will last for the life of the project. Social indicators will also be monitored for the life of the project because these indicators may vary over time. Water temperature and water quality are monitored for a period of 5 years following project commissioning as they are intended to assess the impacts of individual projects. Sediment pollution, zooplankton and benthic invertebrates, and fish community are monitored in year 1, 2, 3, 5, and 10 after the completion of construction. River geomorphology and riparian habitat are monitored in year 3 (or 5) and 10 after project commissioning, but the footprint impacts should be monitored immediately after the completion of construction.

Reporting For management and communication purposes, an annual report should be prepared that documents the findings and recommendations of the monitoring program. Some indicators such as river channel morphology and fish community are not monitored on an annual basis, in which case the results of this monitoring should be included in the relevant annual report. The report should describe the monitoring methods, results, and assessment of the river health components. Meanwhile, the report should detail the compliance and effectiveness of hydropower operation and the mitigation and compensation measures employed. Based on the monitoring results, the annual report should recommend potential improvements to hydropower operations, mitigation and compensation measures, and relevant management actions. The annual report can be used to guide hydropower operations and river health management and can also be used as a basis for communication and dialog with stakeholders in China and lower Mekong countries.

3.4.3.3 Management of the Monitoring Program

The proposed river health monitoring program should incorporate existing monitoring programs that are currently run by different institutions, including government agencies and hydropower project owners on the Lancang River. Major institutions relevant to river health management in the Lancang River Basin include:

1. Water resources monitoring and management departments of Qinghai Province, Xizang (Tibet) Autonomous Region, and Yunnan Province. Large hydropower projects are currently located in Yunnan Province, so the water resources departments of Yunnan Province currently play a more important role in management.
2. Environmental protection monitoring and management departments in the river basin. Departments of Yunnan Province are currently playing a more important role.
3. HydroLancang Corporation. Hydropower is a critical driver of Lancang River health, so the role of HydroLancang Corporation is unique in the management of the river health monitoring program. In addition to conducting monitoring works relevant to hydropower compliance and effectiveness, HydroLancang Corporation should play a critical role in the establishment and management of the river health monitoring program.

In this study, we recommend a framework for river health monitoring and management in the Lancang River Basin that is similar to the management mechanism in the Columbia River Basin in the USA. It is a two-level management and implementation structure. At the regulatory level, the provincial agencies (provincial water resources and environmental protection departments) and river basin agency (Changjiang Water Resources Commission of the Ministry of Water Resources) will guide the river health monitoring and management program by proposing the requirements and implementation plan. At the implementation level, HydroLancang Corporation and relevant technical agencies (e.g., provincial hydrologic bureau and environmental monitoring center) implement the monitoring program. Integrated monitoring and management should be achieved by these agencies through collaboration in conducting the monitoring, compiling and managing the monitoring results, reporting on the outcomes, and implementing management measures to improve river health.

References

Adamson, P.T., I.D. Rutherfurd, M.C. Peel, and I.A. Conlan. 2009. Chapter 4 - The hydrology of the Mekong River. In *The Mekong-Biophysical Environment of an International River Basin*, ed. I.C. Campbell, 53–76. San Diego: Academic Press.

Campbell, I.C. 2007. Perceptions, data, and river management: Lessons from the Mekong River. *Water Resources Research* 43 (2): W02407. https://doi.org/10.1029/2006WR005130.

Chen, X., J. Zhao, T. Zhao, X. Lei, and G. Ni. 2014. Effects of hydropower reservoir operation on natural flow regime and ecosystem – A case study of xiaowan and nuozhadu dams. *Journal of Hydroelectric Engineering* 33 (4): 36–43 (in Chinese).

Chinese Academy of Engineering Project Team (CAE Project Team). 2015. *Greenhouse gases emission of different electricity generation facilities in China*. Beijing: China Atomic Energy Press (in Chinese).

Cochrane, T.A., M.E. Arias, and T. Piman. 2014. Historical impact of water infrastructure on water levels of the Mekong River and the Tonle Sap System. *Hydrology and Earth System Sciences* 18 (11): 4529–4541.

Ding, C. 2015. *The effects of dam removal on fish assemblage, structure and spatial distributions in a tributary of the Lancang River, China*. WLE Mekong WikiSpace. https://wlemekong.wikispaces.com/file/view/The+Effects+of+Dam+Removal+on+Fish+Assemblage+on+Jidu+River.pdf. Accessed 16 Sept 2016.

EIA Department of the MEP. 2006. *Environmental impact assessment guidelines for instream flow, low temperature release, and fish passage of hydro projects*. Beijing: China Environmental Science Press.

Elçi, Ş. 2008. Effects of thermal stratification and mixing on reservoir water quality. *Limnology* 9: 135–142.

Fan, H., D. He, and H. Wang. 2015. Environmental consequences of damming the mainstream Lancang-Mekong River: A review. *Earth-Science Reviews* 146: 77–91.

FCRPS (Federal Columbia River Power System). 2010. *2010–2013 implementation plan*. Portland: Federal Columbia River Power System.

Fu, K., and D. He. 2007. Analysis and prediction of sediment trapping efficiencies of the reservoirs in the mainstream of the Lancang River. *Chinese Science Bulletin* 52 (Suppl 2): 134–140 (in Chinese).

Fu, K., D. He, Y. Chen, C. Ye, and Y. Li. 2007. Impacts of dam constructions on the annual distribution of sediment in Lancang- Mekong River Basin. *Acta Geographica Sinica* 62 (1): 14–21 (in Chinese).

Galy-Lacaux, C., R. Delmas, C. Jambert, J.F. Dumestre, L. Labroue, S. Richard, and P. Gosse. 1997. Gaseous emissions and oxygen consumption in hydroelectric dams: A case study in French Guyana. *Global Biogeochem Cycles* 11 (4): 471–483.

Gao, X., H. Chen, Y. Li, and M. Xu. 2013. Flow patter and effect of withdrawal water from stop log gate with multi-levels in hydropower station. *Journal of Tianjin University* (Science and Technology) 46 (10): 895–900 (in Chinese).

Gile, J. 2006. Methane quashes green credentials of hydropower. *Nature* 444 (7119): 524–525.

Gippel, C.J., and R. Speed. 2010. *River health assessment framework: Including monitoring, assessment and applications*. ACEDP Australia-China Environment Development Partnership, River Health and Environmental Flow in China. Brisbane: International WaterCentre.

Harrison, D., J.J. Opperman, and B. Richter. 2007. Can hydro power be sustainable? *Water Power & Dam Construction* 59 (2007): 22–25.

He, D., Y. Feng, S. Gan, and H. You. 2006. The cross-border hydrological effect of hydropower development along main stream of the Lancang River. *Chinese Science Bulletin* 51 (S1): 14–20 (in Chinese).

Hecht, J., and G. Lacombe. 2014. The effects of hydropower dams on the hydrology of the Mekong Basin. Statement of Knowledge of the WLE Mekong. http://wle-mekong.cgiar.org/download/

state-of-knowledge/hydrology-mekong-basin/Hydrology%20of%20the%20Mekong%20Basin%20_English_web%2016092014.pdf. Accessed 22 May 2015.

Huang, G. 2013. Practices in environmental protection of hydropower development in the Lancang River Basin. In *Proceeding of the Hydropower 2013 – CHINCOLD 2013 Annual Meeting and the 3rd International Symposium on Rockfill Dams*, Kunming, China (in Chinese).

Jin, X., et al. 1995. *Environment of Chinese Lakes*. Beijing: The Ocean Press (in Chinese).

Kang, B. 2013. What do the occurrences of migratory fishes in the Upper Mekong mean? *Ambio* 42 (2013): 877–880.

Kang, B., L. Perreit, Y. Li, and D. He. 2009a. Are the fish of the upper and lower Mekong interconnected? *Chinese Journal of Oceanology and Limnology* 27 (2): 400–407.

Kang, B., D. He, L. Perrett, H. Wang, W. Hu, W. Deng, and Y. Wu. 2009b. Fish and fisheries in the Upper Mekong: Current assessment of the fish community, threats and conservation. *Reviews in Fish Biology and Fisheries* 19: 465–480.

Kummu, M., and O. Varis. 2007. Sediment-related impacts due to upstream reservoir trapping, the Lower Mekong River. *Geomorphology* 85 (3–4): 275–293.

Ledec, G., and J.D. Quintero. 2013. *Good dams and bad dams: Environmental criteria for site selection of hydroelectric projects*, Latin America and Caribbean Region Sustainable Development Working Paper, 16. Washington, DC: World Bank.

Lewis, A., A.J. Harwood, C. Zyla, K.D. Ganshorn, and T. Hatfield. 2013. Long term aquatic monitoring protocols for new and upgraded hydroelectric projects. *DFO Canadian Science Advisory Secretariat Research Document* 2012/166: ix + 88p.

Li, J. 2012. *Ecological changes and relevant impacts in Manwan reservoir in recent 20 years*, Master dissertations. Yunnan University (in Chinese).

Li, D., and J. Zhao. 2016. Hydropower-ecologic benefits tradeoff analysis of Cascade reservoir operation. *Journal of Hydroelectric Engineering* 35 (2): 37–44 (in Chinese).

Li, J., S. Dong, M. Peng, Z. Yang, S. Liu, X. Li, and C. Zhao. 2013. Effects of damming on the biological integrity of fish assemblages in the middle Lancang-Mekong River basin. *Ecological Indicators* 34: 94–102.

Liu, M., D. Chen, X. Duan, K. Wang, and S. Liu. 2011. Ichthyofauna composition and distribution of fishes in Yunnan section of Lancang River. *Journal of Fishery Sciences of China* 18 (1): 156–170 (in Chinese).

Liu, C., Y. He, D.E. Walling, and J. Wang. 2013. Changes in the sediment load of the Lancang-Mekong River over the period 1965–2003. *Science China Technological Sciences* 56 (4): 843–853 (in Chinese).

Liu, S., N. An, S. Dong, H. Zhao, L. Deng, and C. Zhao. 2015. The effects of hydropower stations construction on vegetation dynamics based on NDVI: A case study of Cascade hydropower stations of Lancang River. *Mountain Research* 33 (1): 48–57 (in Chinese).

Lu, X., and R. Siew. 2006. Water discharge and sediment flux changes over the past decades in the lower Mekong River: Possible impacts of the Chinese dams. *Hydrology and Earth System Sciences* 10 (2): 181–195.

Lu, X., S. Li, M. Kummu, R. Padawangi, and J. Wang. 2014. Observed changes in the water flow at Chiang Saen in the Lower Mekong: Impacts of Chinese dams? *Quaternary International* 336: 145–157.

Mei, Z., C. He, and B. Liu. 2006. Analysis of sedimentation of Manwan Reservoir. *Proceedings of the International Symposium on Hydropower 2006*, Kunming, China (in Chinese).

MEP (Ministry of Environmental Protection of China). 2012. *Further enhancing the environmental protection of hydropower development*. http://www.mep.gov.cn/gkml/hbb/bgt/201201/t20120117_222665.htm. Accessed 16 Sept 2016 (in Chinese).

———. 2015a. *Inspection and acceptance of environmental protection of Gongguoqiao HEP*. http://www.mep.gov.cn/gkml/hbb/spwj1/201507/t20150721_306951.htm. Accessed 16 Feb 2016.

———. 2015b. Inspection and acceptance of environmental protection of Xiaowan HEP. http://www.mep.gov.cn/gkml/hbb/spwj1/201509/t20150928_310451.htm. Accessed 16 Feb 2016.

MRC (Mekong River Commission). 2008. *Flood situation report*, August. MRC technical paper No. 21. Vientiane: MRC.
National News Bureau of Thailand. 2015. *China released more water from the dams, hope to increase commercial barge traffic*. http://thainews.prd.go.th/website_th/news/news_detail/WNSOC5901210010013. Accessed 16 Feb 2016.
NOAA Fisheries (NOAA's National Marine Fisheries Service). 2008. *Biological opinion*. Seattle: NOAA's National Marine Fisheries Service (NOAA Fisheries) Northwest Region.
Suter, G.W., II. 1990. Endpoints for regional ecological risk assessments. *Environmental Management* 14: 9–23.
Tennant, D.L. 1976. Instream flow regimens for fish, wildlife, recreation and related environmental resources. *Fisheries* 1 (4): 6–10.
Tharme, R.E. 2003. A global perspective on environmental flow assessment: Emerging trends in the development and application of environmental flow methodologies for rivers. *River Research and Applications* 19 (5–6): 397–441.
Thornton, J., A. Steel, and W. Rast. 1996. Chapter 8 – Reservoirs in water quality assessments. In *A guide to use of biota, sediments and water in environmental monitoring*. 2nd ed. Cambridge: Cambridge University Press.
Wang, Y. 2015. Sustainable hydropower development on the Lancang River. http://www.hydropower.org.cn/showNewsDetail.asp?nsId=15588. Accessed 16 Feb 2016 (in Chinese).
Wei, G., Z. Yang, B. Cui, B. Li, H. Chen, J. Bai, and S. Dong. 2009. Impact of dam construction on water quality and water self-purification capacity of the Lancang River, China. *Water Resources Management* 23: 1763–1980.
Wu, X. 2011. Development of hydropower and protection of eco environment in China. http://english.mep.gov.cn/Ministers/Speeches/201109/t20110907_217061.shtml. Accessed 17 Apr 2017.
Xinhua News. 2011. Eco-environmental design initiatives for hydropower development on the Lancang River. http://www.yn.xinhuanet.com/newscenter/2011-05/12/content_22744101.htm. Accessed 16 Feb 2016 (in Chinese).
Yang, J., X. Chen, and Y. Chen. 2007. On the population status and migration of pangasiid catfishes in Lancangjiang River Basin, China. *Zoological Research* 1: 63–67 (in Chinese).
YEPD (Yunnan Environmental Protection Department). 2014. *Environmental quality communique of Yunnan Province*. Kunming: YEPD.
Yu, X. 2016. International practices and implications of integrated watershed management of fish conservation of hydropower development. *Water Power* 42 (4): 5–9 (in Chinese).
Yunnan News. 2012. Green initiatives of Nuozhadu hydropower project. http://www.yn.chinanews.com/pub/html/special/nuozhadu/. Accessed 16 Sept 2016 (in Chinese).
Zhao, Z., and Y. Li. 2013. Current pollution situation and pollution control in Yunnan section of upper Lancang River. *Environmental Science Survey* 32 (5): 68–70 (in Chinese).
Zheng, L., X. Chen, and J. Yang. 2013. Status and conservation of fishes in the middle and lower Lancangjiang River. *Zoological Research* 34 (6): 680–686 (in Chinese).

Chapter 4
Transboundary Environmental Effects of Hydropower: Hydrology

The Lancang River Basin in China is undergoing extensive hydropower development, and several large dams on the upper Lancang-Mekong River have become operational in the last 20 years. The Lancang-Mekong River flows through Myanmar, Laos, Thailand, Cambodia, and Vietnam after leaving China; thus impacts of the development of the Chinese Lancang cascade of hydropower projects may extend well beyond national boundaries. Transboundary effects may be both positive and negative. For example, although natural ecosystems have the potential to be negatively impacted, water management may alleviate some of the negative effects of drought or flood.

A number of studies have investigated the transboundary environmental effects of the cascade of dams on the mid-lower mainstem of the Lancang River (Lu and Siew 2006; Campbell 2007; Adamson et al. 2009; Lu et al. 2014; Fan et al. 2015). However, only short-term, aperiodic, and sporadic in situ observational data are available (Fan et al. 2015); thus, considerable uncertainties and knowledge gaps remain when evaluating the environmental impacts of the cascade of dams on the Lancang River. In particular, there are few studies of the transboundary effects of China's dams on the Lancang River following the construction of two projects with large reservoirs (i.e., Xiaowan and Nuozhadu).

The study of transboundary environmental effects of hydropower projects on the Lancang River was a key component of this book. Four key transboundary effects were considered: hydrology (discharge and water level), water temperature, sediment transport and geomorphology, and fish community. The approach used for assessing the four key transboundary effects differed. Some effects were investigated through data collection and analysis, some relied on literature review, and others used a combination of approaches.

To assess effects on hydrology, hydrological data from monitoring stations on the Lancang and Mekong rivers from 1960 to 2015 were divided into before and after periods (before and after dam construction), and a comparison of results for hydrological stations located at varying distances downstream was used to evaluate transboundary effects on water level and discharge. Results were also compared to

X. Yu et al., *Balancing River Health and Hydropower Requirements in the Lancang River Basin*, https://doi.org/10.1007/978-981-13-1565-7_4

those from other hydrological studies. Effects on water temperature were assessed by monitoring and/or simulating changes of water temperature within two reaches of the Lancang-Mekong River in China (between Nuozhadu dam and Jinghong dam and between Jinghong dam and the Guanlei gauge station), as well as at a hydrological station in northern Thailand, and linking observed changes in water temperature to dam presence. Analysis of the effects on sediment transport and channel geomorphology was conducted by compiling data on suspended sediment concentrations from the 1960s to 2010 and through a literature review pertinent to the assessment of changes in erosion (sediment yield), sediment transport, and channel geomorphology. Effects of hydropower development on the fish community of the Lancang-Mekong River, such as effects on biodiversity and direct and indirect effects on fish abundance, were investigated primarily through an analysis of information on fish occurrences and captures available from the literature. Conservation efforts directed at fish populations were also summarized.

4.1 Overview

Downstream effects of the Lancang cascade of hydropower projects on hydrology were investigated through comparisons of hydrological data before and after the hydropower projects were constructed. Hydrological data, specifically water level and discharge, were compiled into before and after periods (before and after dam construction and operation) from multiple hydrological stations in the Mekong River. These hydrological stations ranged from Chiang Saen in northern Thailand to Phnom Penh in Cambodia (Fig. 4.1).

The timing of construction of several major dams on the Lancang River defined the before and after periods in the assessment of transboundary effects on hydrology. Manwan, which was initiated in May 1986 and for which the first phase of construction was completed in June 1995, was the first hydropower project on the Lancang River (Fig. 1.2, Table 1.1). Three other dams were completed in the early years of the turn of the century: Dachaoshan in 2003, Jinghong in 2009, and Xiaowan by August 2010. The Nuozhadu project was the most recently completed; its first unit became operational in September 2012 and it became fully operational in June 2014. Thus, for the purposes of this analysis, the time period prior to construction of the Manwan dam (pre-1985) represents the baseline period, and the time period after completion of the Nuozhadu dam (post-2013) represents the after period.

Comparisons among time periods were used to analyze hydrological effects of the dams on water level and discharge, and comparisons among hydrological stations distributed north to south along the Lancang-Mekong River were used to evaluate potential transboundary effects. For both water level and discharge, analyses were conducted from summaries of monthly and daily data. Extreme hydrological conditions were also investigated to determine the influence of the Lancang hydropower projects and evaluate the impacts as well as potential benefits associated with

Fig. 4.1 Hydrological stations along the Mekong River

water regulation. This included investigation of the impact of hydropower water regulation on a flood event in 2013, as well as the effect of water regulation on extreme high and low flows.

Analysis of effects of dam construction on water level and discharge downstream of the dams required that data were also considered seasonally. Two seasons are recognized within the study area that differ by quantity of precipitation: the dry

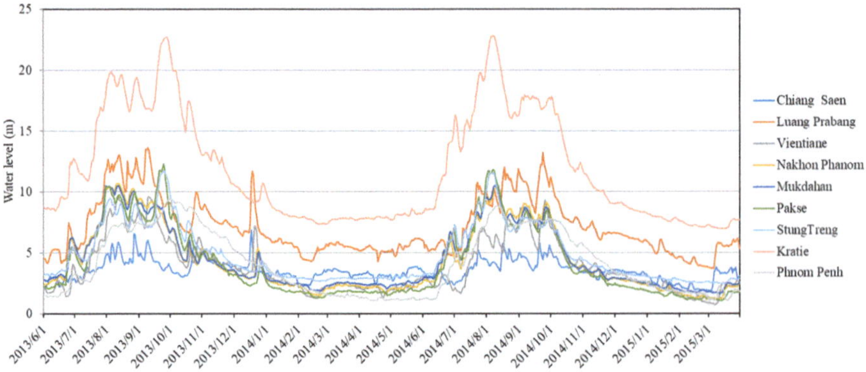

Fig. 4.2 Water level at nine hydrological stations (arranged north to south) in the lower Mekong River from June 2013 to March 2015

season, which extends from November through May, and the flood season (rainy season), which extends from June through October. Water level and discharge vary substantially between seasons, as shown for water level in Fig. 4.2 at nine hydrological stations from June 2013 to March 2015. Thus the impacts of hydropower projects on water level and discharge also have the potential to differ in accordance with the magnitude of seasonal water inputs.

4.2 Methods

Water level and discharge (flow) data as recorded at multiple hydrological stations were compiled from the Mekong River Commission (MRC) from January 1960 to December 2007 and from June 2013 to March 2015. From the hydrological stations distributed along the Mekong River (Fig. 4.1), a subset was selected for each analysis based on data availability and to provide a distribution of data from north to south (upstream to downstream) and thereby allow investigation of effects in relation to distance from dam construction (i.e., transboundary effects). Assessment of effects of the Lancang dams on water level and discharge was conducted by comparing, among hydrological stations, monthly, daily, and seasonal changes, as well as extreme hydrological processes (extreme water levels during a flood event, minimum and maximum flows).

According to the timing of construction and operation of the Manwan, Xiaowan, and Nuozhadu projects, the baseline and impact (post-project) periods were defined as:

- *Baseline period* – up to 1985, prior to the construction of Manwan project. The years for which data were available during this period varied by hydrological station.

- *Impact (post-project) period* – 2013 to 2015, following the start of operations of the Manwan, Xiaowan, and Nuozhadu projects.

4.3 Results and Discussion

4.3.1 Water Level

4.3.1.1 Seasonal Patterns

Monthly Water Levels

A comparison of the monthly average water levels at nine hydrological stations in the lower Mekong River between the baseline and impact periods indicated that water levels post-project were higher during the dry season (November–May) and lower during the flood season (June–October) than during the baseline period (Table 4.1). For example, at the Chiang Saen station (Thailand), which is furthest north (upstream), monthly average water levels were between 0.54 m and 2.02 m higher during the dry season in 2014 than during the baseline period. The effect was least pronounced (and reversed in 1 month (November 2013)), at the beginning of the dry season and most pronounced in March (2.02 m difference) and April (1.56 m difference). During the flood season, monthly average water levels were between 0.60 m and 2.26 m lower post-project (2013 and 2014) relative to the baseline period, with the most pronounced effect observed in July and August (between 1.68 m and 2.26 m difference).

The seasonal pattern observed at the Chiang Saen station is also evident at other hydrological stations; however, the pattern dissipates with distance downstream from the dams, and the speed of this dissipation differs by season (Table 4.1, Fig. 4.3). During the dry season, higher water levels during the impact period relative to the baseline period can be detected as far downstream as the Kratie station (Cambodia), whereas lower water levels during the flood season are not detectable south of Mukdahan (northeast Thailand). The latter can be explained by the influence of important tributaries, such as the Meng River and the 3S River, which provide considerable input to the Mekong River south of Mukdahan, and thereby dilute the stabilizing effect of hydropower projects during the flood season.

Daily Water Levels

Changes in daily water levels since construction of the dams on the Lancang River mirrored the seasonal patterns apparent from monthly averages. Daily water levels during the dry season were generally higher post-project than during the baseline period, and during the flood season, they were lower post-project than during the baseline period (Fig. 4.4). These patterns are evident in spite of substantial

Table 4.1 Monthly average water levels at nine hydrological stations (arranged north to south) in the lower Mekong River during baseline (up to 1985) and impact (2013–2015) periods

Hydrological station	Year(s)	Water level (m) by month											
		Jan	Feb	Mar	Apr	May	Jun	Jul	Aug	Sep	Oct	Nov	Dec
Chiang Saen	1961–1985	1.37	0.97	0.74	0.92	1.49	2.95	4.7	6.07	5.34	4.12	3.04	2
	2013	–	–	–	–	–	2.08	2.83	4.31	3.83	3.22	3.58	3.26
	2014	2.56	2.23	2.76	2.48	2.74	2.35	3.02	3.81	3.95	3.37	2.7	2.76
	2015	1.98	1.44	2.58	–	–	–	–	–	–	–	–	–
Luang Prabang	1961–1985	4.6	3.84	3.32	3.41	4.15	6.47	9.67	12.97	12.14	9.33	7.48	5.76
	2013	–	–	–	–	–	5.19	7.73	11.85	10.74	7.92	7.79	7.41
	2014	5.88	4.89	5.58	5.2	5.64	5.41	7.81	10.14	10.67	8.37	6.74	6.25
	2015	5.21	4.16	5.29	–	–	–	–	–	–	–	–	–
Vientiane	1961–1985	1.59	1.02	0.63	0.67	1.33	3.28	5.96	8.59	8.47	6.04	4.19	2.6
	2013	–	–	–	–	–	2.18	4.35	8.19	7.65	4.95	4.49	4.03
	2014	2.65	1.7	2.32	1.98	2.34	2.59	4.15	6.5	7.54	5.06	3.44	2.88
	2015	2.1	1.2	1.83	–	–	–	–	–	–	–	–	–
Nakhon Phanom	1973–1985	1.81	1.31	1	0.99	1.72	4.39	7.2	9.71	9.71	6.58	4.02	2.57
	2013	–	–	–	–	–	3.44	6.3	10.01	8.52	5.25	4.04	3.43
	2014	2.43	1.82	2.26	2.06	2.17	3.84	6.23	8.91	8.35	5.25	3.23	2.61
	2015	2.01	1.39	1.72	–	–	–	–	–	–	–	–	–
Mukdahan	1960–1985	1.87	1.51	1.29	1.23	1.72	4.4	7.18	9.74	9.53	6.31	3.89	2.59
	2013	–	–	–	–	–	3.6	6.24	9.81	8.41	5.3	4.08	3.41
	2014	2.56	2.11	2.47	2.32	2.49	4.19	6.59	8.83	7.97	5.11	3.28	2.71
	2015	2.21	1.8	2.06	–	–	–	–	–	–	–	–	–
Pakse	1960–1985	1.27	0.92	0.7	0.65	1.14	3.79	6.42	9.23	9.33	6.3	3.53	1.98
	2013	–	–	–	–	–	2.98	5.57	9.28	9.29	6.21	4.23	2.85
	2014	1.98	1.52	1.78	1.71	1.84	3.66	6.88	9.19	7.66	4.8	2.9	2.11
	2015	1.61	1.23	1.4	–	–	–	–	–	–	–	–	–
Stung Treng	1960–1985	2.54	2.22	1.99	1.91	2.36	4.23	6.24	8.37	8.55	6.48	4.34	3.11
	2013	–	–	–	–	–	3.74	5.72	8.46	9	7.03	4.91	3.75
	2014	3.09	2.79	2.88	2.9	3.02	4.36	7.36	8.8	7.55	5.48	3.81	3.2
	2015	2.86	2.6	2.63	–	–	–	–	–	–	–	–	–
Kratie	1960–1985	7.17	6.32	5.8	5.63	6.53	9.99	14.1	18.11	18.73	15.13	10.54	8.4
	2013	–	–	–	–	–	9.6	13.43	18.76	19.3	17	12.38	9.83
	2014	8.41	7.64	7.77	7.88	8.15	10.63	16.61	19.3	17.46	13.95	10.16	8.67
	2015	7.85	7.27	7.25	–	–	–	–	–	–	–	–	–
Phnom Penh	1960–1985	2.6	1.69	1.15	0.88	1.06	2.76	4.93	7.15	8.41	8.07	6.07	4.03
	2013	–	–	–	–	–	1.79	3.7	7.09	7.88	8.68	6.47	4.43
	2014	2.91	1.89	1.46	1.18	1.17	2.09	5.41	7.98	7.59	6.69	4.47	2.89
	2015	2.01	1.65	1.55	–	–	–	–	–	–	–	–	–

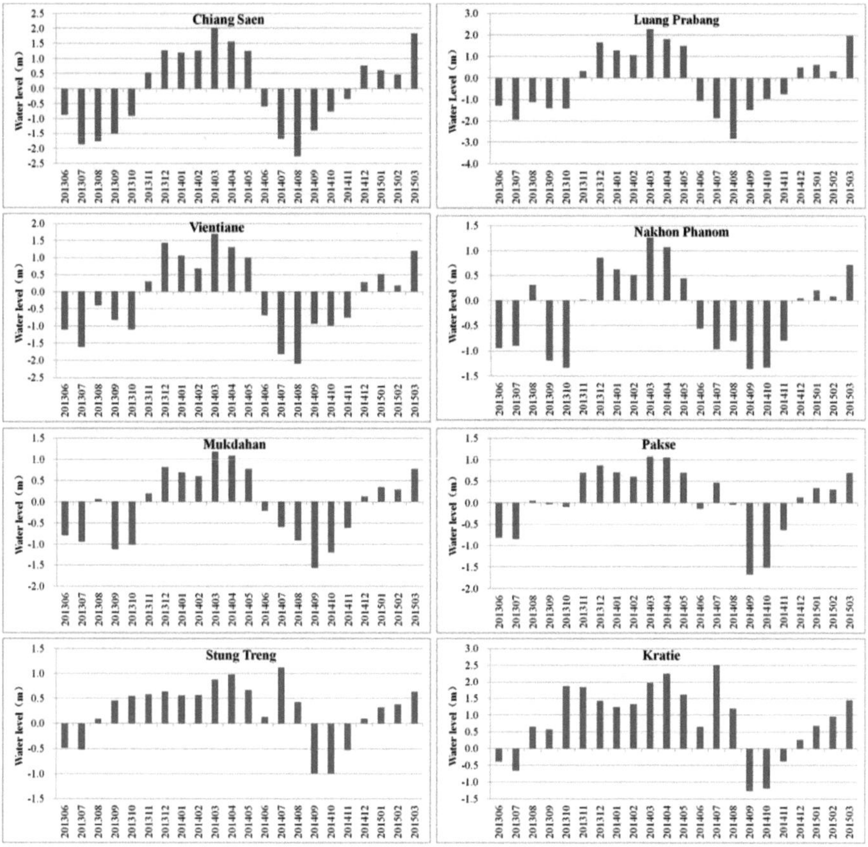

Fig. 4.3 Difference in water levels (positive and negative changes (arranged north to south)) for eight hydrological stations in the lower Mekong River by month between June 2013 and March 2015

inter-year variability resulting from precipitation conditions and differences in operational schedules of the hydropower projects. For example, daily water levels were particularly high during the 2014 dry season: from January to April, they were higher than the highest levels recorded during the baseline period. In contrast, water levels were relatively low during the 2015 dry season (January through March), generally remaining within the range observed during the baseline period.

As was also observed for monthly averages, the difference in daily water levels between periods was evident further downstream during the dry season than during the flood season. For example, at Kratie (the station furthest downstream for which data are presented in Fig. 4.4), daily water levels post-project exceeded all values in the baseline range from February through May (dry season), similar to Chiang Saen,

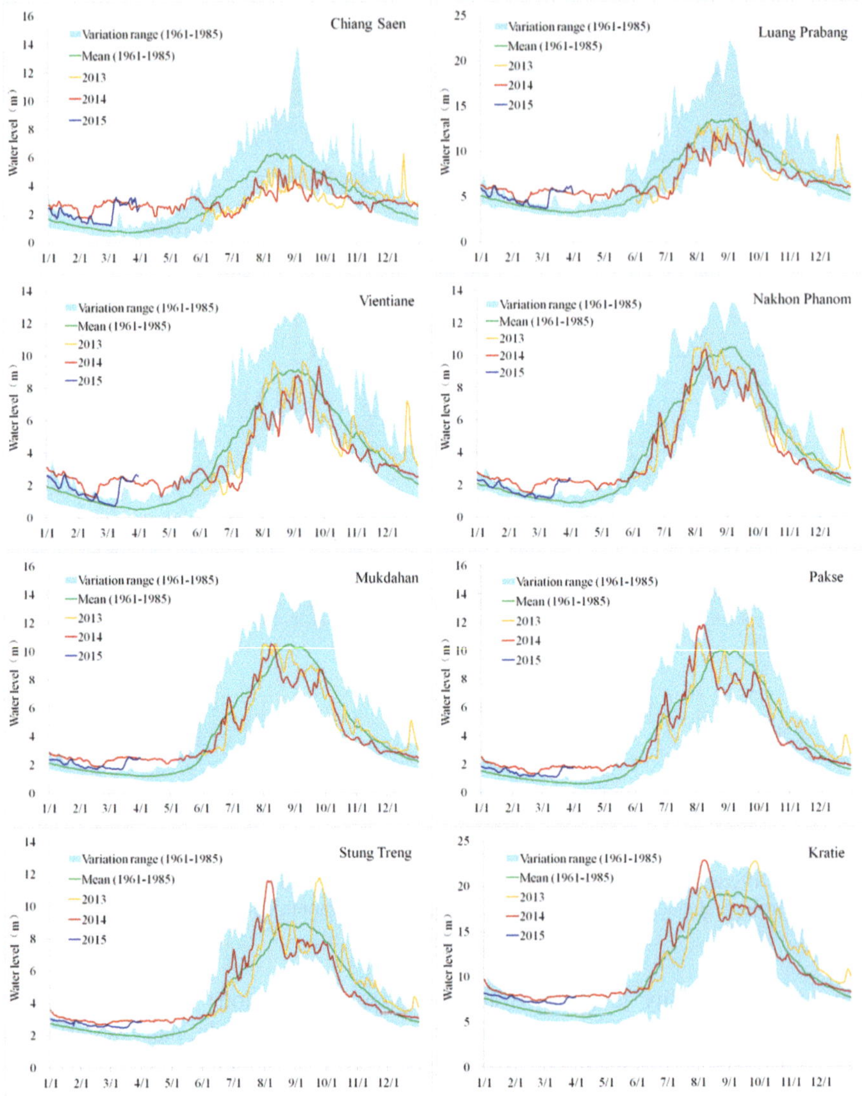

Fig. 4.4 Daily water levels at eight hydrological stations (arranged north to south) on the Mekong River in 2013–2015 compared to the mean and range of baseline (1961–1985) values through the annual cycle

the station furthest upstream. However, during the flood season, water levels at Kratie post-project generally remained within the baseline range, unlike those at Chiang Saen where post-project water levels are less than the baseline range in most cases. In general, during the flood season, the difference in daily water levels between baseline and post-project periods were similar from Chiang Saen to Mukdahan; however, the flood peaks downstream of Mukdahan post-project were more extreme than those in the upstream stations. The timing and intensity of these flood peaks demonstrate that at the southern stations, the influence of regional precipitation and inflow of major tributaries outweigh the stabilizing impacts of the hydropower projects of the Lancang River.

4.3.1.2 Extreme Water Level: Flood Event

A flood event during the dry season in December 2013, which resulted from a regional rainstorm, permitted an analysis of the effect of such an event on water levels during the post-project period. As determined from daily precipitation records from 11 key meteorological stations in the Lancang River Basin, precipitation leading to this event began in the downstream region of the Lancang River (represented by Lancang, Simao, Jinghong, and Mengla meteorological stations in Table 4.2) on December 13, 2013, and reached its maximum value on December 15. In accordance with China's national standard of precipitation intensity, in which total precipitation in 24 h between 25.0–49.9 mm, 50.0–99.9 mm, and 100.0–249.9 mm are categorized as heavy rain, rainstorm, and downpour, respectively, the daily precipitation in portions of the downstream Lancang River during this period reached rainstorm and downpour intensities. This intense rainstorm raised the water level in the lower Mekong River and allowed documentation of the timing of the flood pulse and change in water level with distance downstream.

Examination of water level and timing data at a number of hydrological stations in the Mekong River indicated that the surge in water level during the flood event propagated in a downstream direction over a period of 13 days (Table 4.3 and Fig. 4.5). The highest water level (flood peak) was detected at Chiang Saen on December 17, at Mukdahan on December 24, and at Phnom Penh on December 29. It is also apparent that the magnitude of the water level rise decreased with distance downstream: the rise was greater than 3 m between Chiang Saen and Paksane, greater than 2 m between Nakhon Phanom and Savannakhet, greater than 1.0 m for Khong Chiam and Pakse, and less than 1 m at Kampong Cham and Phnom Penh. Figure 4.5 clearly shows the attenuation of the December 2013 flood peak with distance downstream at eight hydrological stations. Although comparison to a flood event pre-project was not possible, this analysis provides an example of the dissipation of the effects of a severe flood event given the hydropower projects currently in operation.

Table 4.2 Daily precipitation (in mm) at 11 meteorological stations (arranged north to south) in the Lancang River Basin from December 12 to 17, 2013

Date	Daily precipitation (mm) by meteorological station										
	Zaduo	Nangqian	Chengdu	Deqin	Weixi	Dali	Gengma	Lancang	Simao	Jinghong	Mengla
12-Dec-2013	0	0	0	0	0	0	0	0	0	0	0
13-Dec-2013	0	0	0	0.1	0	0	0	0	0	0.5	7.3
14-Dec-2013	0	0	0.2	0	0	0.5	3.5	11.7	0	23.4	34.2
15-Dec-2013	0	0	0	0	1	17.8	1.9	33.5	56.5	91.1	149.4
16-Dec-2013	0	0	0	0	2.7	14.7	4.5	7.6	10.7	0	0.1
17-Dec-2013	0	0	0	0	0	0	0	0	0	0	0

Table 4.3 Timing and magnitude of water level rise associated with a flood event in December 2013 at 16 hydrological stations (arranged north to south) on the Mekong River

Hydrological station	Date of rise in 2013	Rise from the water level (m)	Date of maximum water level in 2013	Maximum water level (m)	Magnitude of rise (m)
Chiang Saen	14-Dec	3.84	17-Dec	6.87	3.03
Luang Prabang	15-Dec	6.66	18-Dec	11.72	5.06
Chiang Khan	17-Dec	6.96	20-Dec	10.68	3.72
Vientiane	17-Dec	3.44	20-Dec	7.20	3.76
Nong Khai	18-Dec	4.08	21-Dec	7.96	3.88
Paksane	19-Dec	5.30	22-Dec	8.50	3.20
Nakhon Phanom	19-Dec	2.98	23-Dec	5.43	2.45
Thakhek	20-Dec	4.30	24-Dec	6.42	2.12
Mukdahan	20-Dec	2.99	24-Dec	5.09	2.10
Savannakhet	20-Dec	1.89	24-Dec	4.00	2.11
Khong Chiam	21-Dec	3.47	25-Dec	5.30	1.83
Pakse	22-Dec	2.55	26-Dec	4.00	1.45
Stung Treng	23-Dec	3.51	27-Dec	4.37	0.86
Kratie	24-Dec	9.25	28-Dec	10.73	1.48
Kampong Cham	25-Dec	5.56	29-Dec	6.51	0.95
Phnom Penh	25-Dec	4.69	29-Dec	5.20	0.51

4.3.2 Discharge

4.3.2.1 Seasonal Patterns

Monthly Flow

Comparison of discharge before and after three large hydropower projects were constructed on the Lancang River indicates that the hydropower projects have a stabilizing effect in both dry and flood seasons such that naturally low discharge during the dry season is augmented, and that naturally high discharge during the flood season is reduced. A comparison between the baseline and impact periods of monthly average flow levels at six hydrological stations in the lower Mekong River indicated that flow post-project was generally higher during the dry season and lower during the flood season relative to the baseline period (Table 4.4, Figs. 4.6 and 4.7). Similar to results from the water level analysis (Sect. 4.3.1), this effect was most pronounced for hydrological stations furthest upstream. Using the northernmost station as an example (Chiang Saen), positive changes in discharge between baseline and impact periods were least pronounced at the beginning of the dry season and most pronounced in March (197.8% change) and April (138.3% change). Negative changes during the flood season were most pronounced in July (43% change) and August (52.4% change). The augmenting effect during the dry season,

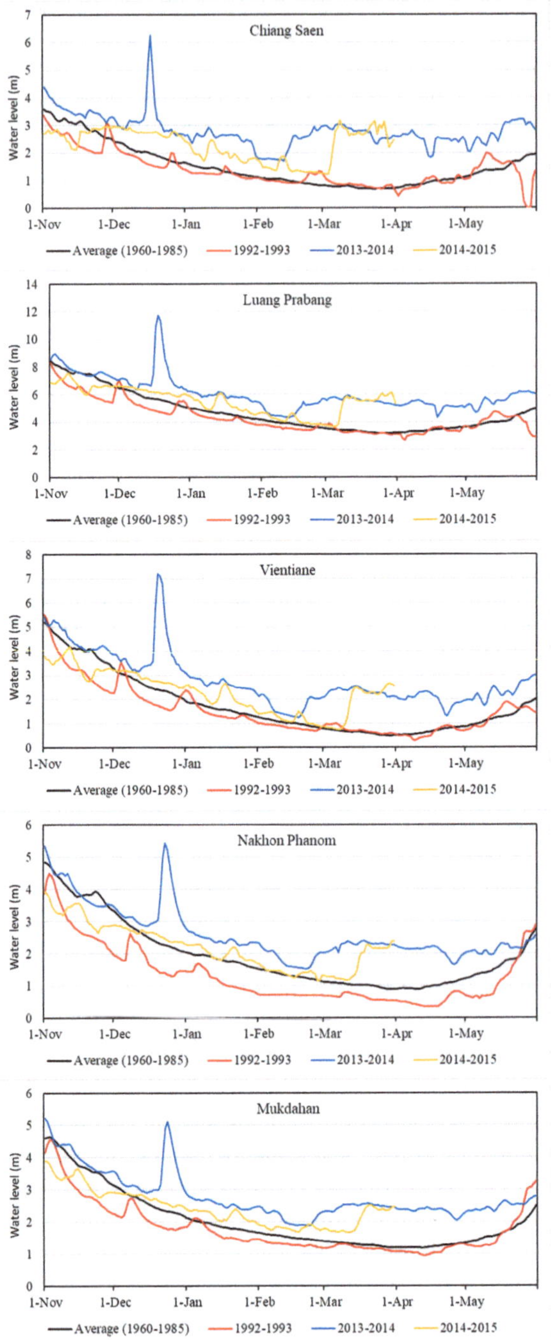

Fig. 4.5 Water level between November 2013 and March 2015 at nine hydrological stations (arranged north to south) on the Mekong River, providing a comparison by station of the magnitude and timing of the December 2013 flood pulse

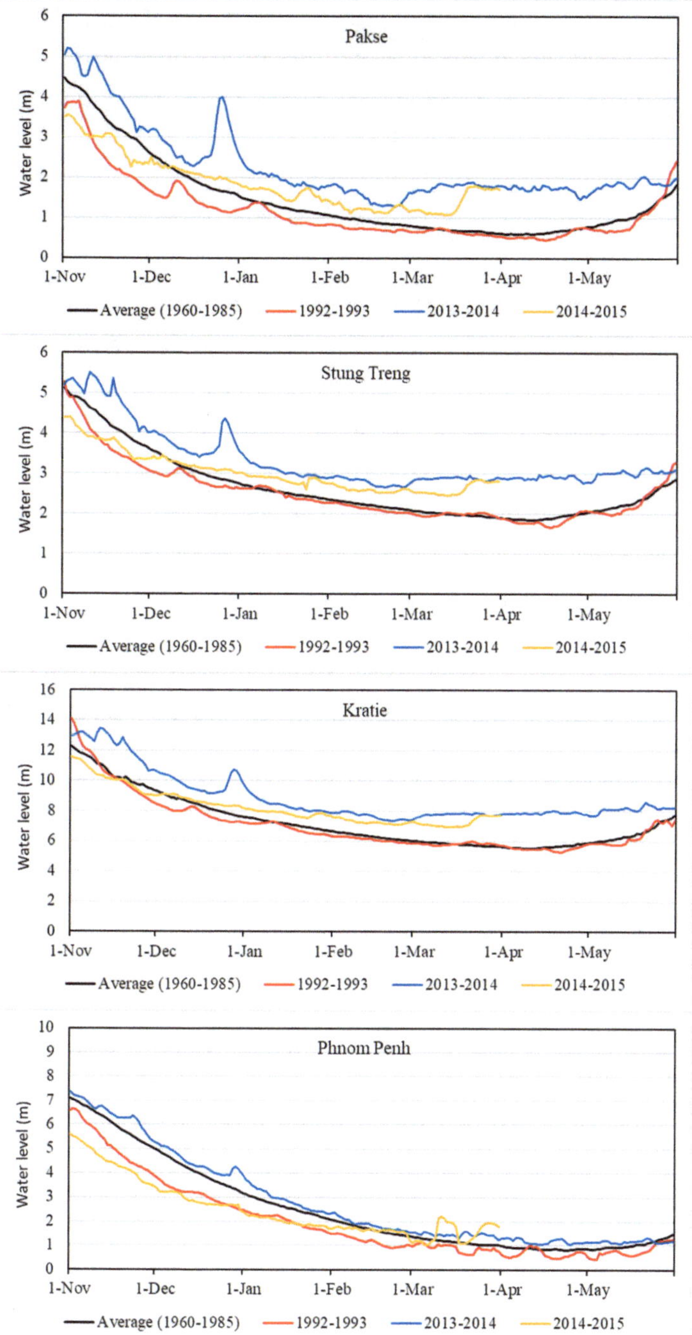

Fig. 4.5 (continued)

Table 4.4 Average monthly discharge at six hydrological stations (arranged north to south) and percent change during baseline (up to 1985) and impact (2013–2015) periods in the lower Mekong River

Hydrological station	Period[a]	Average monthly discharge (m^3/s)											
		Jan	Feb	Mar	Apr	May	Jun	Jul	Aug	Sep	Oct	Nov	Dec
Chiang Saen	Baseline	1146	922	800	885	1239	2507	4719	6934	5690	3819	2568	1611
	2014	2188	1907	2382	2109	2370	1960	2691	3302	3,80	2715	2042	2090
	Change (%)	90.9	106.8	197.8	138.3	91.3	−21.8	−43	−52.4	−38.8	−28.9	−20.5	29.7
Luang Prabang	Baseline	1762	1328	1073	1112	1538	3217	6404	10,511	9381	5900	4028	2561
	2014	–	–	2249	2021	2291	2185	4300	6961	7653	5004	3416	3021
	Change (%)	–	–	109.6	81.7	49	−32.1	−32.9	−33.8	−18.4	−15.2	−15.2	18
Vientiane	Baseline	1848	1438	1204	1215	1673	3517	7073	11,795	11,504	7020	4519	2754
	2014	2448	1749	2178	1937	2205	2430	3997	6635	8458	4915	3173	2696
	Change (%)	32.5	21.6	80.9	59.4	31.8	−30.9	−43.5	−43.7	−26.5	−30	−29.8	−2.1
Mukdahan	Baseline	2419	1893	1583	1515	2211	7098	13,613	21,158	20,400	11,323	5901	3546
	2014	3297	2605	3153	2917	3184	6661	12,386	19,376	16,688	9484	5431	4383
	Change (%)	36.3	37.6	99.2	92.5	44	−6.2	−9	−8.4	−18.2	−16.2	−8	23.6
Pakse	Baseline	2917	2236	1844	1764	2747	9099	16,661	27,027	27,248	16,203	8,135	4384
	2014	4007	3231	3642	3533	3747	8182	17,714	26,522	20,227	11,285	6137	4434
	Change (%)	37.4	44.5	97.5	100.3	36.4	−10.1	6.3	−1.9	−25.8	−30.4	−24.6	1.1
Stung Treng	Baseline	3499	2594	2061	1933	3106	10,737	21,300	35,691	36,722	22,610	10,730	5561
	2014	4878	3902	4147	4198	4595	10,517	28,207	36,975	27,379	15,475	7612	5437
	Change (%)	39.4	50.4	101.2	117.2	47.9	−2	32.4	3.6	−25.4	−31.6	−29.1	−2.2

[a]Baseline flows were calculated based on the average flow during 1960–1985, and the percentage of change was calculated by comparing the flow in 2014 to the previously mentioned natural flow

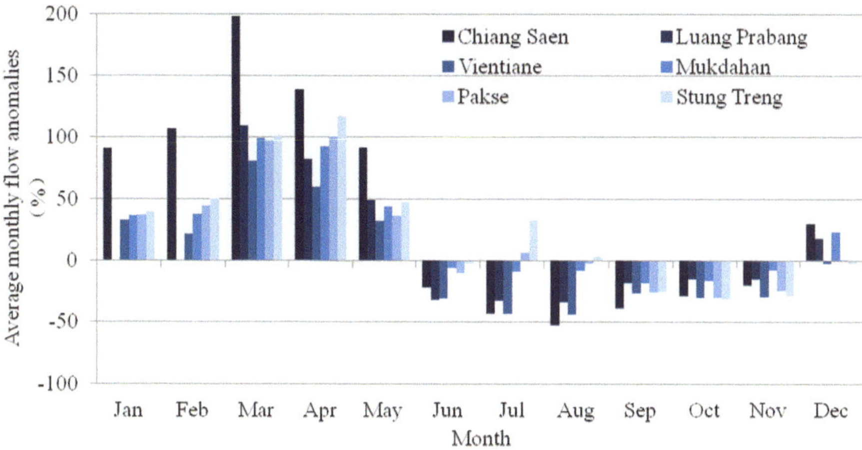

Fig. 4.6 Difference in discharge (positive and negative % change) between the baseline and impact periods for six hydrological stations in the lower Mekong River by month

when flows are low because runoff due to precipitation is reduced, is related to the operational regimes of the hydropower projects. Through water regulation of the large reservoirs associated with Xiaowan and Nuozhadu projects, there is an increase in discharge relative to baseline in the dry season resulting in a substantial increase in flow at locations to the south. Similarly, during the flood season, the reservoirs with high regulatory capacity on the Lancang River reduce discharge to the downstream reach, resulting in reduced flows downstream.

The seasonal difference between periods in the pattern observed at the Chiang Saen station is also evident at other hydrological stations; however, as also observed in the analysis of water levels (Sect. 4.3.1), the pattern dissipates with distance downstream, and the speed of this dissipation differs by season. Figure 4.7 demonstrates the difference in persistence downstream of the positive and negative changes during the dry and flood seasons, respectively. During the dry season, positive changes on discharge during the impact period are detectable as far south as Stung Treng, whereas during the flood season, detectability of negative changes becomes difficult south of Mukdahan.

The reason for the difference in the downstream persistence of the change in discharge post-project relative to baseline between the dry and flood seasons is related to the proportion of total discharge in the Mekong River that originates in the Lancang River. During the dry season, owing to the low levels of precipitation in the lower Mekong River Basin, the proportion of runoff from the Lancang River Basin (i.e., from the basin above the Chiang Saen hydrological station) relative to the total runoff of the Mekong River is large. In contrast, during the flood season, when precipitation levels in the Mekong River Basin are higher, the proportion of discharge in the Mekong River that originates in the Lancang River is smaller. Thus the positive change in discharge between baseline and impact periods during the dry season persists downstream, whereas in the flood season, when the discharge from the

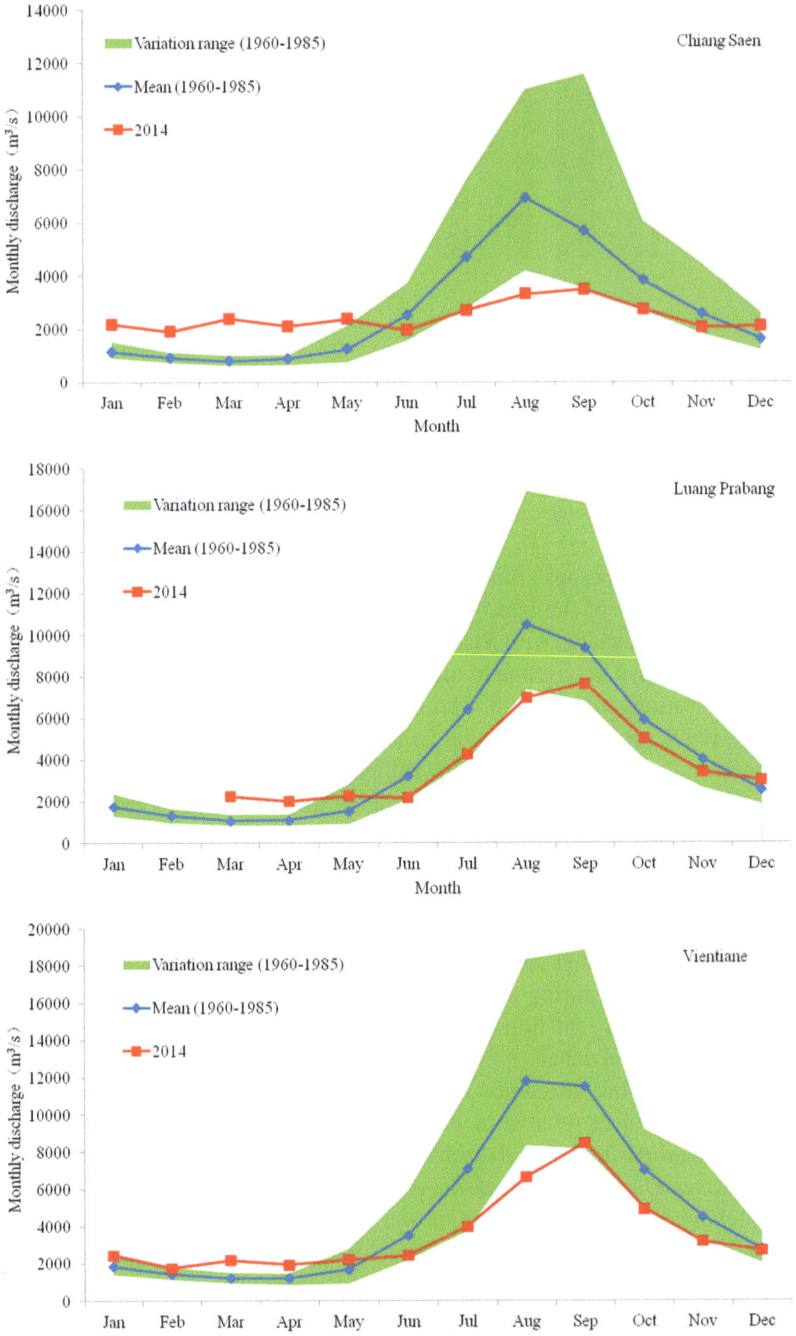

Fig. 4.7 Average monthly flow at six key hydrological stations (arranged north to south) of the lower Mekong River in 2014 in comparison to the mean and range of baseline values through the annual cycle

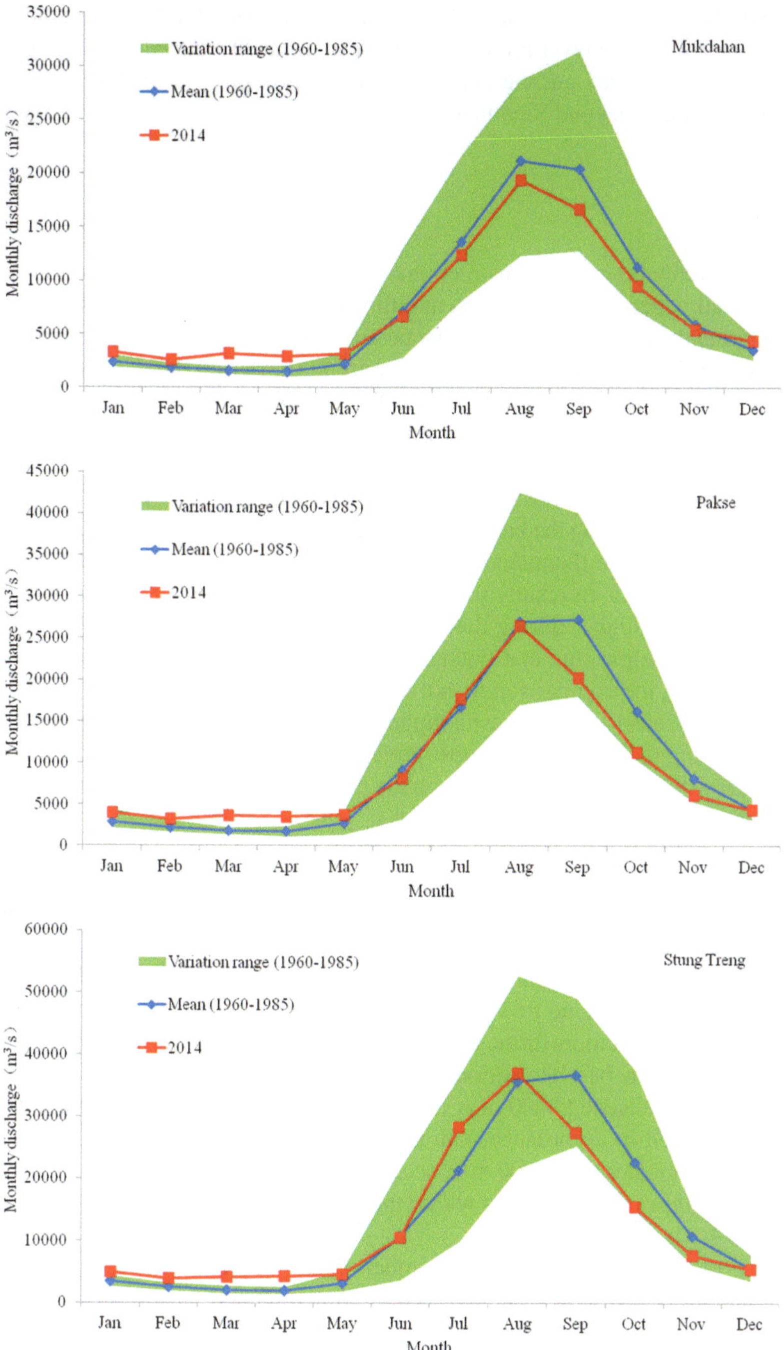

Fig. 4.7 (continued)

Lancang River represents a relatively small proportion of total discharge, the effect of the Lancang hydropower projects downstream is much reduced. During the flood season, impacts of the Lancang cascade of hydropower projects on discharge generally do not extend beyond Vientiane (in Laos).

Daily Flow

Changes in daily flow since construction of the dams on the Lancang River mirrored the seasonal patterns apparent from monthly averages. Figure 4.8 shows that daily flows during the dry season were higher post-project than during the baseline period, especially in March and April, and flows were lower during the flood season post-project compared to the baseline period. The dissipation of these effects with distance downstream is also apparent in Fig. 4.8, as is the difference in the persistence of the effect between the dry and flood seasons.

Comparison of 2014 post-project flows to extreme (1975, 1985) and normal (1963) flow years during the baseline period at the Chiang Saen hydrological station were used to evaluate the magnitude of the effect of hydropower projects on flow downstream. Similar to results from previous analyses, Fig. 4.9 shows that post-project flows are higher during the dry season, and lower during the flood season compared to baseline, and that fluctuations throughout the annual cycle are dampened. However, the comparison to extreme years also indicates that during the dry season, post-project flows are even higher than those in a baseline high flow year (1985). During the flood season, post-project flows tend to be even lower than in a baseline low flow year (1975), although the effect in the flood season is weaker than in the dry season.

The effect of distance from the Lancang hydropower projects on the comparison of discharge rates between a post-project year (2014) and the three baseline years (representative of normal (1963), low (1975), and high (1985) flows) was further explored by examining the annual proportions of flow of different rates at five hydrological stations distributed various distances downstream. Figure 4.10 demonstrates the loss of extreme flows (flows of highest and lowest rates) post-project at Chiang Saen (the station furthest north). At this station, all flow categories were recorded during the baseline period (with the exception of the low flow year (1975) when no flows exceeded 8000 m^3/s), although the proportion of highest and lowest flow rates varied in accordance with the type of year (e.g., in the high flow year (1985), flows greater than 2000 m^3/s occurred for more than half the year). In 2014, however, no flows were recorded within the highest and the lowest flow rate categories (Fig. 4.10).

The post-project reduction in the number of days during the dry season with low flows, and the decrease in number of days during the flood season with high flows, is also apparent at the other four hydrological stations. This is evident from the increased occurrence of the intermediate flow categories post-project compared to pre-project and less frequent occurrences of the extreme categories (Figs. 4.11, 4.12, 4.13, and 4.14). However, as also observed in other analyses, although the post-project reduction in the proportion of low flows (i.e., flows during the dry season) is observed at all stations and therefore persists downstream, the reduction of high flows is not. At the Stung Treng station (which is the furthest south of those considered; Fig. 4.14), there is little difference in the proportion of extremely high flows (>20,000 m^3/s) in 2014 relative to baseline years. However, no extreme low flows (<2000 m^3/s) were observed in 2014, whereas such low flows were observed between 12% and 24% of days during baseline years. Thus, a substantial reduction in high flows during the flood season post-project was only observed as far south as Vientiane.

4.3.2.2 Extreme Discharge: Minimum and Maximum Flows

Minimum Daily Flow

Changes between baseline and impact periods during periods of extreme flow are also important considerations during analysis of transboundary effects on flow. The difference in the range of daily minimum discharge between baseline and post-project periods in the lower Mekong River demonstrates how extreme low flows have been augmented following dam construction. As shown in Table 4.5, the daily minimum flow in 2014 at the Chiang Saen hydrological station is substantially greater than the daily minimum flow during the baseline period (including the greatest values of minimum flow). This relationship extends to the other four stations located further downstream shown in Table 4.5 (Vientiane, Mukdahan, Pakse, and Stung Treng). For these four stations, the minimum daily flows in 2014 were all the largest in the historical minimum daily flow series.

The timing of occurrence of minimum flows has also changed following dam construction. As shown in Fig. 4.15, minimum daily flow in 2014 occurred on June 24 during the flood season. In contrast, during the 48 years included in the baseline period (1960–2007), the minimum daily flows most commonly occurred in March (28 years) and April (14 years) and occasionally in February (3 years) and May (3 years). However, the minimum daily flow never occurred during the flood season (June through October) as observed in 2014. Thus minimum daily flows post-project (in 2014) were both greater, and occurring later, than during the baseline period.

As discussed under seasonal patterns above, the Lancang hydropower reservoirs have high regulatory capacity and therefore have the potential to mitigate extreme discharge situations downstream. For example, the release of water from the Lancang dams supplemented the low discharge and eased the regional drought of 2016 in the lower Mekong River Basin. A total of 12.65 billion cubic meters of water was discharged from the Jinghong hydropower reservoir during the period of March to May 2016. These releases amounted to between 40 and 89% of flows along various sections of the Mekong River. This emergency water supplement increased water level and discharge along the Mekong mainstem by 0.18–1.53 m and 602–1010 m^3/s, respectively. If these emergency releases had not occurred, flows would have been 47% lower at Jinghong, 44% lower at Chiang Saen, 38%

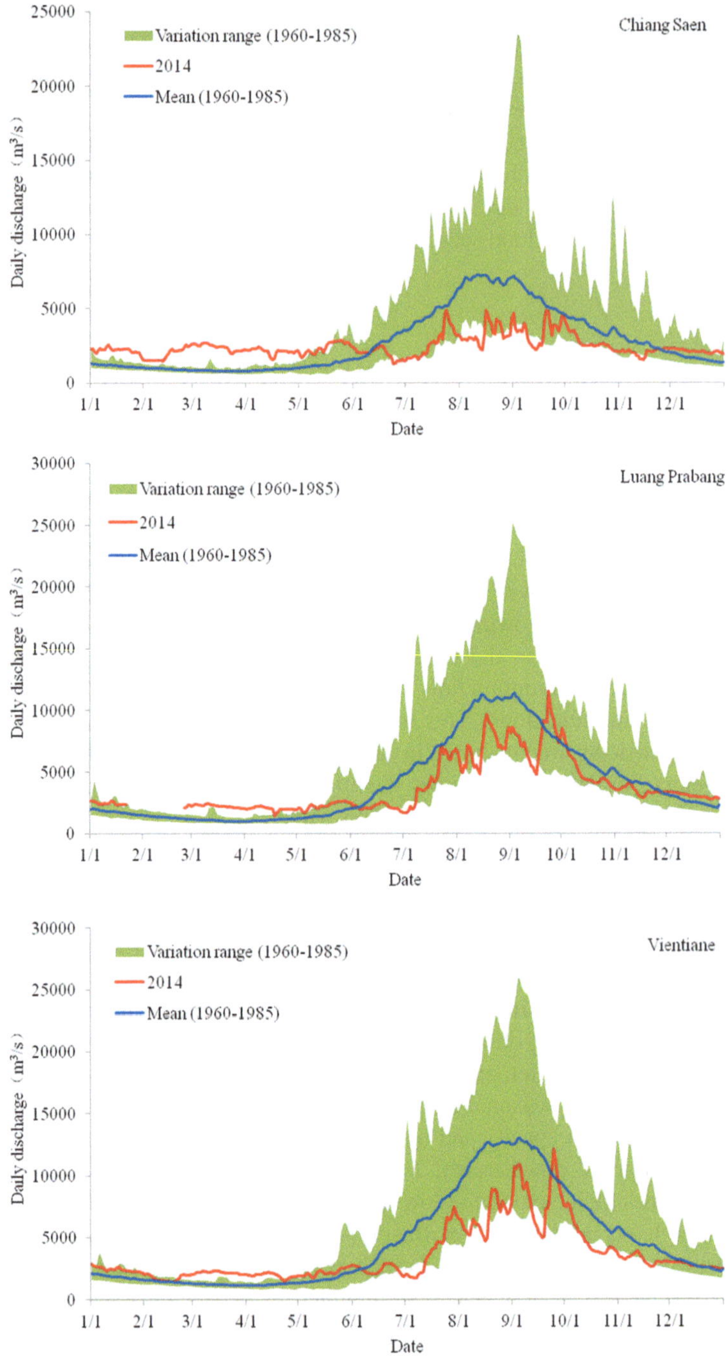

Fig. 4.8 Daily discharge at six hydrological stations (arranged north to south) in the lower Mekong River in 2014 in comparison to the mean and range of baseline values through the annual cycle

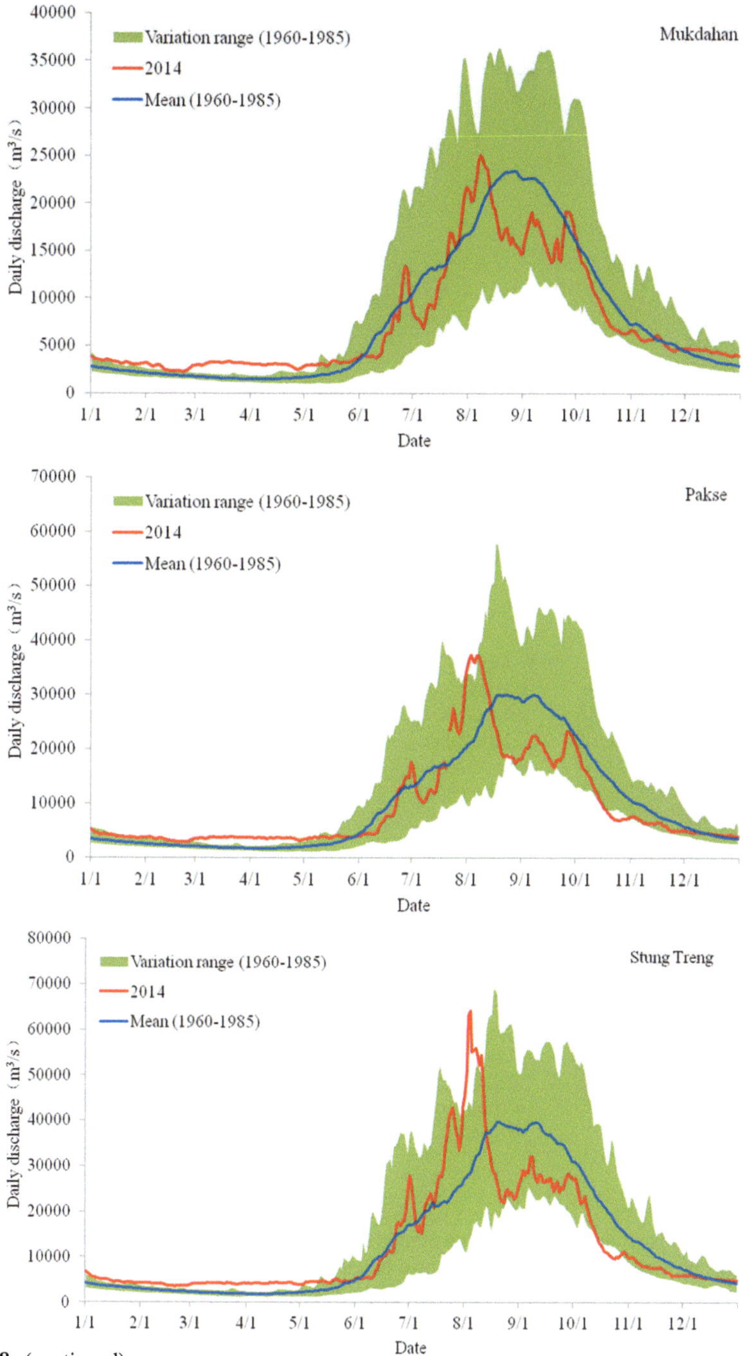

Fig. 4.8 (continued)

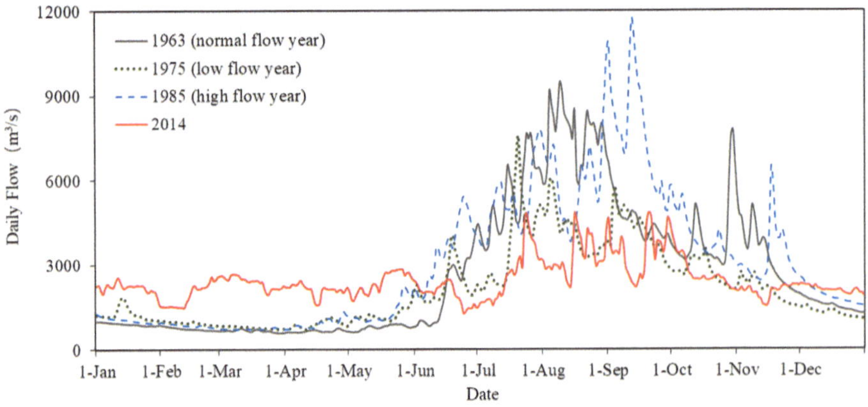

Fig. 4.9 Comparison of daily flow over the annual cycle at the Chiang Saen hydrological station between 2014 and three representative years of normal, low, and high flow during the baseline period

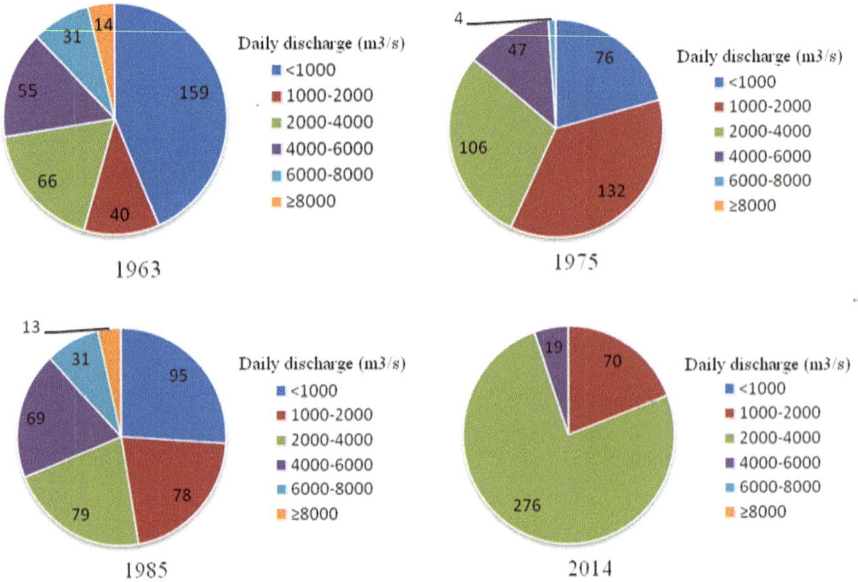

Fig. 4.10 Comparison of the proportion of the year (number of days) with various discharge rates among one post-project year (2014) and three baseline years representative of normal (1963), low (1975), and high (1985) flow at the Chiang Saen hydrological station

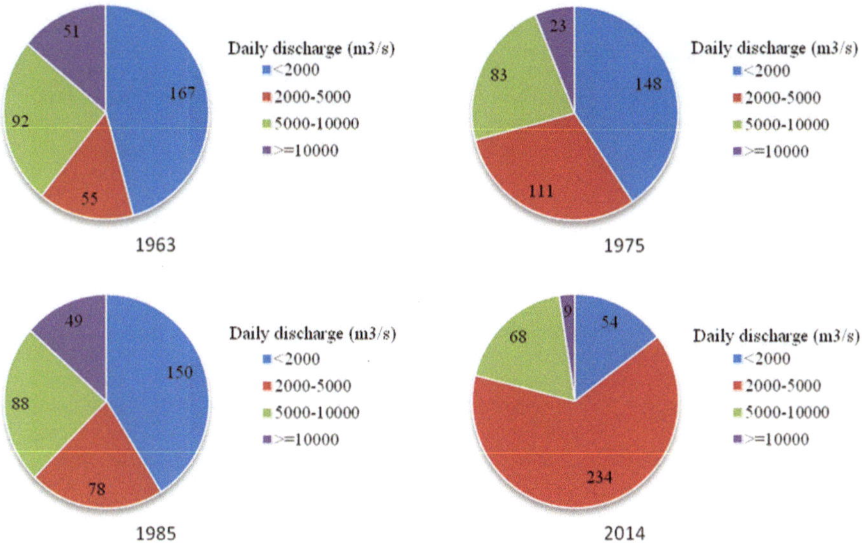

Fig. 4.11 Comparison of the proportion of the year (number of days) with various discharge rates among one post-project year (2014) and three baseline years representative of normal (1963), low (1975), and high (1985) flow at Vientiane hydrological station

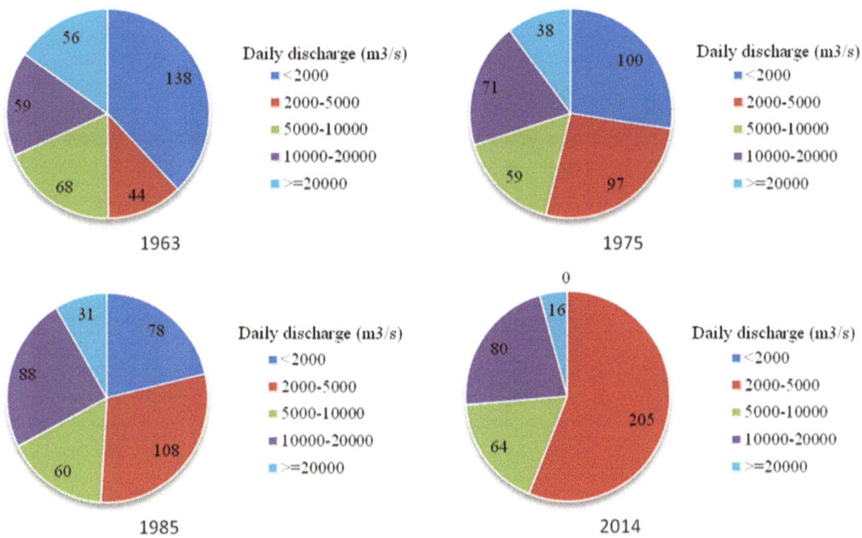

Fig. 4.12 Comparison of the proportion of the year (number of days) with various discharge rates among one post-project year (2014) and three baseline years representative of normal (1963), low (1975), and high (1985) flow at the Mukdahan hydrological station

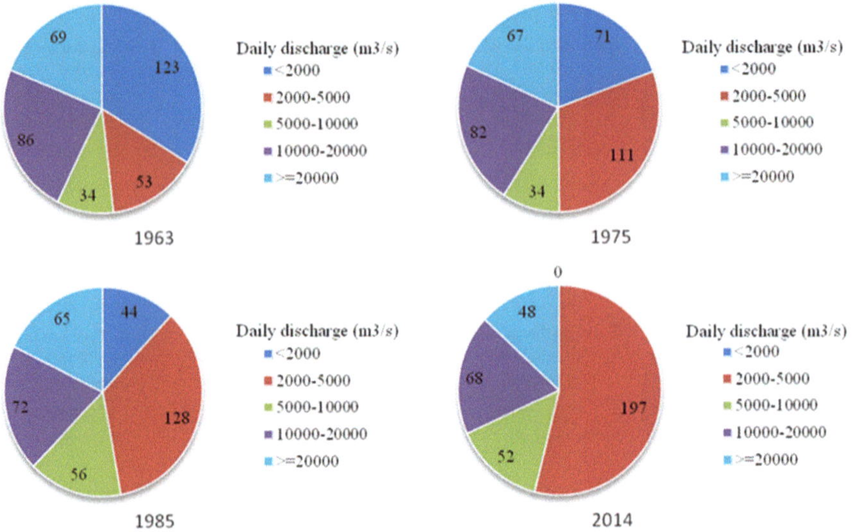

Fig. 4.13 Comparison of the proportion of the year (number of days) with various discharge rates among one post-project year (2014) and three baseline years representative of normal (1963), low (1975), and high (1985) flow at the Pakse hydrological station

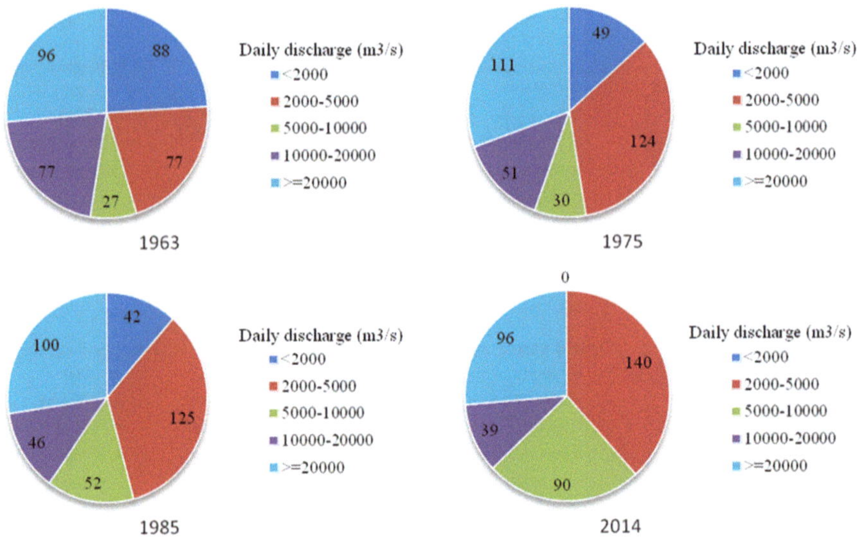

Fig. 4.14 Comparison of the proportion of the year (number of days) with various discharge rates among one post-project year (2014) and three baseline years representative of normal (1963), low (1975), and high (1985) flow at the Stung Treng hydrological station

Table 4.5 Comparison of minimum daily flow at five hydrological stations in the lower Mekong River

Hydrological station	Number of years of data	Flow (m^3/s) before 2008				Flow (m^3/s) in 2014
		Mean	Median	Maximum	Minimum	Minimum
Chiang Saen	49	692	720	1010	338	1274
Vientiane	95	1036	1035	1380	678	1444
Mukdahan	85	1399	1401	2240	958	2283
Pakse	84	1553	1570	2220	661	2847
Stung Treng	96	1557	1540	3030	855	3546

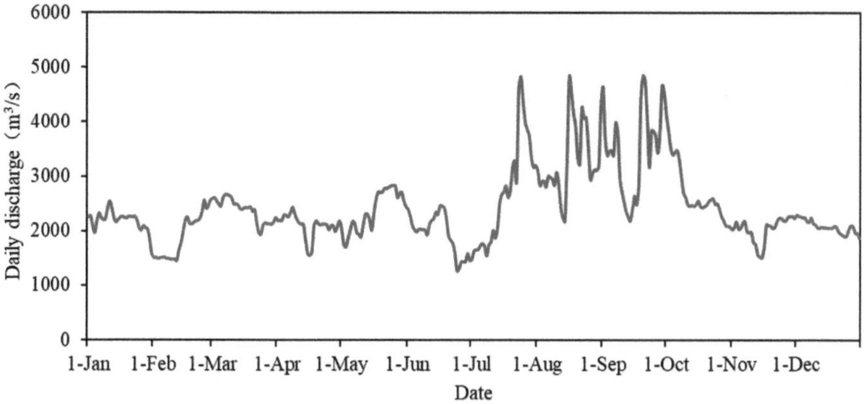

Fig. 4.15 Daily flow at Chiang Saen hydrological station in 2014

lower at Nong Khai, and 22% lower at Stung Treng. The supplemental flow also alleviated salinity intrusion in the Mekong Delta (MRC 2017). The ability of the Lancang dams to regulate water also has implications for potential water issues that may arise during the dry season as a result of climate change.

Maximum Daily Flow

Parallel to the results from the minimum daily flow analysis, the difference in the range of daily maximum discharge between baseline and post-project periods in the lower Mekong River demonstrates how extreme high flows have been dampened following dam construction. As shown in Table 4.6, the daily maximum flow in 2014 at the Chiang Saen hydrological station is lower than in any year of the baseline period. Thus, the regulation and storage of floodwaters, which occur during operation of the cascade of hydropower projects on the Lancang River, have the potential to lower the peak flow at the downstream stations and thereby provide

Table 4.6 Comparison of maximum daily flow at five hydrological stations in the lower Mekong River

Hydrological station	Number of years of data	Flow (m^3/s) before 2008				Flow (m3/s) in 2014
		Mean	Median	Maximum	Minimum	Maximum
Chiang Saen	49	10,890	10,700	29,300	5370	4855
Vientiane	95	16,809	16,450	25,900	7650	12,111
Mukdahan	85	28,476	28,475	36,800	16,100	24,975
Pakse	84	37,600	38,087	57,800	24,600	37,299
Stung Treng	96	54,007	53,907	78,093	30,532	64,171

flood control and have positive impacts downstream. However, the ability of these hydropower projects to regulate maximum flow downstream is more limited than their ability to regulate minimum flows owing to the large amount of precipitation that can occur downstream of the cascade of hydropower projects, in the lower Mekong River, during the flood season. Thus, in contrast to the ability of the cascade of hydropower projects to manage extreme low flows in the dry season, during the flood season extreme high flows are mainly regulated upstream of Vientiane.

4.3.3 *Comparison to Other Studies*

Other studies have assessed transboundary effects of the Lancang cascade of hydropower projects on water level and flow downstream into the Mekong River, the results and predictions of which can be compared to results from this study. Räsänen et al. (2012) reported significant dry season hydrological changes, especially in March and April, detectable as far south as Kratie, and concluded that the Mekong's hydrological regime has been significantly altered. These results are generally consistent with those presented here, although the magnitude of the changes differ, potentially because the Nuozhadu hydropower project had not yet become operational at the time of the Räsänen et al. (2012) study. Räsänen et al. (2012) reported that the monthly average flow at Kratie in March increased by 49% and water level increased by ~1.1 m. Results from our study, which was conducted after the Nuozhadu hydropower project became operational, indicated that the water levels at Kratie during March of 2013 and 2014 increased by 1.97 and 1.45 m, respectively, which is substantially greater than reported by Räsänen et al. (2012). A more recent study (Räsänen et al. 2017) assessed discharge changes using observed river discharge data and a distributed hydrological model over the period of 1960–2014. Results indicated that hydropower operations have considerably modified discharge since 2011, and that the largest changes were observed in 2014. According to observed and simulated discharge data, the most notable changes occurred in northern Thailand (Chiang Saen) in March to May 2014, when discharge increased by 121% to 187%, and in July to August 2014, when discharge decreased by 32% to

46% compared to average. In contrast, in Cambodia (Kratie), flow increased from 41% to 74% in March–May 2014 and decreased from 0 to 6% in July–August 2014. The authors noted that discharge impacts are expected to vary from year to year depending on hydropower operations.

Through simulation modeling, Piman et al. (2013) predicted future changes to the hydrological regime after the construction of hydropower facilities on the Mekong River, including the Lancang cascade of hydropower projects. The "Definite Future" scenario (next 5 years) of Piman et al. (2013) predicted that there would be a 20–30% increase in flow in the Mekong Delta during the dry season and a 4–14% flow reduction during the flood season. Although our results consider only current operational hydropower projects, they already have demonstrated that in 2014 there was an 18% increase in flow at Stung Treng station during the dry season and a 7% flow reduction during the flood season.

Lu et al. (2014) used the Indicators of Hydrological Alteration (IHA) to examine the impacts of the Lancang dams on river discharge. The authors investigated the difference in discharge between pre- and post-dam periods and found that the monthly mean values averaged over the entire post-dam period were higher in July (an increase of 15%), but lower in August (a decrease of 9%), than in the pre-dam period. This suggests that the reservoirs released more water for hydropower generation in July than would naturally be flowing, but began to store water in August, thereby reducing discharge relative to baseline. Further, the study reported that although flows at Chiang Saen were dominated by precipitation upstream, the existing reservoirs (Manwang, Dachaoshan, Jinhong, and Xiaowan (up to 2010)) nevertheless altered discharge in this location to a certain degree.

Cochrane et al. (2014) indicated that hydropower operations and irrigation development in the Mekong may have already caused observable alterations to natural water levels along the Mekong mainstem and the Tonle Sap River beginning as early as 1991. Increases in water levels during the dry season (March, April, and May) of 35% to 20% post-1991 from Chiang Saen downstream to Stung Treng were documented, and such alterations, although relatively minor, were thought to be likely caused by water infrastructure development in the basin. It was reported that the effect of the upper Mekong hydropower development tributary operations was clearly observable, in terms of water level fluctuations and fall rates, downstream to the Mukdahan station. It was further noted that alterations observed in Pakse and downstream are likely a result of irrigation development, flood control, and hourly/daily hydropower operations (at Pak Mun dam in particular) in the Chi-Mun basin. Alterations observed downstream from Stung Treng will be exacerbated by the ongoing development in the 3S basin.

4.4 Conclusions

Comparisons of changes to water level and discharge between baseline and post-project periods, and the distances for which changes are apparent downstream, revealed the magnitude and extent of effects of the Lancang cascade of hydropower projects on downstream reaches of the Mekong River during dry and flood seasons. Results from comparative water level and discharge analyses indicated that operation of the cascade of hydropower projects (i.e., those that were operational by 2013) had a stabilizing effect on natural water fluctuations in both seasons for some distance downstream. This effect was particularly pronounced in the dry season when, due to water regulation within the reservoirs, water levels and flows were increased post-project relative to baseline, especially in March and April, and effects extended to the entire lower Mekong River. For example, in 2013 and 2014, the average water levels between Chiang Saen and Kratie increased relative to baseline between 1.67 m (at Kratie) and 0.68 m (at Nakhon Phanom). However, downstream of Kratie, water level changes were much smaller (e.g., water level at Phnom Penh station increased only by 0.29 m relative to baseline). Similarly, in March 2014, flows increased between Chiang Saen and Stung Treng relative to baseline between 197.8% (at Chiang Saen) and 80.9% (at Vientiane). Positive effects of the hydropower projects identified during the dry season included more effective utilization of water resources in the lower Mekong River Basin, which included alleviation of drought and prevention of the invasion of seawater into the Mekong Delta.

The effect of the hydropower projects during the flood season, when naturally high water levels and flows were diminished by the hydropower projects, was less pronounced than dry season effects and did not extend as far downstream. For example, in the flood season of 2013, monthly average water levels between Chiang Saen and Mukdahan decreased relative to baseline between 1.43 m (at Luang Prabang) and 0.76 m (at Mukdahan). Post-project flows also decreased relative to baseline, with the percentage of monthly average flow decrease at Chiang Saen from July to October 2014 ranging between 43.0% (in July) and 28.9% (in October). Although the reservoirs also regulate extreme high flows, the attenuation of high flows is more limited than effects on low flows, owing to the high levels of precipitation that occur downstream of the Lancang hydropower projects during the flood season, and the inflow of large tributaries in Thailand, Laos, and Cambodia that became the main sources of flow in the Mekong River at this time of the year. Thus, during the flood season, effects of water regulation were observed to only extend to Mukdahan and Vientiane for water level and flow, respectively. Positive effects of water regulation during the flood season include providing flood control to downstream areas through the storage of floodwaters within the reservoirs.

Xiaowan and Nuozhadu projects have only been in operation for a short period of time. As such, this study mainly analyzed changes to water level and flow from June 2013 to March 2015. Differing climate conditions and varying operational approaches of the cascade of hydropower projects will affect water level and flow in future years. Thus an in-depth understanding of changes to water level and flow in the lower Mekong River requires a longer timeframe and further study.

References

Adamson, P.T., I.D. Rutherfurd, M.C. Peel, and I.A. Conlan. 2009. Chapter 4 – The hydrology of the Mekong River. In *The Mekong-biophysical environment of an international river basin*, ed. I.C. Campbell, 53–76. San Diego: Academic.

Campbell, I.C. 2007. Perceptions, data, and river management: Lessons from the Mekong River. *Water Resources Research* 43 (2): W02407. https://doi.org/10.1029/2006WR005130.

Cochrane, T.A., M.E. Arias, and T. Piman. 2014. Historical impact of water infrastructure on water levels of the Mekong River and the Tonle Sap System. *Hydrology and Earth System Sciences* 18 (11): 4529–4541.

Fan, H., D. He, and H. Wang. 2015. Environmental consequences of damming the mainstream Lancang-Mekong River: A review. *Earth-Science Reviews* 146: 77–91.

Lu, X., and R. Siew. 2006. Water discharge and sediment flux changes over the past decades in the lower Mekong River: Possible impacts of the Chinese dams. *Hydrology and Earth System Sciences* 10 (2): 181–195.

Lu, X., S. Li, M. Kummu, R. Padawangi, and J. Wang. 2014. Observed changes in the water flow at Chiang Saen in the Lower Mekong: Impacts of Chinese dams? *Quaternary International* 336: 145–157.

MRC (Mekong River Commission). 2017. *The effects of Chinese dams on water flows in the Lower Mekong Basin*. http://www.mrcmekong.org/news-and-events/news/the-effects-of-chinese-dams-on-water-flows-in-the-lower-mekong-basin/. Accessed 12 June 2017.

Piman, T., T. Lennaerts, and P. Southalack. 2013. Assessment of hydrological changes in the lower Mekong Basin from Basin-Wide development scenarios. *Hydrological Processes* 27: 2115–2125.

Räsänen, T.A., J. Koponen, H. Lauri, and M. Kummu. 2012. Downstream hydrological impacts of hydropower development in the Upper Mekong Basin. *Water Resources Management* 26 (12): 3495–3513.

Räsänen, T.A., P. Someth, H. Lauri, J. Koponen, J. Sarkkula, and M. Kummu. 2017. Observed river discharge changes due to hydropower operations in the Upper Mekong Basin. *Journal of Hydrology* 545: 28–41.

Chapter 5
Transboundary Environmental Effects of Hydropower: Water Temperature

5.1 Overview

Downstream effects of the Lancang cascade of hydropower projects on water temperature were investigated through temperature data comparisons and modeling. Similar to the approach for hydrology (Chap. 4), comparisons of water temperature were conducted in part by comparing water temperature between baseline (historical) and post-project periods. Other comparisons were also made, such as comparing post-project water temperature upstream and downstream of a hydropower project and comparing temperature gradients within different river segments and during different time periods. Modeling of actual and virtual development scenarios, in which the effect of a hydropower project could be included or removed, was also used to investigate the consequences of the presence of a particular hydropower project on water temperature. Water temperature data were compiled from existing hydrological stations along the Lancang-Mekong River Basin (MRC data) and additional data collected from gauges installed specifically for the study.

The Lancang-Mekong River flows in a southerly direction and crosses temperate, sub-tropical, and tropical latitudes; thus, natural temperature changes must be considered when analyzing temperature effects. Under natural conditions, water temperature in the Lancang-Mekong River increases with decreasing latitude, with the rate of this increase becoming more rapid with distance south (Zhang et al. 2007). Thus, increases in temperature with distance downstream would be expected without the influence of hydropower projects, and this effect must be explicitly considered in the analysis of potential hydropower project effects on water temperature.

Water temperature analyses were conducted for three locations along the Lancang-Mekong River, two of which are within China (the Nuozhadu-Jinghong reach and the Jinghong-Guanlei reach) (Fig. 5.1) and one of which is in northern Thailand (Chiang Saen hydrological station; Fig. 4.1 in Sect. 4.1). Reaches within China were defined in accordance with the locations of existing hydropower proj-

X. Yu et al., *Balancing River Health and Hydropower Requirements in the Lancang River Basin*, https://doi.org/10.1007/978-981-13-1565-7_5

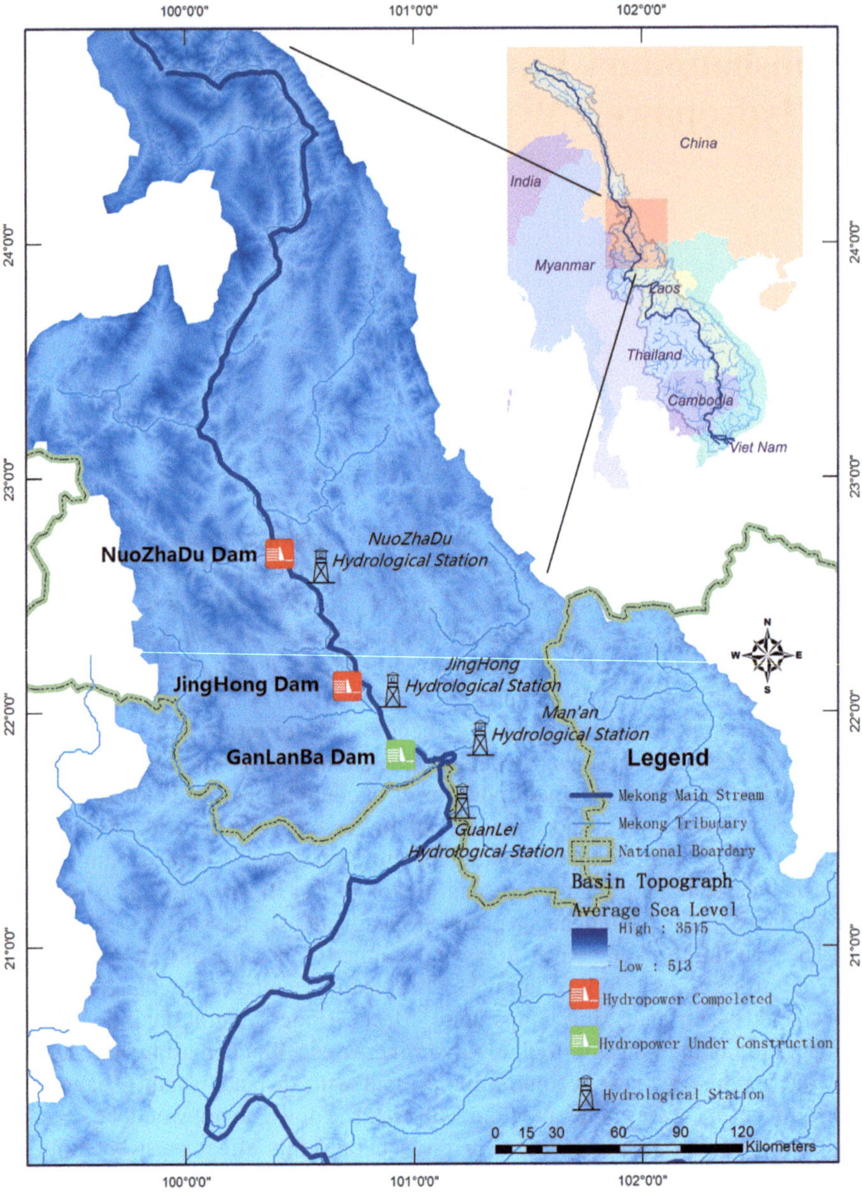

Fig. 5.1 Locations of hydrological stations and hydropower projects in the Nuozhadu-Jinghong and Jinghong-Guanlei reaches

ects and hydrological stations, as reflected in the analyses conducted for each. Key features of the three locations relevant to this study are described below.

5.1.1 Nuozhadu-Jinghong Reach

After the Jinghong hydropower project was constructed, the Jinghong reservoir was formed with water stored along the Nuozhadu-Jinghong reach of the river (i.e., what used to be a naturally flowing river became a reservoir). The river channel in this reach is long and narrow: it is 105 km long (measured along the course of the river), but only 400 m wide, on average. When water stored in the Jinghong reservoir reaches or exceeds normal water level, the conditions of the river segment downstream of the Nuozhadu dam become lake-like. However, during the data collection period of this study, the water level in the Jinghong reservoir stayed below the maximum operating level, and river flow was maintained along this reach. The impoundment of the Nuozhadu reservoir began in December 2011, and the project was completed in 2012. Thus post-project water temperatures used to evaluate temperature changes were compiled from 2012 through 2016. However, 2015 was the first year the reservoir operated at a normal level. The baseline, or historical, period was defined as 1985–2001.

Within the Nuozhadu-Jinghong reach, hydrological stations exist at Nuozhadu and Jinghong (Fig. 5.1). The Nuozhadu hydrological station (Nuozhadu) is located approximately 5 km downstream of the Nuozhadu dam on the right side of the channel and ~5 km upstream of the operational backwater region of the Jinghong reservoir. The Jinghong hydrological station (Jinghong) is located 4 km downstream from the Jinghong dam on the right side of the channel.

Water temperature analyses focused on the Nuozhadu-Jinghong reach investigated post-project temperature changes by comparing historical values to those following construction of the Nuozhadu hydropower project (completed in 2012) and also considered post-project temperature differences upstream and downstream of the Jinghong hydropower project (completed in 2009). Nuozhadu and Jinghong post-project water temperatures were compared not only to historical values for the same locations but also to historical values for southern areas where temperatures are expected to be warmer and rates of change are expected to be greater under natural conditions. The historical data and the 2015 data from Jinghong were measured at exactly the same location, thus allowing effective comparisons between periods.

5.1.2 Jinghong-Guanlei Reach

The Jinghong-Guanlei reach is ~100 km long and includes the confluences of three level 1 tributaries to the Lancang River: the Liusha River, the Buyuan River, and the Nanla River. The amount of natural runoff from the Liusha and Nanla rivers is small, and the influence of these tributaries on water temperature in the Lancang River is considered negligible. However, the Buyuan River is a large tributary with the potential to influence temperature in the Lancang River, and, as such, its contributions to temperature effects were considered in this study. Hydrological stations in the Jinghong-Guanlei reach are the Jinghong, Man'an, and Guanlei hydrological stations (Figs. 5.1 and 5.2).

The Buyuan River is also called the Luosuo River or Nanban River. It is ~300 km long and originates in Mohei Township, Ninger County, Yunnan Province. The confluence of the Buyuan River with the Lancang River is located between Jinghong (~60 km downstream) and Guanlei (~30 km upstream) hydrological stations. The Buyuan River is the largest tributary to the Lancang River in terms of flow input. It provides abundant habitat for fish spawning, has numerous shoals of fish, and is China's only confirmed spawning ground for migratory fish in the Lower Mekong River Basin. Man'an hydrological station is located in the Buyuan River tributary on the left side of the channel, ~22 km upstream of the Lancang-Buyuan River confluence. Approximately 1.5 km upstream of the station, there is a run-of-the-river hydropower project. However, this hydropower project is small, and the Man'an runoff conditions can be considered natural.

Water temperature analyses for the Jinghong-Guanlei reach investigated differences in water temperature between historical and post-project periods in the mainstem below the outlet of the Jinghong dam, which is currently the last mainstem dam in China, and at Guanlei, which is near the border of China. Further, modeling permitted evaluation of the effect of the existence of a hydropower project on water temperatures up to 90 km downstream.

Fig. 5.2 Schematic diagram of the Jinghong-Guanlei reach and location of temperature gauges

5.1.3 *South of the Chinese Border (Chiang Saen)*

The Chiang Saen hydrological station (Chiang Saen) is located on the Mekong River in northern Thailand, 330 km downstream from Jinghong (Fig. 4.1). Comparison of monthly average water temperatures at Chiang Saen in the mainstem of the Mekong River before, during, and after construction of the Jinghong hydropower project allowed direct analysis of the potential for water temperature effects to extend south of China's border. Because the Jinghong hydropower project was completed in 2009, water temperatures were compared between baseline (1986–2002), during (2003–2008), and after (2009–2014) construction of the Jinghong project.

5.2 Methods

There were two components to the analysis of the transboundary effects on water temperature due to construction and operation of the Lancang cascade of hydropower projects: field data collection and hydrodynamic modeling. Field data were used to provide summary statistics and define boundary conditions to calibrate and verify information for the hydrodynamic model. The hydrodynamic model was then adopted to integrate flow and water temperature data and to analyze the effect of the presence of hydropower projects on water temperature.

5.2.1 *Data Collection*

Data from existing hydrological stations along the Lancang-Mekong River Basin were used to provide summary statistics for historical (baseline) conditions, and MRC data from the Chiang Saen hydrological station in Thailand were used to investigate transboundary effects given that it is the closest downstream hydrological station outside of China. Existing data have limitations, however, given that only surface water temperature is measured and this occurs only twice daily (at 8:00 and 20:00). As such, automatic temperature monitoring equipment that take hourly temperature readings was installed in key locations in China to better record post-project water temperatures for this study.

Three kinds of instruments (gauges) were used to record hourly temperature data, all of which are reliable, precise, and portable (Fig. 5.3): the Onset HOBO® U20-001-2 automatic water temperature and water pressure gauge (hereinafter referred to as U20), the Onset HOBO® UTBI-001 water temperature logger (UTBI), and the Onset HOBO® U26-001dissolved oxygen water temperature logger (U26). Several U20 were used to collect data in this study. The U20 loggers are widely used for research in water environments and function in water up to 30 m deep. When

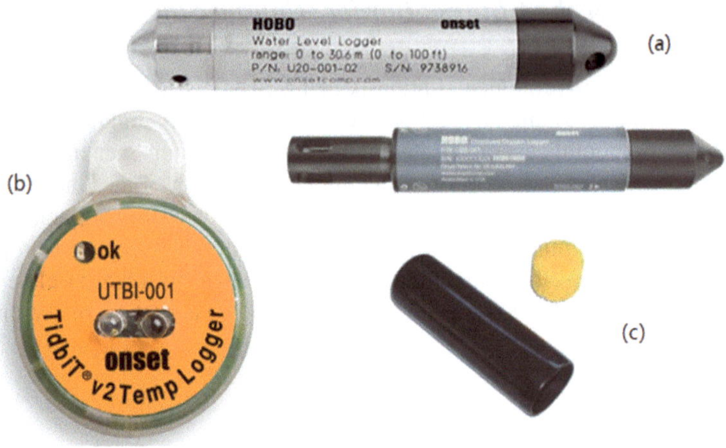

Fig. 5.3 Water temperature recording equipment used in this study: (**a**) U20 water temperature and pressure logger, (**b**) UTBI temperature logger, and (**c**) U26 dissolved oxygen water temperature logger

deployed below the surface of the water, they measure water temperature and combined air and water pressure, and when deployed above the surface of the water, they measure air temperature and air pressure. These devices have a stainless steel, sealed tube design, and are easy to install, reliable, robust, small, and precise. UTBI devices were used where the main body of the Mekong River crosses the Chinese border (the farthest point downstream within the temperature study area). In this location, water depth measurements could not be taken, thus the UTBI device, which functions in up to 300 m of water, was used. In addition to providing adequate depth allowance, the measurable range and accuracy (error of less than 0.02 °C at 25 °C) satisfied study requirements. Three U26 devices were deployed (at Nuozhadu, Jinghong, and Man'an, see below). This device measures dissolved oxygen and water temperature, and data collected were used to verify and validate U20 and UTBI measurements.

The gauges deployed in this study were installed at the existing hydrological stations, specifically, at Nuozhadu, Jinghong, Man'an, and Guanlei hydrological stations (Fig. 5.1). This decision was made based on three primary considerations: equipment security, achieving representation of the area of interest, and ease of access. Loss of equipment due to environmental conditions (e.g., high flows) or theft was a key concern for equipment placement given that lost equipment would represent a significant cost in terms of data availability. Further, due to restrictions on the number of measurement devices that could be deployed in secure locations, each logger needed to be representative of the area into which it was deployed (i.e., within flowing water). Finally, locations had to be easy to access. All of these requirements were met by installing the equipment at existing hydrological stations. Jinghong, Man'an, and Guanlei stations are located in the mainstem river channel, thus measurements taken here are representative of the river segment. As described in Sect. 5.1.1, although conditions at the Nuozhadu station can be lake-like if the Jinghong reservoir reaches or exceeds normal water level, during the data collection period, the water level in the Jinghong reservoir stayed below the maximum operating level; thus, equipment deployed at the Nuozhadu station remained within flowing river conditions. Further, because the Jinghong hydropower project is the downstream-most project constructed on the Lancang River mainstem, and the Jinghong hydrological station is located downstream of the dam, water temperature measurements taken at the Jinghong hydrological station were entirely within natural river conditions and are representative of this section of the river.

Measurements recorded during this study included water temperature and pressure and air temperature and pressure. Equipment used to measure air temperature and pressure was deployed inside the existing hydrological stations. Equipment used to measure water temperature and combined air and water pressure were deployed in monitoring wells at the stations. The gauges were suspended in the hydrological station wells 20 cm below the 2013 low-water mark, and their positions did not change with rising and falling water levels. The 2013 low-water mark was chosen as the standard reference depth for measurements, and water-level data were calculated based on water pressure changes measured by the equipment. To ensure that water temperature measurement data were representative of the water in the river, water from the center of the river channel was directed to the monitoring wells at hydrological stations with pipes. The rope used to secure the equipment inside the monitoring wells was specially designed Kevlar rope that had high tensile strength and high resistance to wear and corrosion and was designed for extreme environments.

Water temperature data recorded by the gauges installed at the hydrological stations were collected at Jinghong and Man'an from October 2014 through December 2015, at Nuozhadu for the entire year of 2015, and at Guanlei from September 2015 through December 2015. All water temperature data (1986–2014) for the Chiang Saen hydrological station was MRC data.

During data analysis, outlying data points were occasionally identified as being untrustworthy. When such an outlier was identified, the measurements from the same point, at the same time of day, from the previous 5 days and the following 5 days, were averaged together and substituted for the outlying data point. The number of outlying data points that were substituted in this manner composed less than 0.5% of the total data set.

5.2.2 *Modeling*

Modeling based on Delft3D was used to study the effects of different development scenarios on the water temperature of the Lancang River. By analyzing temperature differences at Jinghong and at the Chinese border (Guanlei) under different hydropower development scenarios, it was possible to investigate the effects of the Jinghong hydropower project on the Jinghong-Guanlei reach. Analysis of the

Table 5.1 Key features of development scenarios used to compare of the actual development scenario with a virtual scenario in which the Jinghong hydropower project and reservoir do not exist

Development scenario	Existence of Jinghong power station	Conditions of Nuozhadu-Jinghong reach	Conditions of Jinghong-Guanlei reach
Actual	Yes	Jinghong reservoir	Natural river
Virtual	No	Natural river	Natural river

influence of the hydropower project on water temperature downstream was conducted by setting up two development scenarios (Table 5.1): a "virtual" scenario without a hydropower plant and reservoir at Jinghong and an "actual" scenario (based on actual water temperatures measured at the Chinese border) with a hydropower plant and reservoir at Jinghong. Thus, in the actual development scenario, the Nuozhadu-Guanlei reach of the Lancang River contains the Jinghong hydropower plant and reservoir and the Jinghong-Guanlei reach of the river. In the virtual development scenario, the Nuozhadu hydropower plant exists, but the Jinghong plant and reservoir do not, and thus the Nuozhadu-Guanlei reach is composed entirely of free-flowing river. The Jinghong-Guanlei reach includes the confluence of the Buyuan River in both scenarios. We chose the months from December 2014 through May 2015 for the simulation period. The first month (December 2014) was the model's initiation phase, and simulation results were provided from January to May 2015.

The hydrodynamic model used for this study was constructed using the Delft3D platform. Delft3D is a hydrodynamic and water quality modeling platform developed by the Delft University of Technology and Deltares Institute in the Netherlands. It is mainly used to study surface water dynamics and non-steady flow and transport phenomena of water environments, such as rivers, oceans, and lakes. It has flexible architecture and comprehensive features. It can precisely simulate the 2D and 3D flow field in the research zone, including the characteristics of waves, water quality, mud and silt transport, and changes in riverbed topology, as well as interactions among various types of physical processes.

The Delft3D model is composed of six main analysis modules (Fig. 5.4): FLOW is the hydrodynamic calculation module, WAVE is used to simulate the propagation of waves, WAQ is the water quality calculation module, PART is the water quality simulation and particle tracking model, ECO is the ecological simulation module, and SED is used to simulate cohesive and non-cohesive sediment transport. The Delft3D system has seen widespread use internationally and is commonly applied in China for water environment research, such as storm surge modeling in the Yangtze River estuary and hydrodynamic characteristics and water quality research in Hangzhou Bay and Poyang Lake. This study mainly adopted the Delft3D-Flow module to simulate 2D and 3D hydrodynamics combined with water surface heat transfer modeling to simulate water temperature.

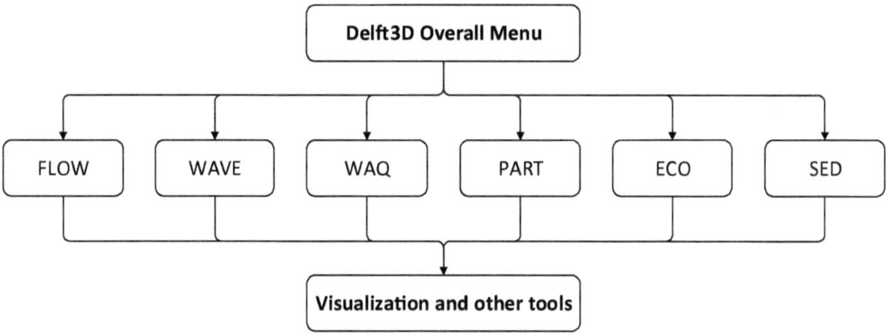

Fig. 5.4 Architecture of the Delft3D model

5.3 Results and Discussion

5.3.1 Nuozhadu-Jinghong Reach

Comparison of water temperature between baseline and post-project periods was used to investigate monthly and annual changes in water temperature following construction of the Nuozhadu hydropower project. Monthly average water temperature before and after the operation of the Nuozhadu project is presented in Table 5.2 and Fig. 5.5. After operation of the Nuozhadu project, water temperature in winter (November–January) at the gauge station downstream of the dam increased by 1.0 °C–6.6 °C, and water temperature in summer (May–July) decreased by 0.4 °C–3.0 °C (Table 5.2). The decrease during summer was most pronounced in June and July, when, averaged over all post-project years, it was 2.4 °C and 1.7 °C, respectively. Averaged over the entire year, there was an increase in temperature during all post-project years relative to baseline. Further, 2015 was the first year the reservoir operated at a normal level, and the increase in yearly average water temperature relative to baseline was greatest in this year (a 2.1 °C increase). The maximum change in water temperature over the annual cycle occurred in December, when it was 6.9 °C higher in 2015 than during baseline.

The timing of temperature changes also differed post-project relative to baseline (Fig. 5.5). Prior to project development, water temperature began to increase in January, reached a peak between June and August, and decreased gradually from September to December. Post-project, water temperature was relatively stable between January and April, began to increase in April or May, peaked in August through October, and decreased from October through December. Thus, following project construction, the period during which water temperature increased was shortened by 3 months, and there was a decrease in the magnitude of the temperature rise (from 9.2 °C to 4.5 °C). Overall, downstream of the Nuozhadu dam following project construction, there was an overall increase in water temperature and a significant reduction in the magnitude of temperature change over the annual cycle,

Table 5.2 Monthly average water temperature before and after the operation of the Nuozhadu project and change relative to baseline as recorded at the Nuozhadu hydrological station

	Monthly average water temperature (°C)													
Period/change	Year	Jan	Feb	Mar	Apr	May	Jun	Jul	Aug	Sep	Oct	Nov	Dec	Average
Baseline		13.2	15.2	17.3	18.4	20.2	22.5	22.7	22.4	21.2	19.4	17.0	13.9	18.6
Post-project	2012	17.7	17.1	17.5	19.0	19.7	20.3	20.6	21.2	21.9	22.1	21.8	20.5	19.9
	2013	19.1	18.8	18.7	18.4	17.9	19.5	20.7	21.2	21.5	21.5	20.6	19.8	19.9
	2014	18.6	18.0	18.4	17.9	20.8	20.3	20.4	22.6	21.9	21.9	21.0	20.0	20.2
	2015	13.9	18.4	18.7	18.7	19.1	20.4	22.2	23.1	23.7	23.6	21.1	20.8	20.7
	2016	19.5	18.7											
Change	2012	4.5	1.9	0.2	0.6	−0.5	−2.2	−2.1	−1.2	0.7	2.7	4.8	6.6	1.3
	2013	5.9	3.6	1.4	0.0	−1.3	−3.0	−2.0	−1.2	0.3	2.1	3.6	5.9	1.3
	2014	5.4	2.8	1.1	−0.5	0.6	−2.2	−2.3	0.2	0.7	2.5	4.0	6.1	1.5
	2015	5.7	3.2	1.4	0.3	−1.1	−2.1	−0.5	0.7	2.5	4.2	4.1	6.9	2.1
	2016	6.3	3.5											

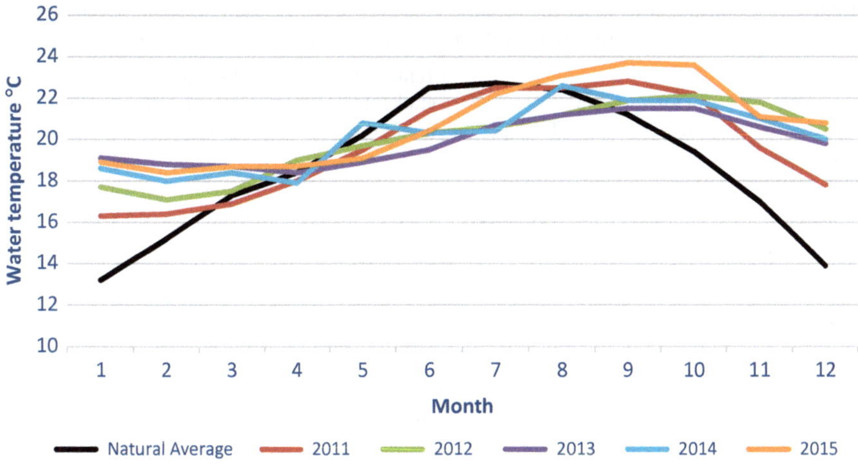

Fig. 5.5 Monthly average water temperature before (natural) and after the operation of the Nuozhadu hydropower project, as recorded at the Nuozhadu hydrological station

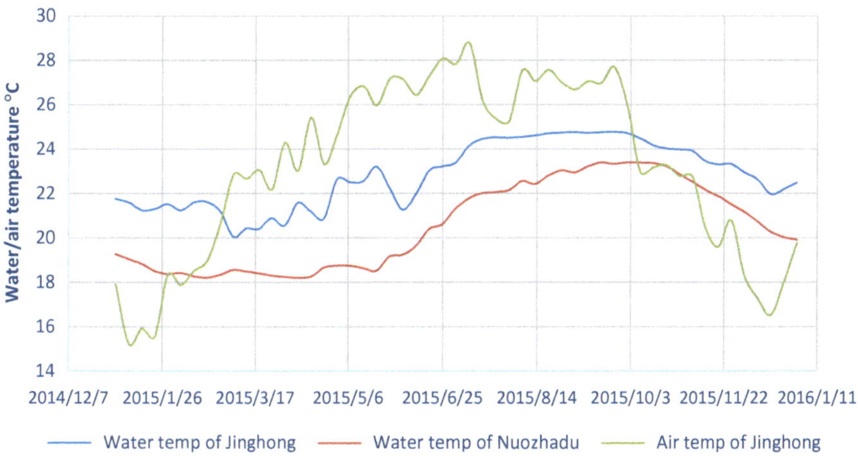

Fig. 5.6 Water temperature at Jinghong and Nuozhadu hydrological stations and air temperature at Jinghong (weekly average values) between Dec 2014 and January 2016

temperature rises and drops were not as steep in spring and fall, and there was a delayed temperature peak, which resulted in lower temperatures from May to July.

A comparison of post-project water temperatures between Nuozhadu and Jinghong hydrological stations was used to more directly investigate the effects on temperature of water flowing through the Jinghong hydropower project. As shown in Fig. 5.6, weekly average values in 2015 indicate that water temperature at Nuozhadu is always lower than it is at Jinghong, and that water temperatures reached

Table 5.3 Comparison of water temperature gradients in the Lancang-Mekong River between Nuozhadu-Jinghong (post-project) and between Jinghong-Chiang Saen (baseline)

Reach of Lancang-Mekong River	Difference in latitude	Time period	Temperature gradient (°C /degree of latitude)	Length of river reach (km)	Temperature gradient (°C / 100km)
Nuozhadu-Jinghong	0.5523	2015	3.98	100	2.20
Jinghong-Chiang Saen	1.7972	1985–2001	2.23	330	1.21

maximum and minimum values in both locations at about the same time, which lags approximately 2 months behind air temperature. Because the Nuozhadu hydrological station is above the Jinghong dam (but within flowing river conditions during the data collection period; see Sect. 5.2.1) and the Jinghong hydrological station is below the Jinghong dam, this demonstrates that water temperature increases as the water flows through the Jinghong hydropower project.

Given the direction of flow of the Lancang-Mekong River and the latitudes it flows through, water temperature naturally increases as one proceeds downstream, and this rate of temperature increase becomes more rapid with latitude (Zhang et al. 2007). Thus, the natural warming of the river as it flows south must be taken into consideration when evaluating downstream effects of hydropower projects on water temperature. A comparison of the recent temperature gradient in the northern portion of the river (Nuozhadu-Jinghong) with the historical temperature gradient in a southern portion of the Lancang-Mekong River (Jinghong to Chiang Saen) demonstrates the magnitude of temperature changes following dam construction. As shown in Table 5.3, the historical average rise in water temperature between Jinghong and Chiang Saen hydrological stations was 1.21 °C/100 km, whereas the 2015 rise in water temperature between Nuozhadu and Jinghong was 2.20 °C/100 km. Thus, following construction of the Jinghong hydropower project, the water temperature gradient along the higher latitude Nuozhadu-Jinghong reach was higher than it historically was in the lower latitude Jinghong-Chiang Saen reach by a factor of 1.73. Using latitude gradient as a metric for comparison of the rate of water temperature increase, and given that the latitude differences between Jinghong and Chiang Saen and Nuozhadu and Jinghong are 1.7972° and 0.5523°, respectively, the latitudinal temperature gradient between Jinghong and Chiang Saen is 2.23 °C per degree of latitude and that between Nuozhadu and Jinghong is 3.98 °C per degree of latitude (Table 5.3). Thus, the recent (2015) latitude gradient of the northern portion of the river is substantially greater than the southern portion was historically, and the reversal in the trend in latitude gradient relative to what would be expected to exist naturally suggests that the temperature gradient along the Lancang River has increased after the construction of the Jinghong hydropower project and is therefore indicative of hydropower project effects downstream.

Water temperatures measured at Jinghong and Nuozhadu in 2015 were also compared to historical averages (1985–2001) from Jinghong and Chiang Saen, the latter of which is located 330 km downstream of Jinghong. As shown in Fig. 5.7, the 2015

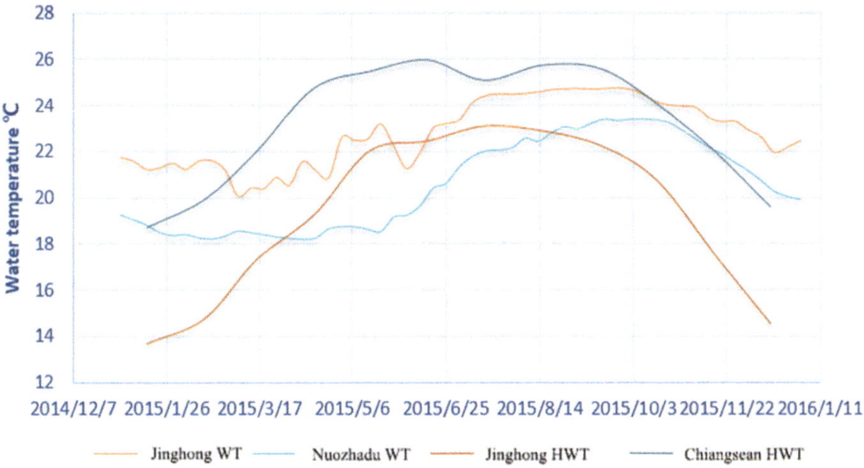

Fig. 5.7 Comparison of water temperatures (WT) in 2015 (5-day averages) with historical (baseline) water temperatures (HWT) (monthly averages) at Jinghong, Nuozhadu, and Chiang Saen

water temperatures at Jinghong were higher than the historical temperatures in the same location by an average of ~3.6 °C, and the 2015 Jinghong water temperatures were more similar to the historical temperatures at far away Chiang Saen than to the historical Jinghong temperatures. It is also apparent in Fig. 5.7 that the greatest differences between historical and 2015 Jinghong temperatures were the minimum temperatures, and that this is associated with a difference in the range of temperatures. The minimum water temperature at Jinghong in 2015 was ~6 °C higher than the historical minimum from the same location, and there was a roughly 40% decrease in the range of 2015 temperatures relative to the historical range. Further, timing of maximum and minimum temperatures differed between 2015 and historical values, with 2015 maximum and minimum temperatures occurring approximately 2 months later in the year.

5.3.2 *Jinghong-Guanlei Reach*

Comparison of recent water temperature measurements among hydrological stations, and correlations between water and air temperature in the Jinghong-Guanlei reach, were used to further investigate effects of the Lancang hydropower projects on mainstem Lancang River water temperature, including the portion of the river south of the Buyuan River tributary (where the Lancang River leaves China). Data from the Man'an gauge station, which is located on the Buyuan River tributary, allowed comparison of water temperature between the tributary and the river, both north (Jinghong) and south (Guanlei) of the tributary input (see Sect. 5.1 and

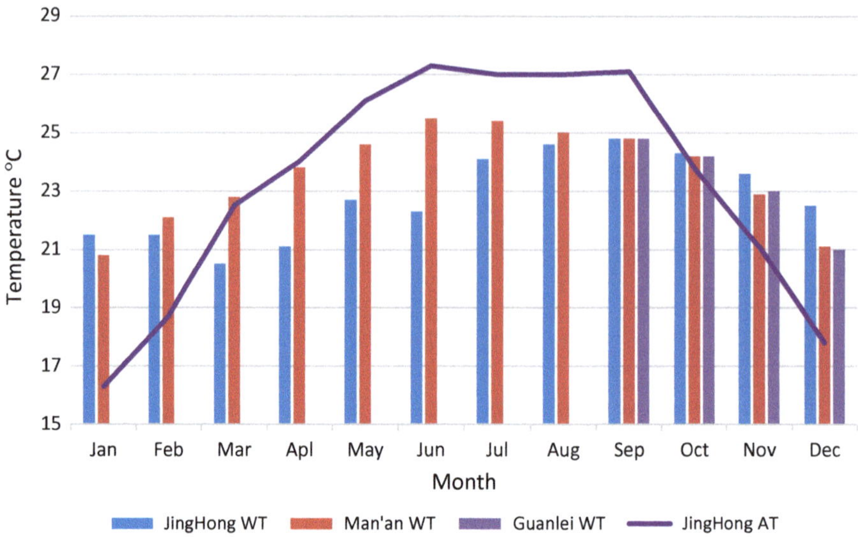

Fig. 5.8 Water temperature (WT) at Jinghong, Man'an, and Guanlei and air temperature (AT) at Jinghong, by month in 2015

Fig. 5.1), and model simulation allowed comparison of effects on water temperature for scenarios with and without a hydropower project.

In 2015, water temperature in the Buyuan River tributary (at Man'an) was higher than that in the mainstem of the Lancang River north of the Buyuan River tributary (Jinghong) from February through September (Fig. 5.8). From September through December, water temperatures in the tributary (Man'an) were similar to those in the Lancang River south of the Buyuan River tributary (Guanlei), and in November and December, they were both slightly lower than that in the mainstem of the Lancang River to the north (Jinghong). As previously demonstrated in Sect. 5.3.1, and as is also evident in Fig. 5.8, water temperature at Jinghong reached maximum and minimum values approximately 2 months after air temperature peaked. Thus the correlation between water temperature and air temperature at Jinghong, which is related to the hydropower plant upstream (see Sect. 5.3.1), is poor ($R^2 = 0.30$; Fig. 5.9). At Man'an, however, air temperature and water temperature are relatively well correlated ($R^2 = 0.84$; Fig. 5.9). Since there are only a number of small run-of-river hydropower projects upstream of Man'an, but there are large dams upstream of Jinghong, the difference in the correlational relationship between air temperature and water temperature in these two locations may reflect the impact of the upstream dam on water temperature.

Comparison of the two hydropower development scenarios (actual scenario with the Jinghong hydropower project and reservoir, and virtual scenario without the Jinghong project and reservoir) at Man'an and at Guanlei allowed investigation of the effects of a reservoir on downstream water temperatures in the tropics (i.e., on the Mekong-Lancang River) and on tributary effects (i.e., effects of the Buyuan River on the Lancang River with and without a reservoir). Model results (Fig. 5.10) indicate that temperatures at Guanlei between January and May are consistently

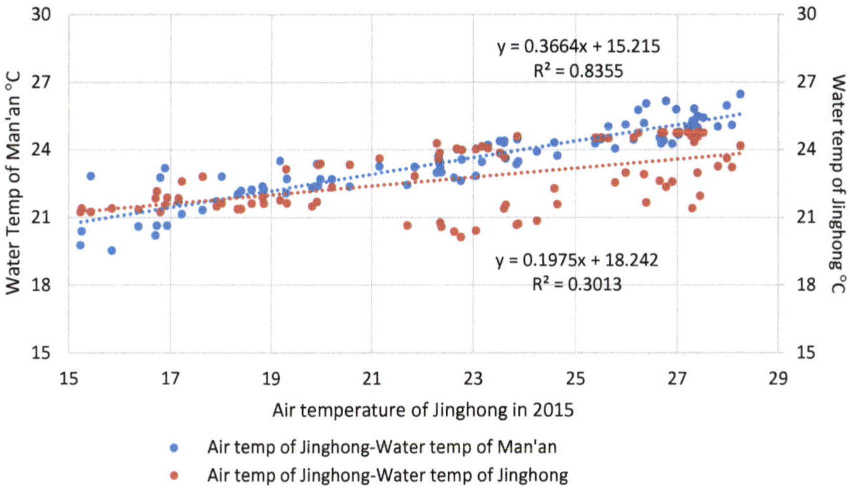

Fig. 5.9 Linear relationships between air temperature at Jinghong and water temperature at Jinghong and Man'an in 2015 (data points represent 5-day average values)

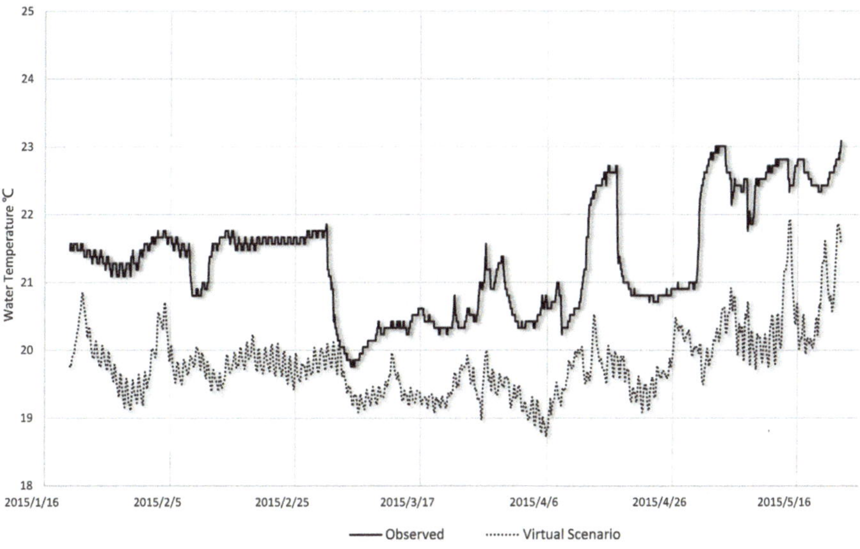

Fig. 5.10 Comparison of water temperature at Guanlei for the actual (observed) and virtual development scenarios (data points represent daily average water temperature)

higher in the actual than the virtual development scenario. On average, over these months, modeling results indicate a difference of 1.2 °C (averages of 20.4 °C and 21.6 °C for virtual and actual scenarios, respectively). In addition to a higher average temperature, the range of temperatures at Guanlei is also greater in the actual scenario than in the virtual scenario (Fig. 5.10). This difference is due to the thermoregulatory characteristics of large bodies of water (i.e., reservoir) that leads to temperature stabilization. Temperature stabilization occurs because the heat

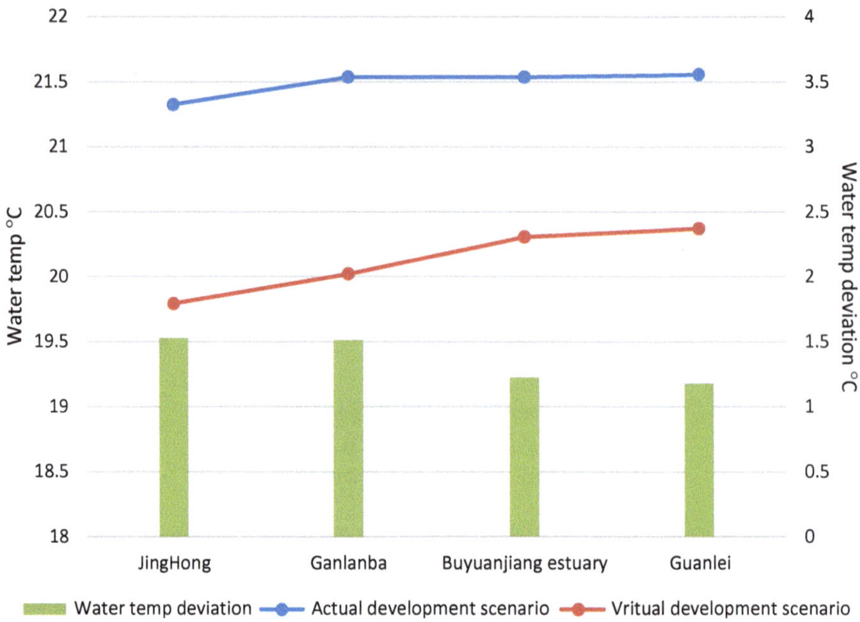

Fig. 5.11 Water temperature in the Lancang River at Jinghong, Ganlanba, Buyuanjiang estuary, and Guanlei under the actual and virtual development scenarios and the temperature difference between scenarios (deviation)

exchange rate between the water and the surroundings is reduced when water is contained within a large body. Without the reservoir (virtual scenario), the Jinghong reach of the river does not have a temperature regulating mechanism, and thus temperature ranges are greater.

Comparison of the difference between the two development scenarios in a number of locations along the Lancang River demonstrates the effect of distance from the nearest dam on water temperature changes in the Lancang River (Fig. 5.11). Among the locations considered, the planned Ganlanba dam is located ~40 km downstream from the Jinghong hydropower project, the Buyuan River confluence is located another ~20 km downstream from the planned Ganlanba dam, and the distance between the Buyuan River confluence (Buyuanjiang estuary) and Guanlei is ~30 km (see Figs. 5.1 and 5.2). Thus, the distances between the Jinghong dam and Ganlanba, the Buyuanjiang estuary, and Guanlei are approximately 40, 60, and 90 km, respectively. Modeling results indicate that the overall pattern in both scenarios is for water temperature to increase between Jinghong and Guanlei and for water temperatures to start higher in the actual compared to the virtual scenario. However, in the virtual scenario (no dam), the magnitude of water temperature change is greater, and the water temperature gradually increases along the course of the river. In the actual scenario, the temperature increase is most pronounced between Jinghong and Ganlanba, after which it stabilizes. In the actual scenario, the temperature gradient between Jinghong and Guanlei is 0.3 °C/100 km, whereas in the virtual scenario, it is approximately 0.7 °C/100 km.

Modeling results also indicated that the effect of the dam on water temperature decreases with increasing distance from the dam (Fig. 5.11), with the difference between development scenarios being greater at Jinghong and Ganlanba (1.5 °C) than at Guanlei (1.2 °C). Although other factors could play a role in the water temperature differences between scenarios, temperature increases in the Nuozhadu-Jinghong reach are mainly due to solar radiation and heat exchange with the air. Temperature inputs are identical in each scenario, flow rates are similar, and the difference in the surface area of the body of water due to differences in flow rate is insignificant. Thus, the amount of solar radiation received by the water in each scenario is the same. As such, the difference in water temperature between scenarios reflects the existence of the upstream hydropower project (Jinghong dam and reservoir) which has a moderating effect on water temperature variation downstream.

5.3.3 South of the Chinese Border (Chiang Saen)

Comparison of monthly average water temperatures in the Mekong River at Chiang Saen (Thailand) among the three time periods (before (baseline), during, and after construction of the Jinghong dam) indicates that an increase in water temperature in the Mekong River south of the Chinese border followed construction of the Jinghong dam (Fig. 5.12). More detailed analysis revealed that yearly average water temperature in the construction phase was 0.5 °C lower than during the baseline period, and post-project water temperature was 1.1 °C higher than baseline (Table 5.4, Fig. 5.13). The biggest water temperature increase was between July and December, when the

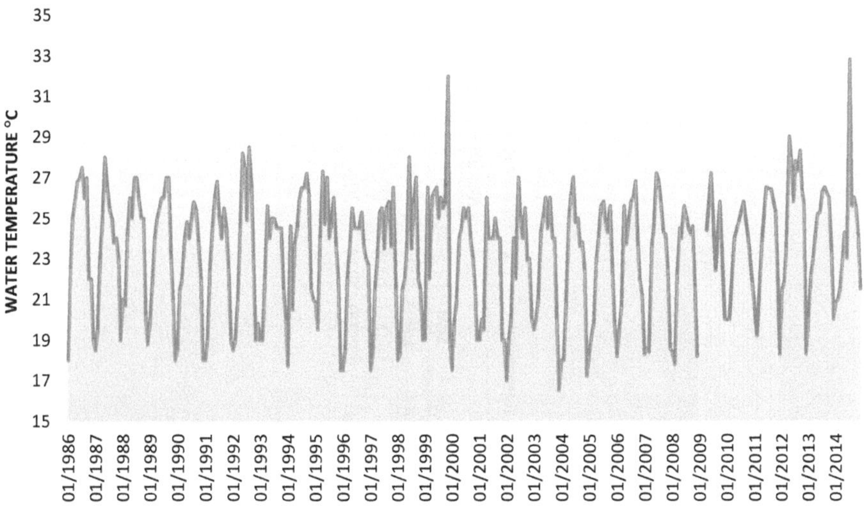

Fig. 5.12 Monthly average water temperature in the Mekong River at Chiang Saen from 1986 to 2014

Table 5.4 Water temperature in the Mekong River at Chiang Saen before, during, and after construction of the Jinghong dam

Period	Year	Monthly average water temperature (°C)												
		Jan	Feb	Mar	Apr	May	Jun	Jul	Aug	Sep	Oct	Nov	Dec	Average
Before construction	1986–2002	18.6	20.4	22.1	24.7	25.2	26.1	25.1	25.5	25.6	23.8	22.4	19.4	23.2
construction	2003–2008	18.4	18.9	20.5	24.3	24.6	26	25.5	25.4	24.5	24.3	21.8	18.6	22.7
Post construction	2009–2014	20.8	20.8	22.7	24.7	25.8	25.4	28.9	25.7	26.8	25.5	24.3	20	24.3

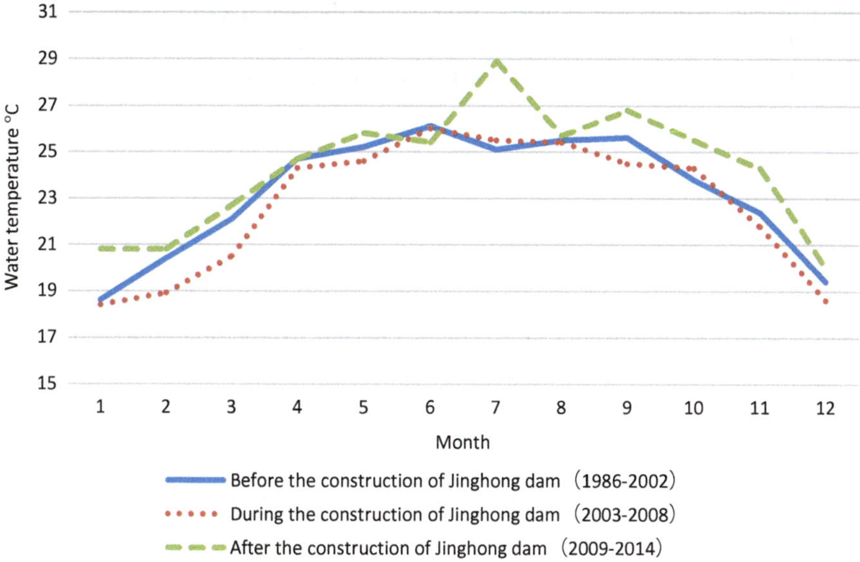

Fig. 5.13 Monthly average water temperature in the Mekong River at Chiang Saen before, during, and after construction of the Jinghong dam

temperature difference between pre- and post-project was 1.6 °C. In contrast, the difference between February and June was less than 0.1 °C. Differences pre- and post-project were pronounced during the months when lowest and highest temperatures were recorded. The lowest water temperature recorded during the baseline period was in January, when baseline (18.6 °C) and post-project (20.8 °C) averages differed by 2.2 °C. The highest water temperature recorded during the baseline period was in June (26.1 °C), when post-project water temperature (28.9 °C) was 2.8 °C higher. Temperature variability also differed among the three periods, with greatest variability observed during construction: the coefficient of variation (CV) of monthly average water temperature in the construction phase (2.91) was higher than that during baseline (2.61) and post-project (2.69) periods.

Existing data indicate that water temperatures at Chiang Saen increased after the operation of Jinghong project; however, the Jinghong and Nuozhadu projects were completed in 2009 and 2012, respectively; thus post-project water temperature data for Chiang Saen were only available for a few years. In addition, water temperature at Chiang Saen may have been impacted by factors other than hydropower projects on the Lancang River (e.g., hydrology, climate); thus observed changes in water temperature may not be entirely attributed to the presence of hydropower projects. Nevertheless, analyses conducted above, including those for the Nuozhadu-Jinghong reach (Sect. 5.3.1) that investigated historical and post-project latitude gradients of northern and southern river segments and found a relationship with latitude that is the reverse of that observed in natural river systems, provide confidence that the effects observed at Chiang Saen are due to the construction and operation of the Lancang hydropower projects. However, additional data in future years are needed to fully investigate the effect of hydropower project-related water temperature effects over large distances downstream, which may be complex and variable.

5.4 Conclusions

Comparisons of historical and recent water temperature data at selected hydrological stations, in conjunction with modeling of actual and virtual development scenarios, have demonstrated the effects of the Lancang hydropower projects on water temperature downstream. Results indicate that relative to baseline, there is a general increase in water temperature downstream of the Lancang projects, a reduction in the range of water temperature, and a shifting in the timing of peak temperatures, such that maximum and minimum values are delayed relative to historical values and relative to air temperature. For example, after the Nuozhadu project began operating at a normal level (2015), the yearly average water temperature relative to baseline increased by 2.1 °C downstream of the Nuozhadu dam. However, although yearly average temperatures increased relative to baseline downstream of the Nuozhadu dam in all post-project years, seasonal differences were apparent in monthly average temperatures, as was the decrease in temperature range: relative to baseline, water temperature in winter (November to January) increased by up to 6.6 °C and in summer (May–July) decreased by up to 3.0 °C.

Water temperature may be impacted by factors other than hydropower projects on the Lancang River, and the post-project period was limited to a few years of data; thus additional analyses were conducted to improve confidence in the link between observed temperature changes and the presence of hydropower projects on the Lancang River. Firstly, comparison of 2015 temperatures upstream and downstream of the Jinghong dam indicated that water temperature increases as the water flows through the Jinghong hydropower project. Secondly, a modeling comparison of actual and virtual development scenarios (in which only the presence of the Jinghong power station and reservoir is different between scenarios) indicated that temperatures at Guanlei between January and May are consistently lower (by 1.2 °C) in the virtual scenario, which excluded the effects of the Jinghong power station and reservoir, than the actual scenario. Thirdly, an investigation of historical and post-project latitude gradients of northern and southern river segments indicated that the post-project northern gradient was substantially greater than the historical southern gradient, even though this is contrary to the relationship between latitude and temperature change under natural conditions.

A key component of the assessment of transboundary water temperature effects is to evaluate changes to temperature effects with distance downstream. Comparison of water temperature changes from Jinghong and Guanlei using actual and virtual development scenarios and data from gauges located at various distances from Jinghong indicated that, although starting water temperatures are higher in the actual than the virtual scenario, an effect of the Jinghong dam and reservoir is to reduce the magnitude of temperature change that naturally occurs through southerly flow. Modeling results also indicated that in the actual scenario, the rate of temperature change decreases rapidly with increasing distance from Jinghong, whereas in the virtual scenario it decreases more gradually. Further, the difference between development scenarios decreased with distance from the dam. Analysis of temperature effects at Chiang Saen (Thailand) explicitly investigated the potential for transboundary water temperature effects. This analysis showed that following construction of the Jinghong dam, yearly average water temperature at Chiang Saen was 1.1 °C higher than baseline and that pre- and post-project differences were most pronounced during the months when lowest and highest temperatures were historically recorded. Nevertheless, confidence in these conclusions is limited owing to the short period of time that the Lancang hydropower projects have been operational, and further analyses documenting temperature patterns and identifying causes are required.

Reference

Zhang, Y., F. Gao, D. He, and L. Li. 2007. Comparison of spatial-temporal distribution characteristics of water temperatures between Lancang River and Mekong River. *Chinese Science Bulletin* 52 (Suppl II): 141–147.

Chapter 6
Transboundary Environmental Effects of Hydropower: Sediment Transport and Geomorphology

6.1 Overview

Downstream effects of the Lancang cascade of hydropower projects on sediment load, sediment transport, and river channel geomorphology were investigated through a combination of literature review and analysis of suspended sediment data recorded at hydrological stations on the Lancang-Mekong River. For the analysis of sediment transport and channel morphology, it was first important to establish the background necessary to put potential changes due to hydropower projects into perspective. An understanding of sediment transport and channel morphology, and the potential for hydropower projects to affect these processes, is dependent on an understanding of the physical, biological, and anthropogenic factors that affect them. These include sediment yield of the adjacent land base through soil erosion, sediment load and transport within the river system, and key aspects that govern channel morphology. Providing background material on these topics required a thorough literature review and compilation of information relevant to our analysis of transboundary effects.

The sections below introduce key concepts relevant to the investigation of sediment transport and geomorphology including topography, climate, soil erosion, sediment loading, and channel morphology in the Lancang-Mekong Basin. Literature review and analysis of sediment concentration data from the 1960s to 2010 in the Lancang-Mekong River Basin were then used to quantify sediment yield and load and investigate the potential for transboundary effects of the Lancang hydropower projects on sediment transport. A semiquantitative model along with a summary of results of a key study was then used to evaluate potential transboundary effects on channel geomorphology.

X. Yu et al., *Balancing River Health and Hydropower Requirements in the Lancang River Basin*, https://doi.org/10.1007/978-981-13-1565-7_6

6.1.1 Topography, Climate, Soil Erosion, and Sediment Load

Sediment load of a river basin is affected by a variety of natural and social factors. These include climate, rainfall intensity, vegetation cover, soil type, slope, slope direction, land use, and land use practices. Because there is potential for these factors to vary greatly, there is also variability in the amount of regional soil erosion that occurs (sediment yield) and the type of sediment that is transported into the river system (e.g., sediment source, composition, content, means of erosion). Sediment transport within the river system can also be affected by the construction of hydropower projects that affect the river's natural flow regime.

In general, the Lancang River Basin has high potential for erosion, partly due to topography and partly due to vegetation cover and land practices. Topography is variable, and mountains can be steep. River valleys, which are generally characterized as "V" in shape, are strongly incised and have tributaries and gullies that have formed a pinnate drainage pattern. Hilly fields with steep slopes make up a substantial proportion of the land base. Slopes ≤8° account for only 4.8% of the total land area, whereas slopes >25° account for 41.8% of the total land area. There are many barren hills and slopes, and the arable land is limited. The forest cover is 46.1%, but the distribution of forest is uneven (He 1995). The altitude in the watershed decreases from north to south, while temperature and precipitation increase from north to south. Climate and topographical conditions of the study area that affect soil erosion and river sediment load, as described by He (1995) and summarized here, can be roughly delineated into two sections; the middle and lower reaches of the Lancang River.

The land surrounding the middle reach of the Lancang River, from Liutongjiang to Gongguoqiao, has a transitional climate that varies from frigid to subtropical. The average annual rainfall is 650–1100 mm, and the annual average temperature is 5–16 °C. Topography is described as rolling terrain, and elevation is highly variable, with an average altitude of 2520 m. The river is fast flowing, and deep canyons have been incised by its erosionary forces. Bedrock is exposed on both banks of the river where vegetation cover is sparse. Owing to strong mechanical and chemical weathering forces, slope erosion, and sediment transport capacity, combined with the geological structure and substantial metamorphism, river sediment concentration is high. Human activities are frequent.

The lower Lancang River, from Gajiu to the China-Myanmar border, has a subtropical climate and low terrain, with an average altitude of 1540 m. The annual average temperature is 17–22 °C, and the annual average rainfall is 900–1700 mm. Vegetation cover (e.g., forest cover) is greater than in the middle reaches of the Lancang River, and human activities are also frequent. Water is abundant, and there are frequent rainstorms. As such, surface runoff resulting from frequent precipitation washes soil into the Lancang-Mekong River.

6.1.2 *Channel Morphology*

Water stored and sediment trapped by reservoirs result in physical, chemical, and ecological effects on the downstream watercourse. Sediment loading in a river can be divided into three components: erosion, equilibrium, and deposition (Kondolf 1997). If the continuity of a river's natural sediment transport regime is interrupted by a dam, the balance of these three components is likely to shift. When a dam causes creation of a reservoir, in which flow is slowed, the sediment load of the water upstream of the dam tends to be deposited in the reservoir. Downstream of the reservoir, the clear, sediment-free water released from the dam tends to erode the channel beds and banks, producing channel incision (downcutting) and causing coarsening of bed material as smaller gravels are transported without replacement from upstream (Brandt et al. 2000; Grant et al. 2003; Kondolf 1997). The clear water released from dams that expends excess energy through erosion is often referred to as "hungry water" (Kondolf 1997). The alteration of stream substrate may be, in turn, associated with negative biological effects (e.g., loss of spawning gravels, changes to the habitat of other aquatic species). Given the potential for changes in flow to affect channel morphology through erosion and deposition and to thereby result in physical and biological consequences, it is important to consider the potential impacts of the Lancang cascade of hydropower projects on downstream channel morphology.

6.2 Methods

Methods of investigating transboundary effects of sediment load and transport and channel geomorphology involved summarizing relevant information from literature reviews, compiling data from studies identified during literature review, and compiling discharge and sediment concentration data from hydrological stations along the main stem of the Lancang River.

6.2.1 *Sediment Load and Transport*

A literature review was undertaken to describe topography, climate, and land use which affect sediment yield and load, characterize and quantify sediment yield and load in the Lancang River Basin, and demonstrate changes of sediment yield and load in space and time. Changes in space and time are important considerations because they may occur for many reasons such as a change in land use and/or land use practices, not only as a result of hydropower project development. Key topics investigated through literature review included differences in erosion rate by region of the Lancang River Basin, changes in erosion rate over time, changes in sediment transport downstream along the Lancang-Mekong River, and temporal changes in sediment discharge in relation to periods of economic development. Assessment of transboundary effects was conducted by including time periods following construction of the Manwan dam and quantifying sediment discharge at varying locations downstream.

6.2.2 *Channel Geomorphology*

Investigation of transboundary effects of the Lancang hydropower projects on channel geomorphology was done for two components, as described in the sections below: longitudinal changes and cross-sectional changes.

6.2.2.1 Longitudinal Changes

The potential for the Lancang hydropower projects to affect channel morphology longitudinally was investigated through a semiquantitative model that analyzes channel morphology changes in the downstream reach of a dam (Grant et al. 2003). The conceptual and analytical framework for predicting geomorphic response of rivers to dams developed by Grant et al. (2003) is based on two dimensionless variables: ratio of sediment supply below the dam to supply above the dam (S*) and fractional change in frequency of sediment-transporting flows (T*). Predicted downstream effects of dams in relation to these two variables can be presented as a bivariable plot of T* and S* (Fig. 6.1).

As described in Sect. 6.1.2, suspended sediment loads decline dramatically when water enters a reservoir, and clear water released from the reservoir can erode the downstream channel. Because fine sediments suspended in the water become replaced with coarser particles downstream of the dam (sediment loading downstream of the dam is supplied by bed load), the ratio of fine to coarse sediment decreases. Further, because sediment particle size increases, the transporting distance of each particle is reduced relative to baseline conditions. Assuming that other hydrological and hydraulic conditions remain the same, this will lead to a more frequent exchange between the suspended sediment and the bed load (Han and Yang 2003). The distance for which this effect persists depends on the discharge rate and frequency of water released from the reservoir, which together define their capacity to transport sediment (Qian et al. 1987; Grant et al. 2003). Grant et al. (2003) applied modifications of discharge and sediment load caused by dam operation to assess the morphology changes of downstream channels semiquantitatively. The two variables, S* and T*, are defined in the model to predict the magnitude and trend of the longitudinal downstream response to dammed rivers.

Longitudinal changes to river morphology in the Lancang River were analyzed in relation to the cascade of hydropower projects using the model of Grant et al. (2003) by predicting responses to S* and T* for seven locations: two in the upper

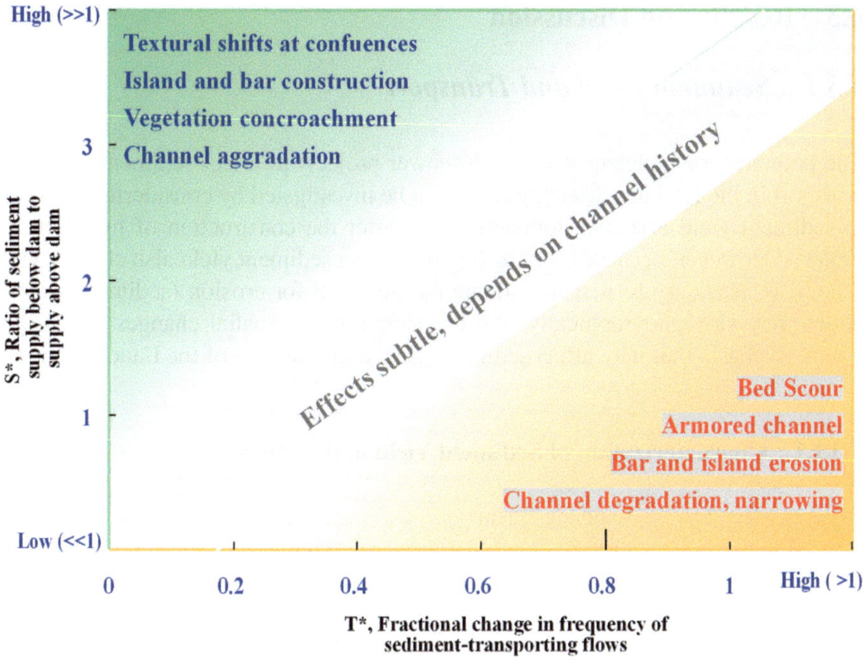

Fig. 6.1 Response domain for predicted channel adjustment in relation to the change in T* and S*. (Reproduced from Grant et al. 2003)

Mekong (Yunjinghong which locates at 3.5 km downstream of Jinghong dam, China-Myanmar border) and five in the lower Mekong (Chiang Saen, Luang Prabang, Mukdahan, Khong Chiam, and Pakse; see Fig. 4.1 in Sect. 4.1). Effects were analyzed for a series of eight cascade dams. For these analyses, T* was estimated from the seasonal storage capacity of the reservoir and the ability of the hydropower projects to regulate the hydrological and hydraulic conditions of the river. Values of S* were taken from sediment loading data from the hydrological stations for three time periods distinguished by the hydropower projects operational at the time.

6.2.2.2 Cross-Sectional Changes

Assessment of the potential for cross-sectional changes to channel morphology caused by the Lancang hydropower projects was conducted through a literature review that identified one pivotal study. Results and conclusions of Fu et al. (2008), who investigated downstream channel erosion triggered by the Manwan dam, were described and summarized.

6.3 Results and Discussion

6.3.1 Sediment Load and Transport

The potential for Lancang River hydropower projects to affect sediment load and transport in the Lancang-Mekong River can be investigated by considering changes in sediment yield and transport before and after the construction of hydropower projects. However, because other factors that affect sediment yield also change with time, it is necessary to first investigate the potential for erosion (sediment yield), which may vary geographically, and the temporal and spatial changes in erosion within the basin that may affect sediment yield and transport of the Lancang River.

6.3.1.1 Characterization of Sediment Yield and Load

Sediment Yield in the Lancang Basin

Four sections of the Lancang River were identified that differ in sediment inputs to the Lancang River: upstream and downstream of the Changdu hydrological station in the upper Lancang River, between Gongguoqiao and Nuozhadu in the middle reach of the Lancang River, and downstream of Nuozhadu.

Upstream of the Changdu station there is little sediment input into the Lancang River. In this area, the Lancang River flows through the Qinghai-Tibet Plateau where both the population density and the cultivated land area are small. Land reclamation rate is low, there is relatively little impact from human activity, and vegetation cover is generally intact and native. The annual rainfall of most areas is about 400 mm, and the surface of the land is frozen for long periods of the year; thus, the water erosion capacity is weak. These factors result in low sediment input, and that which is transported into the river is in the form of suspended sediment load. According to He (1995), the average annual sediment discharge in Tibet is about 8.50×10^6 t.

Downstream of the Changdu station to Gongguoqiao, sediment input into the Lancang River is moderate and mostly composed of bed load. The topography is generally steep with many high mountains and canyons, rainfall is abundant, and there is little arable land. The main external forces that affect soil erosion are freezing, weathering, mass movement (erosion due to gravity), and flowing water (runoff). Human activities mainly include the construction of hydropower projects and riverside roads, as well as the reclamation of steep slopes. Given these erosional characteristics, sediment input into the river is mainly composed of large particles of bed load rather than suspended load. The average annual sediment concentration is only 0.73 kg/m^3. However, flow from tributaries is greater than that of the mainstem, and the soil erosion of Bijiang basin is more prominent. For example, the maximum average sediment concentration of Tangshan station on the Yongchun

River from 1963 to 1979 is about 608 kg/m^3, which is more than twice as much as values at Jiuzhou station (303 kg/m^3).

In Yunnan Province, soil erosion and the potential for sediment input into the Lancang River are greatest in the middle reach of the Lancang River Basin between Gongguoqiao and Nuozhadu. The topography of this area is characterized by broken terrain and deep, narrow river valleys with elevation ranging between 1000 and 2000 m. Here the climate is generally dry and hot, and precipitation levels are low. High rates of soil erosion are caused by poor quality loose soil, poor vegetation cover, and a large population. The area with the highest erosion is the tributary of the Weiyuan River. The erosion modulus of this area is 919 t/km^2, which is the highest in the whole Lancang River Basin (second only to Pijia Station (1080 t/km^2) on the Yili River, which is the tributary of the Jinsha River in Yunnan Province).

In the lower reach of the Lancang River downstream of Nuozhadu, soil erosion and sediment input to the Lancang River are lower than in the upper and middle reaches. In this area, the river valleys are relatively open, and the precipitation and heat are moderate. Soil reclamation conditions are good, and the surface forest coverage is high, which reduces the potential for soil erosion.

Sediment Load in the Lancang River

Current data provide comparatively quantitative summaries on area-specific sediment loading in the Lancang River. Data compiled from 13 hydrological stations in the Lancang River Basin indicate that the sediment concentration in rivers in Yunnan Province ranges between 0.057 and 1.45 kg/m^3. The greatest sediment concentrations were recorded for the Weiyuan River (1.35 to 1.45 kg/m^3), and the smallest concentrations were recorded for the Xi'er River and the Liusha River (0.054 and 0.14 kg/m^3).

Sediment concentration at key hydrological stations on the Lancang-Mekong River was compiled from He (1995) and Lu et al. (2006) (Table 6.1 and Fig. 6.2). Results indicate that sediment concentrations at nine stations from Liutong River to Base range between 0.43 and 1.23 kg/m^3, that sediment discharge at the Yunjinghong station on the lower Lancang River is larger than that at the Chiang Saen station on the upper Mekong River, that the majority of the annual sediment discharge occurs during the flood season, and that sediment discharge increases with distance downstream (Fu et al. 2006) (Fig. 6.2).

Changes in Sediment Yield in the Lancang Basin over Time

Information on changes in soil erosion over time was taken from results of the remote sensing survey of soil erosion in Yunnan in 2000 (Yunnan Province Development Planning Commission 2004). Results from this survey (Table 6.2) indicated that in 1999, the soil erosion area of the Lancang basin was 2.58×10^4 km^2, which is equivalent to 29.15% of the total area of the basin. Comparison to values

Table 6.1 Sediment concentration, transport rate, discharge, and erosion modulus at nine hydrological stations in Lancang-Mekong River

Hydrological station	Sediment concentration (kg/m^3)	Maximum sediment concentration (kg/m^3)	Sediment transport rate (kg/s)	Annual sediment discharge (10^6t)	Annual sediment discharge: Percentage in Jun–Sep (%)	Erosion modulus (t/km^2)
Liutong River[a]	0.8	15	642	20.3	95	296
Jiuzhou[b]	0.73	30.3	682	24.8	95	247
Gajiu[c]	1.06	33.9	1,341	45.6	95	393
Yunjinghong[b]	1.23	19	2,298	90.8	96	514
Chiang Saen[d]	0.645	–	1,719	54.2	–	287
Luang Prabang[d]	0.483	–	1,906	60.1	–	224
Langkai[d]	0.521	–	2,388	75.3	–	251
Mudahan[d]	0.461	–	3,561	112.3	–	287
Base[d]	0.431	–	4,198	132.4	–	243

[a]1965–1992 (Lu X. 2006)
[b]1965–2005 (Lu X. 2006)
[c]1956–1985 (He 1995)
[d]1962–2000 (Lu X. 2006)

for 1987 indicated that the total soil erosion area in the basin has decreased by 217.93 km^2 (0.8%) in the last 12 years. Classification of erosion area by intensity further indicated that the reduced erosion area in 1999 relative to 1987 was due entirely to a change in light erosion (1118.60 km^2) and that erosion area within moderate, intensive, and very intensive erosion intensity categories all increased between 1987 and 1999. Increases in erosion area for moderate and intensive erosion were similar (411.63 km^2 and 487.79 km^2, respectively), but the rate of change since 1987 differed substantially (4.95% for moderate, 80.44% for intensive).

Region-specific differences in changes in erosion amounts and intensities were also observed (Table 6.2). Diqing had the largest decrease in light erosion (734.43 km^2) and the second largest decrease in total erosion (536.95 km^2). This result is largely attributable to the long period during which the soil is in a frozen state in this area. Puer and Lincang also had relatively large decreases in light erosion (394.22 km^2 and 378.48 km^2, respectively), although their total erosion areas increased slightly. Regions with large increases in erosion of moderate intensity were Puer (493.02 km^2), Lincang (386.56 km^2), Dali (267.10 km^2), and Diqing (177.05 km^2). Regions with large increases in intensive erosion areas were Nujiang (226.32 km^2), Lincang (110.98 km^2), and Dali (81.55 km^2). Very intensive erosion was only recorded for Dali (1.25 km^2) and only in 1999. Xishuangbanna had the greatest decrease in erosion area over all intensities combined (540.47 km^2), and Dali had the largest increase (388.62 km^2).

Due to the increase of moderate and intensive erosion areas from 1987 to 1999 in all regions combined, in combination with the reduction in total basin area, erosion

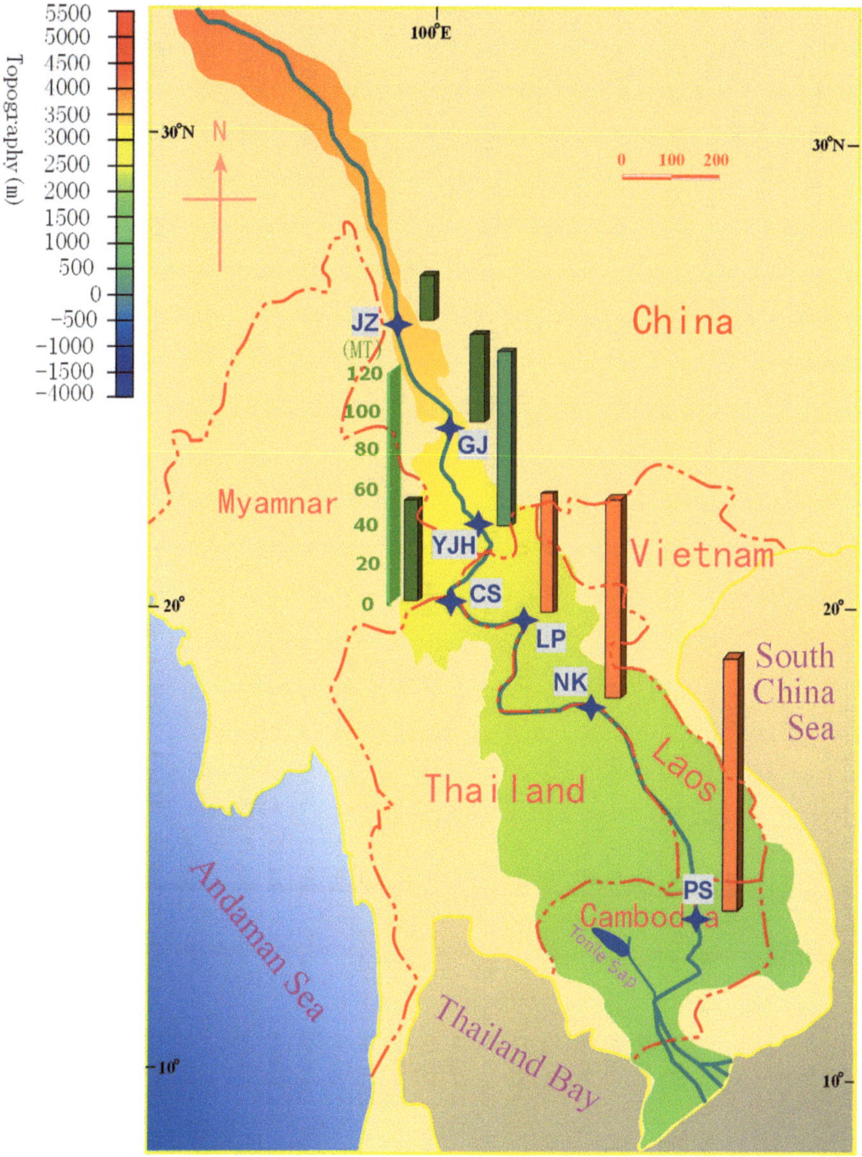

Fig. 6.2 Sediment discharge (10^6 t) of seven hydrological stations in the Lancang-Mekong River. (Reproduced from Fu et al. 2006)

depth and erosion modulus (soil erosion amount per unit area) also increased. From 1987 to 1999 the total average annual erosion amount increased from 0.905×10^8 t to 0.9337×10^8 t, the erosion modulus increased from 1022 to 1055 t/km^2, and the annual erosion depth increased from 0.76 to 0.78 mm. Thus, results indicate that although total erosion area has decreased from 1987 to 1999, erosion within the

Table 6.2 Comparison of soil erosion remote sensing survey results in Lancang River Basin in 1987 and 1999

Region	Year	Soil erosion areas and change		Soil erosion intensity category										Survey results in 1999		
				Light		Moderate		Intensive		Very intensive		Fierce				
		Area (km^2)	Change	Area (km^2)	Change	Area (km^2)	Change	Area (km^2)	Change	Area (km^2)	Change	Area (km^2)	Change	Erosion modulus (t/ km^2•a)	Erosion amount (10^{4}t)	Erosion depth (mm)
Diqing	1999	1,474.65	−536.95	697.45	−734.43	699.47	177.05	77.73	20.43	0	0	0	0	948	628	0.7
	1987	2011.6		1,431.88		522.42		57.3		0		0				
Lijiang	1999	95.64	−56.19	88.38	−54.58	7.26	−1.61	0	0	0	0	0	0	759	29	0.56
	1987	151.83		142.96		8.87		0		0		0				
Nujiang	1999	1,750.56	173.45	1,095.03	97.12	356.2	−149.99	299.33	226.32	0	0	0	0	1,487	650	1.1
	1987	1,577.11		997.91		506.19		73.01		0		0				
Dali	1999	5,781.22	388.62	2,979.13	38.72	2,319.71	267.1	481.13	81.55	1.25	1.25	0	0	1,313	2,189	0.97
	1987	5,392.60		2,940.41		2,052.61		399.58		0		0				
Baoshan	1999	905.72	93.68	377.02	28.13	469.5	59.11	59.2	6.44	0	0	0	0	1,545	344	1.14
	1987	812.04		348.89		410.39		52.76		0		0				
Lincang	1999	4,666.82	119.06	2,010.84	−378.48	2,522.47	386.56	133.51	110.98	0	0	0	0	1,423	1,729	1.05
	1987	4,547.76		2,389.32		2,135.91		22.53		0		0				
Puer	1999	6,101.08	140.87	5,000.50	−394.22	1,057.27	493.02	43.31	42.07	0	0	0	0	768	2,084	0.57
	1987	5,960.21		5,394.72		564.25		1.24		0		0				
Xishuang banna	1999	5,028.29	−540.47	3,729.82	279.14	1,298.47	−819.61	0	0	0	0	0	0	887	1,684	0.66
	1987	5,568.76		3,450.68		2,118.08		0		0		0				
Total	**1999**	**25804**	**−217.93**	**15978.17**	**−1118.6**	**8,730.35**	**411.63**	**1,094.21**	**487.79**	**1.25**	**1.25**	**0**	**0**	**1,055**	**9,337**	**0.78**
	1987	**26,021.91**		**17,096.77**		**8,318.7**		**606.42**		**0**		**0**				

Reproduced from Yunnan Provincial Development Planning Commission (2004) decreased area has become more intense, and there is substantial variability among regions in the intensity of erosion and the direction of change in erosion area.

6.3.1.2 Evaluation of Transboundary Effects on Sediment Load Due to Hydropower Projects

Before the first dam (Manwan) was commissioned on the Lancang River in 1993, transport of suspended sediment was relatively unimpeded within the Lancang-Mekong River, and human activities were the main causes of the alteration of natural sediment transport mechanisms. However, the construction of the dams in the Lancang River Basin had the potential to have substantial impacts on sediment transport, given that dams and reservoirs affect key characteristics of the river's flow regime, which in turn affect factors that influence the amount of sediment that is suspended within the river, such as deposition, transport capability, and channel erosion.

Many studies of sediment transport in the Lancang-Mekong River were based on sediment load data recorded at the hydrological stations (Roberts 2001; Wolanski et al. 1996; You 1999, 2001; Kummu and Varis 2007; Lu et al. 2006). Most of the previous work estimated that ~80 million tons (MT) of sediment is drained from the Chinese Lancang River (upper Mekong) into the lower Mekong Basin. Further, previous authors concluded that the sediment loads to the Mekong Delta and Tonle Sap system were reduced by as much as a half relative to previous loads (160 MT) owing to sediment trapping in China by the cascade of hydropower projects on the Lancang River. These conclusions assumed that sediment loads from China to the lower Mekong Basin would have been fully transported to the Mekong estuary and would not have settled along the way.

Comparison of sediment loads at Yunjinghong and Chiang Saen hydrological stations suggests that the assumption made regarding sediment transport in previous studies may be incorrect. Rather than finding a greater sediment load at downstream Chiang Saen than at Yunjinghong, which would be predicted from previous assumptions of complete sediment transport, results indicated that the load of suspended sediment was greater at Yunjinghong (~80 MT) than at Chiang Saen (~50 MT). Thus, although prevalent opinion was that half of the sediment loads of the Mekong River (160 MT) originated from China, this conclusion is based on a dataset that has low comparability and that may exaggerate China's contribution to sediment load. The cross-sectional sediment concentration was always replaced by surface water sediment concentration for sediment monitoring in the Mekong regions, so it is believed that the downstream sediment discharge was underestimated. The dataset need to be revised according to the established relationship between surface and cross-sectional sediment concentration and be compared to the sediment fluxes from the upper Mekong Basin.

In addition, annual suspended sediment load (SSL) distribution curves also show that SSL at Yunjinghong and Chiang Saen are in two different groups (Fig. 6.3). The next section provides discharge and SSL data at Yunjinghong and Chiang Saen for assessing the impact of the Manwan dam on downstream sediment loads.

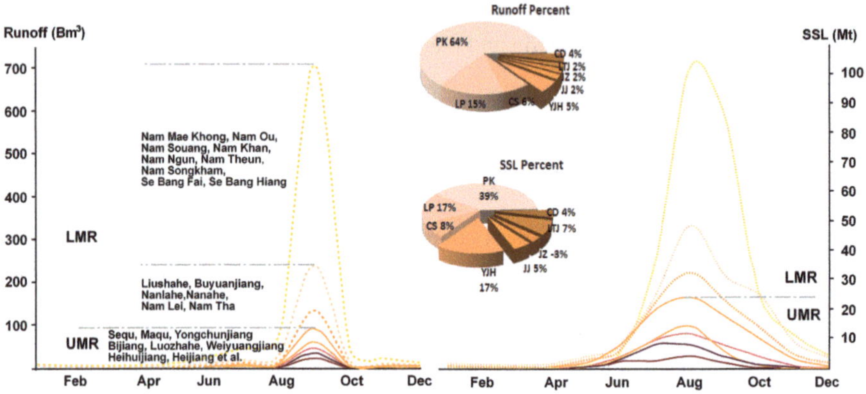

Fig. 6.3 Annual suspended sediment load (SSL) distributions along the Lancang-Mekong River. (Data from He 1995) and Lu and Siew 2006)

Eight cross sections on the mainstem Lancang-Mekong River were selected from Changdu in China to Pakse in Laos for the comparison of sediment fluxes. The annual water discharge measured in the river reaches of Changdu, Liutongjiang, Jiuzhou, Gajiu, and Yunjinghong were 4%, 2%, 2%, 2%, and 5% of the water discharge measured at Pakse, while sediment discharges were 4%, 7%, −3%, 5%, and 17% of the sediment discharge of Pakse. Accumulated water discharge and sediment discharge from Changdu to Yunjinghong are 15% and 30%, respectively. If taking the reach from Yunjinghong to Myanmar-Laos boundary into account (including the Liusha River, Nana River, Buyuan River, Nanla River, and Nam Lwe River), the ratio of water and sediment became 18% and 35%. The reach between Yunjinghong and Chiang Saen contributed around 6% and 8% of the water and sediment flux of Pakse, and the reach between Chiang Sean and Luang Prabang provided 15% and 18%. Since most of the large tributaries are concentrated in the reach between Luang Prabang and Pakse, the water and sediment made up 64% and 42% of the discharge at Pakse.

Comparison of sediment discharge values at four hydrological stations over the last 60 years, including before and after the construction the Manwan dam, allowed more specific assessment of impacts of the Lancang hydropower projects on sediment transport downstream (Table 6.3 and Fig. 6.4). From the 1960s to the 1970s, when the Lancang River Basin was relatively undeveloped, sediment discharge was relatively low. However, as land use changed due to economic development after Chinese economic reform, forest coverage decreased due to deforestation, and this exacerbated soil erosion which caused more sediment to be flushed into the river system. This increase in sedimentation is evident from the substantial increases in sediment discharge seen at all stations shown in Table 6.3 and Fig. 6.4 between 1978 and 1992. After the impoundment of Manwan reservoir in 1993, sediment discharge at downstream stations declined steeply. At Gajiu (also referred to as Jiajiu), which is several kilometers downstream of the Manwan dam, the observed

Table 6.3 Average sediment discharge at three hydrological stations prior to (1964–1977 and 1978–1992) and following (post 1993) impoundment of the Manwan reservoir. Note that Jiuzhou and Gajiu are also referred to as Jiuzho and Gajiu, respectively

Hydrological station	Location relative to Manwan reservoir	Sediment discharge (MT per year) and percent change from preceding period		
		1964–1977	1978–1992	Post 1993
Jiuzhou	Immediately upstream	21.96	26.94 (+22.7%)	27.06 (+0.4%)
Gajiu	~4 km downstream	38.74	54.34 (+40.3%)	19.38 (−64.3%)
Yunjinghong	~400 km downstream	73.05	108.45 (+48.6%)	26.66 (−75.4%)

change in sediment discharge indicates that 64.3% of the sediment that had been discharged during 1978 to 1992 (54.34 MT) was now being retained by the Manwan reservoir. At Yunjinghong, which is ~400 km downstream of the Manwan dam, the decrease in sediment discharge post-project was even greater. However, the sediment yield between Gajiu and Yunjinhong increased by 18.1 MT (40%) in 1993 (Fu 2006), possibly due to increased sediment input from tributaries or due to the "hungry water" effect that results from the release of sediment-free water downstream of a dam (Kondolf 1997). In contrast, since the Jiuzhou hydrological station is immediately upstream of the Manwan reservoir, little change in sediment discharge was evident between the post-1993 and 1978–1992 periods.

Data from Chiang Saen hydrological station in northern Thailand showed a slightly different pattern of sediment discharge over time than the other stations. Sediment discharge results at this station identified three distinctive sediment discharge periods between 1986 and 2004 (Fig. 6.4 and Fig. 6.5). Sediment discharge was high between 1986 and 1992 (average of 131.26 MT per year), and this dropped substantially after 1991 (to an average of 69.46 MT per year from 1993 to 1997) owing to impoundment of the Manwan reservoir. Sediment discharge then increased slightly again during 1998–2004 period (to an average of 83.16 MT per year). To allow comparison with the Yunjinghong station, sediment discharge at Chiang Saen in the 1986–1992 period (131.26 MT per year) was also compared to the 1993–2004 period (76.31 MT per year). The difference between these periods, a decrease of 54.95 MT per year (41.8%), is almost twice the absolute decrease observed at Yunjinghong (26.66 MT per year) during a similar time period (Table 6.3). This suggests that the low sediment discharge observed at Chiang Saen as of 1992 was likely due to other environmental factors that aggravated the sediment reducing effects of the Manwan dam.

The loss of sediment discharge observed at Yunjinghong after 1993 could also be related to the Dachaoshan and Jinghong dams, which were completed in 2003 and 2009, respectively. Data up to 2010 show that there was a steep decline in sediment discharge at Yunjinghong in 2003 (when the Dachaoshan dam was completed) and another one in 2008 (when the Jinghong dam was completed) and that sediment discharge after 2008 was very small (Fig. 6.5). Figure 6.5 also shows that there was

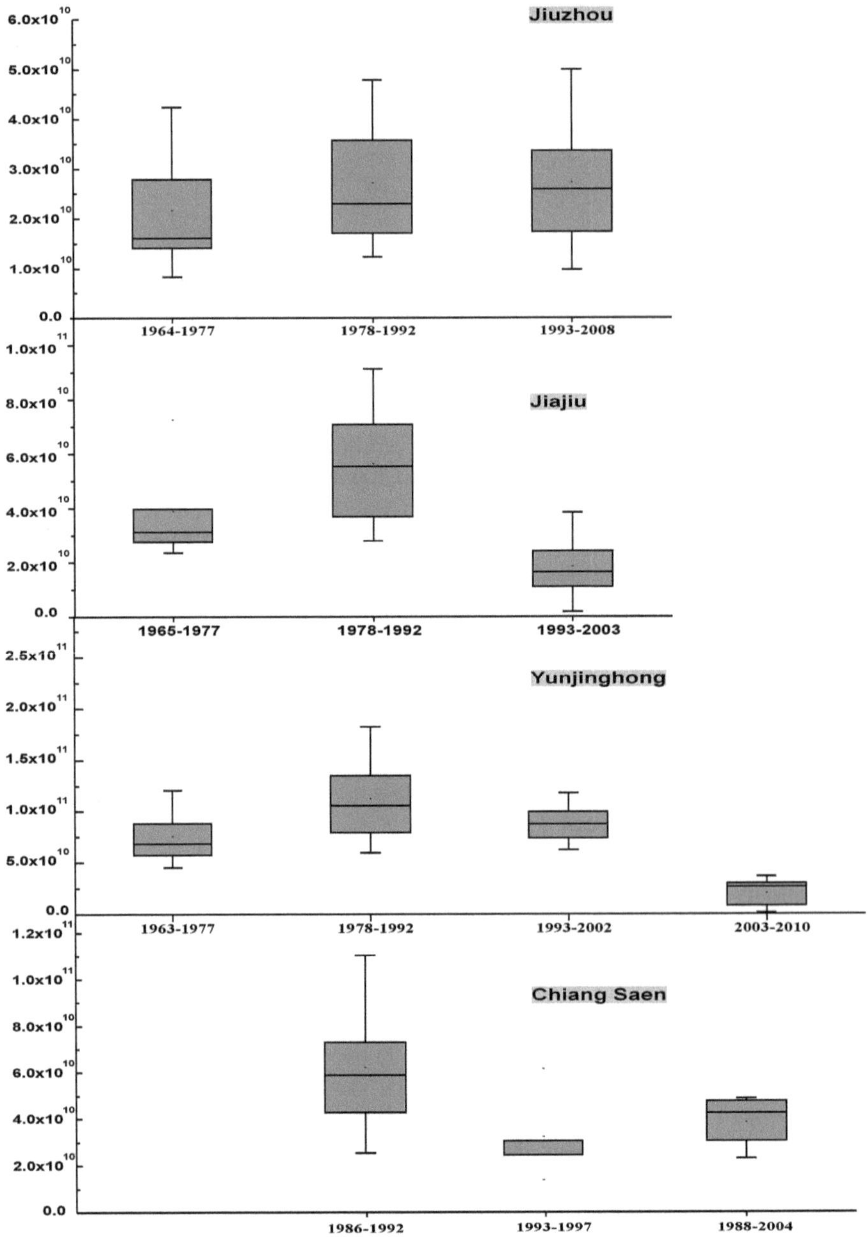

Fig. 6.4 Sediment discharge at four hydrological stations (arranged north to south) along the Lancang-Mekong River prior to (1964–1977 and 1978–1992) and following (1993–2010) impoundment of the Manwan reservoir. Note that Jiuzho and Jiajiu are also referred to as Jiuzhou and Gajiu, respectively

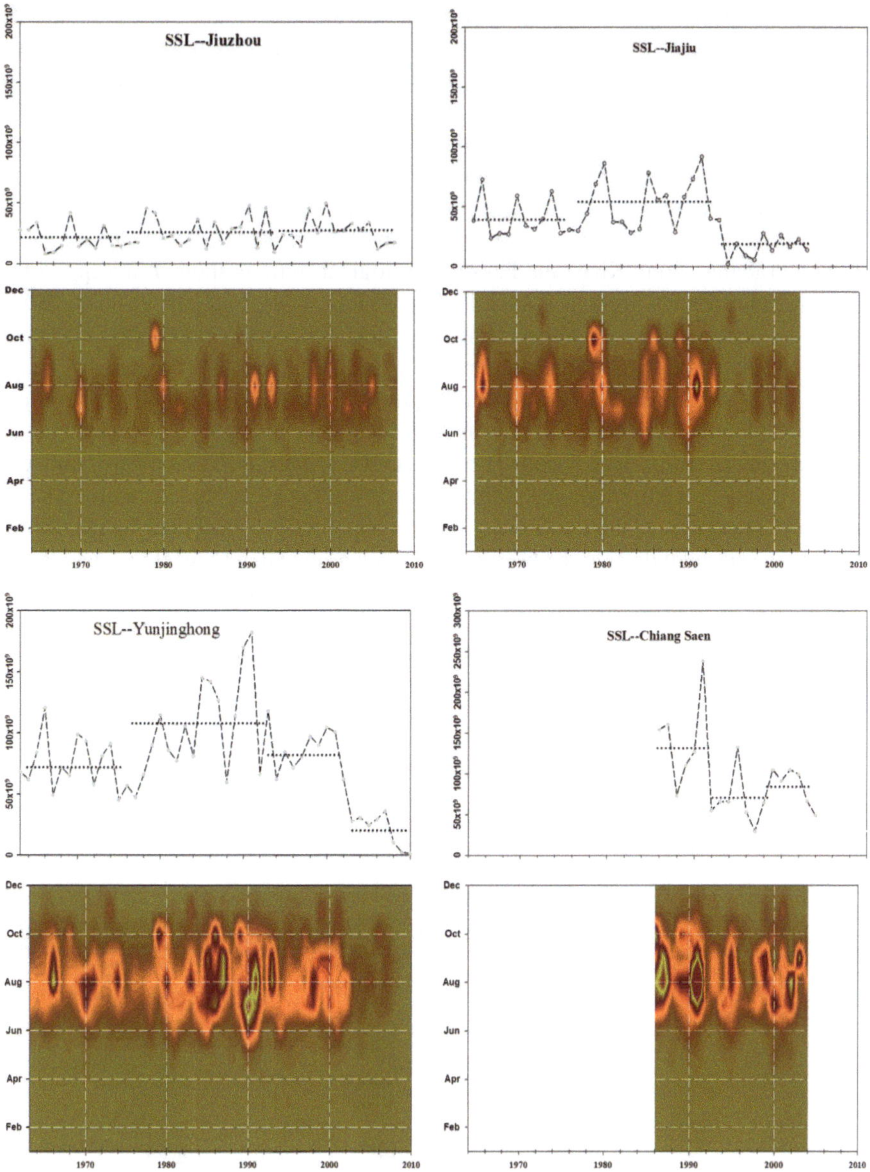

Fig. 6.5 Sediment discharge at Jiuzhou, Yunjinghong, Gajiu (Jiajiu), and Chaing Saen in MT from 1960 to 2010 and by month during this period

a similar decline observed in 1993 when the Manwan dam was completed (but that sediment discharge increased again somewhat after this and prior to the decline noted in 2003) and that annual sediment discharge peaks occur during the flood season.

Suspended sediment data at four hydrological stations in the lower Mekong River (see Fig. 4.1 in Sect. 4.1 for station locations) were summarized to evaluate the potential for sediment discharge impacts from the Lancang cascade of hydropower projects to extend downstream past Chiang Saen. Results from Chiang Saen, Luang Prabang, Nong Khai, and Pakse hydrological stations show little support for declines in sediment loads at Luang Prabang, Nong Khai, and Pakse stations after impoundment of Manwan reservoir in 1993 (Fig. 6.6). It is evident, however, that sediment loads are higher at Luang Prabang, Nong Khai, and Pakse than at Chiang Saen. These results suggest, firstly, that the impact of the Lancang hydropower projects on sediment loading in the Lancang-Mekong River are not easily detectable at great distances from the hydropower projects. The effects of the Manwan dam were clearly detected at Yunjinghong, which is about 400 km away, and were also detectable at Chiang Saen, although other factors apparently contributed to the strong decline in sediment discharge observed at this location at that time. However, effects do not appear to persist hundreds of km into the southern Mekong River. Secondly, the increase in sediment yields with distance downstream in the Mekong River likely reduces the potential for the Lancang hydropower projects to have significant effects over great distances downstream.

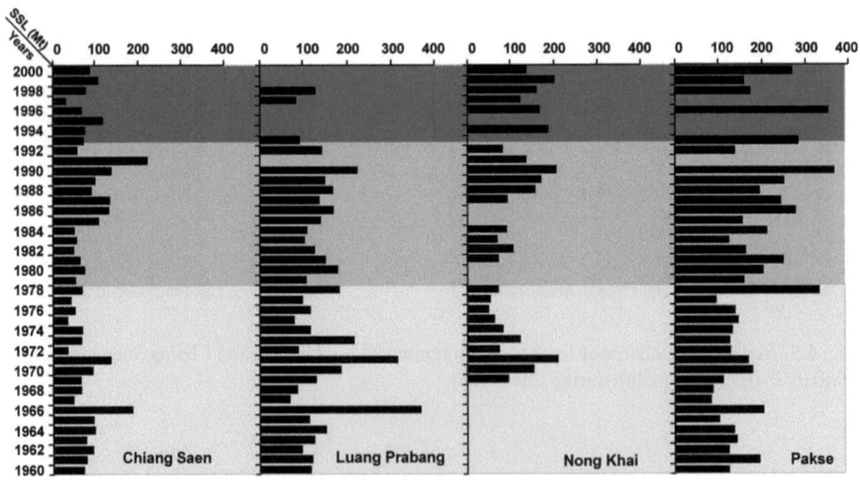

Fig. 6.6 Sediment loads at four hydrological stations (arranged north to south) along the Lancang-Mekong River from 1960 to 2000. Gray shading is used to identify periods prior to 1978, from 1978 to 1993, and from 1993 to 2000

6.3.2 *Channel Geomorphology*

6.3.2.1 Longitudinal Changes

In the analysis of the response of channel morphology changes to the eight cascade dams within three time period scenarios, it was assumed that, because the cascade dams all have a strong ability to regulate the hydrological and hydraulic conditions of the river, the value of T* from the model of Grant et al. (2003) would be between 0.80 and 1.00. Table 6.4 provides the values of S* at the seven locations in the lower and upper Mekong River for the three time period scenarios.

Results of the model indicate that the value of S*, which is the ratio of sediment supply below to above the dam, decreases at each location with each period scenario and falls below 1 for scenarios after 2005 (Table 6.4, Fig. 6.7). This means that there is more sediment below than above the dam for all locations in 2005 except Chiang Saen (where the ratio is never greater than 1) but that this pattern is reversed in the next two time periods and becomes more extreme at Chiang Saen, and the ratio universally continues to decrease between 2010 and 2015. However, owing to the proximity of the hydropower projects to each other, effects are complex. The channel at the Gajiu station, which is immediately downstream of the Manwan dam, were inundated by the backwater of the Dachaoshan reservoir. The Yunjinghong station is close to the Jinghong dam, so the channel and nearby channels was eroded in 2010 and 2015. It is unclear why the channel response to dam construction in the lower Mekong is different from that in the upper Mekong. This phenomenon may be caused by the impenetrable relationship between the sediment loads and discharge described in the previous section. Petts and Gurnell (2005) suggested that after dam closure, there is a migration of channel changes (erosional front) downstream, which may have extreme migration rates of more than 30 km per year in some sand-bed channels. Thus, it will likely take a long time (e.g., tens of years) for the erosion front to arrive in the lower Mekong River and the process of change will be complicated.

Table 6.4 Values of S* at seven hydrological stations along the Mekong River under the 2005, 2010, and 2015 scenarios

	S* values		
Hydrological station	2005	2010	2015
Yunjinghong	1.849	0.32138	0.0036
Border	2.51126	0.5854	0.10364
Chiang Saen	0.85507	0.10715	0.0556
Luang Prapang	1.13357	0.20792	0.10976
Mukdahan	2.75779	0.44033	0.31052
Khong Chiam	2.52334	1.24337	1.07492
Pakse	2.39766	0.75251	0.60143

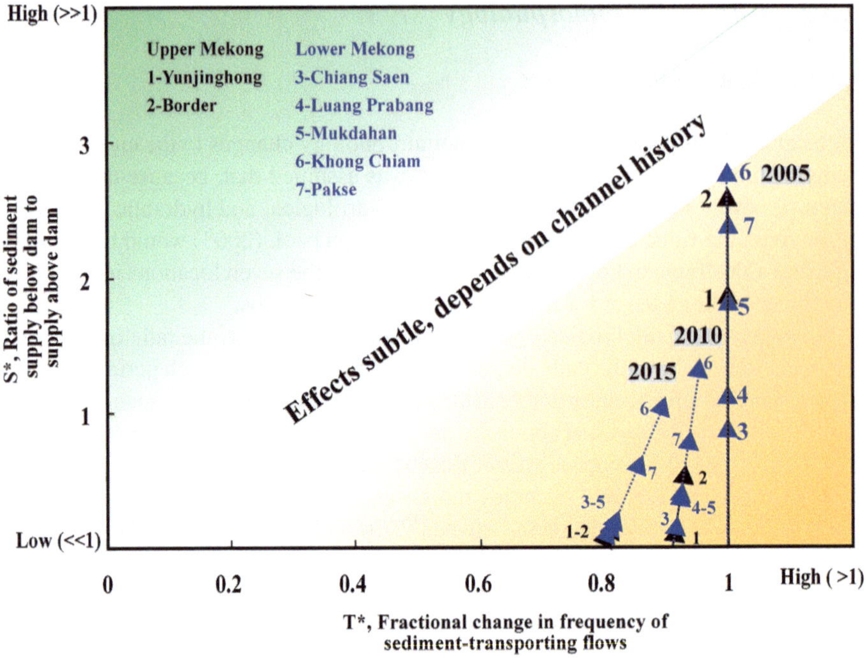

Fig. 6.7 Response of long profiles along the Mekong River to dam constructions under the scenarios of 2005, 2010, and 2015 at seven locations in the upper and lower Mekong River

6.3.2.2 Cross-Sectional Changes

During investigation of downstream channel erosion triggered by the Manwan dam, Fu et al. (2008) concluded that the sediment trapping of the Manwan dam has not caused channel erosion downstream in the mountainous river course at Gajiu (located ~4 km downstream of the Manwan reservoir). This conclusion was based on three key factors. First, the channel substrate in the Gajiu location is largely bedrock, which is not easily eroded (Fig. 6.8). Second, the relationship between water level and discharge did not change after the Manwan dam was constructed (Fig. 6.9). Third, although the relationship between water level and discharge differed at Yunjinghong following construction of the Manwan dam (Fig. 6.10), these changes were not likely to result in increased erosion. For the same discharge rate, water level was higher post-dam than pre-dam, and conversely, for the same water level, discharge was lower post-dam than pre-dam, which suggests that erosion would not be higher post-project relative to pre-project. Based on these results, the authors suggested that no visible erosion of the channel of Gajiu and Yunjinghong reaches had taken place and that erosion had not occurred in the lower Mekong up to 2003 as a result of Manwan dam construction.

Fig. 6.8 Photograph showing the geomorphology at the Gajiu gauging station. In this location the channel near the cross section has a bedrock substrate that is not easily eroded

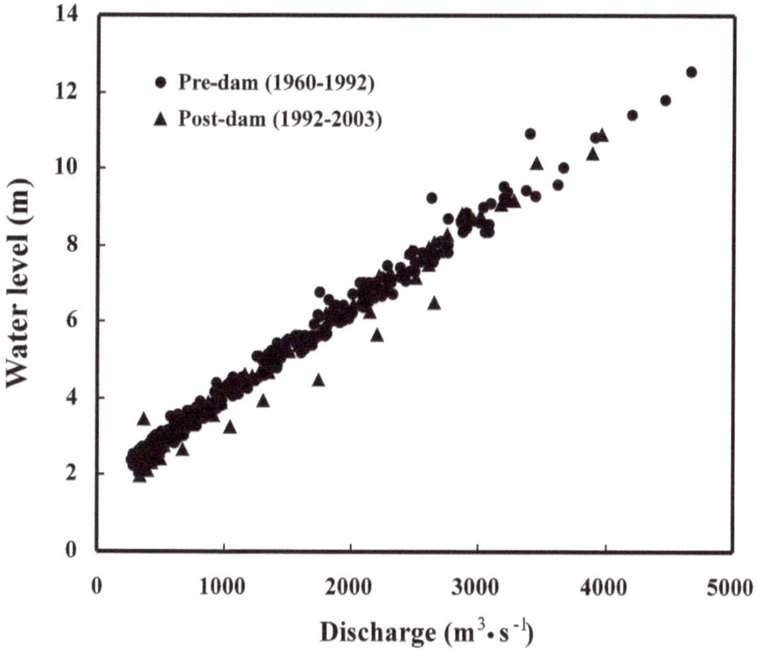

Fig. 6.9 Relationship between monthly mean water level and monthly mean discharge at Gajiu station before and after construction of the Manwan dam. (Reproduced from Fu and He 2007)

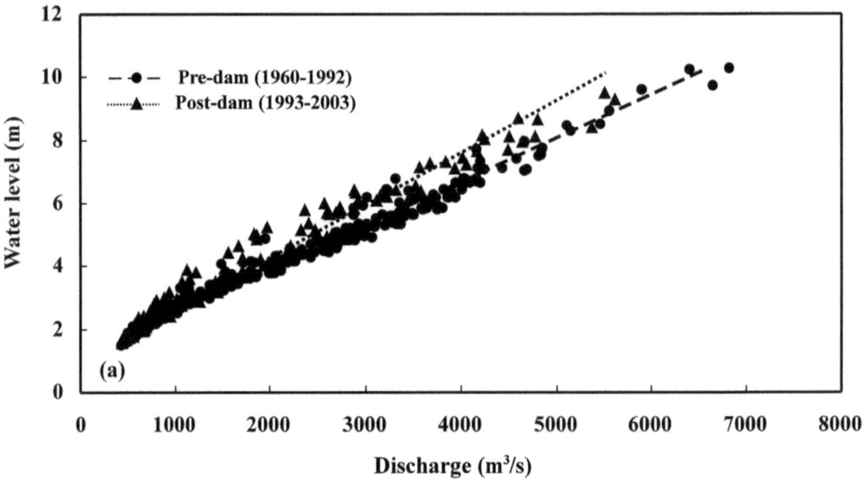

Fig. 6.10 Relationship between monthly mean water level and monthly mean discharge at Yunjinghong station

6.4 Conclusions

Results of literature review and analysis of suspended sediment data recorded at hydrological stations were used to characterize sediment yield in the Lancang River Basin and sediment load in the Lancang River. Sediment yield in the Lancang River Basin was documented to vary in accordance with topography, precipitation, land use, and human population and therefore to vary across space and time. Comparisons among locations indicated that in general, sediment discharge increases with distance downstream in the Lancang-Mekong River and that the majority of the annual sediment discharge occurs during the flood season. Soil erosion surveys conducted in the Lancang River Basin in 1987 and 1999 indicated that changes in erosion rates (which are directly related to sediment yield) over time differ by region and have generally decreased in area but increased in intensity. Such spatial and temporal changes have the potential to confound interpretation of the impacts of the Lancang hydropower projects on sediment transport and channel geomorphology.

Comparison of sediment discharge values over the last 60 years (including before and after the construction the Manwan dam) allowed assessment of transboundary effects of the Lancang hydropower projects on sediment transport, in spite of temporal changes due to factors unrelated to the hydropower projects (e.g., changes in land use associated with economic development). Although there were substantial increases in sediment discharge after 1977 at four hydrological stations that were associated with a general increase in economic development, specific changes in sediment discharge could be linked to hydropower development. After the impoundment of Manwan reservoir in 1993, sediment discharge at downstream stations declined steeply due to retention of sediment in the reservoir. For example, at Gajiu (several kilometers downstream of the Manwan dam), there was a 64.3% decrease in sediment discharge, and at Yunjinghong (~400 km downstream of the Manwan dam), the decrease in sediment discharge post-project was even greater. In contrast, little change in sediment discharge was evident at the Jiuzhou hydrological station, which is immediately upstream of the Manwan reservoir and therefore served as a control site for this analysis. The initial decrease in sediment discharge following impoundment of the Manwan reservoir was also apparent at the Chiang Saen hydrological station in northern Thailand, although other factors were also involved. Effects of construction of the Dachaoshan and Jinghong dams were also apparent in the steep declines in sediment discharge observed at Yunjinghong that coincided with the timing of dam completion. However, comparison of sediment data at four hydrological stations in the lower Mekong River before and after impoundment of Manwan reservoir in 1993 indicated that impacts of the Lancang hydropower projects on sediment loading in the Lancang-Mekong River are not easily detectable at great distances from the hydropower projects (i.e., into the lower Mekong River). Further, the increase in sediment yields with distance downstream in the Mekong River likely reduces the potential for the Lancang hydropower projects to have significant effects over great distances downstream.

Potential effects of the Lancang hydropower projects on channel morphology, due to changes in sediment loading above and below a reservoir and to erosional forces of the water released from reservoirs, were also investigated. Results and conclusions of Fu et al. (2008), who investigated downstream channel erosion triggered by the Manwan dam, concluded that no visible erosion of the channel at Gajiu and Yunjinghong reaches (cross-sectional changes) had taken place and that erosion had not occurred in the lower Mekong up to 2003 as a result of Manwan dam construction. Further, although erosional changes within the Lancang-Mekong River downstream of the hydropower projects are predicted, it is anticipated that movement of channel changes downstream of the hydropower projects will be a slow and complicated process.

References

Brandt, S.A., et al. 2000. Classification of geomorphological effects downstream of dams. *Catena* 40: 375–401.

Fu, K., and D. He. 2007. Analysis and prediction of sediment trapping efficiencies by dams on mainstream of Lancang River. *Chinese Science Bulletin* 52: 134–140.

Fu, K., D. He, and S. Li. 2006. Response of downstream sediment to water resource development in mainstream of the Lancang River. *Chinese Science Bulletin* 51 (Suppl): 100–105.

Fu, K., D. He, and X. Lu. 2008. Sedimentation in the Manwan reservoir in the upper Mekong and its downstream impacts. *Quaternary International* 186: 91–99.

Grant, G.E., J.C. Schmidt, and S.L. Lewis. 2003. A geological framework for interpreting downstream effects of dams on rivers. In *A Peculiar River*, Water science and application, ed. J.E. O'Connor and G.E. Grant, vol. 7, 203–219. San Francisco: American Geophysical Union.

Han, Q., and X. Yang. 2003. A review of the research work of reservoir sedimentation in China. *Journal of China Institute of Water Resources and Hydropower Research* 11 (13): 169–178 (in Chinese).

He, D. 1995. Analysis of hydrological characteristics of Lancang-Mekong River. *Yunnan Geographical and Environmental research* 7 (1): 58–74.

Kaidao, F., D. He, and S. Li. 2006. Response of downstream sediment to water resource development in mainstream of the Lancang River. *Chinese Science Bulletin* 51 (Suppl): 100–105.

Kondolf, G.M. 1997. Hungry water: Effects of dams and gravel mining on river channels. *Environmental Management* 21 (4): 533–551.

Kummu, M., and O. Varis. 2007. Sediment-related impacts due to upstream reservoir trapping, the lower Mekong River. *Geomorphology* 85 (3–4): 275–293.

Lu, X., and R. Siew. 2006. Water discharge and sediment flux changes over the past decades in the lower Mekong River: Possible impacts of the Chinese dams. *Hydrology and Earth System Sciences.* 10 (2): 181–195.

Petts, G.E., and A.M. Gurnell. 2005. Dams and geomorphology: Research progress and future directions. *Geomorphology Dams in Geomorphology* 71 (1–2): 27–47.

Qian, N., R. Zhang, and Z. Zhang. 1987. *Fluvial process*, 453–496. Beijing: Science Press (in Chinese).

Roberts, T. 2001. Downstream ecological implications of China's Lancang Hydropower and Mekong Navigation project. http://www.irn.org/programs/lancang/.

Wolanski, E., Nguyen Ngoc Huan, Le Trong Dao, et al. 1996. Fine-sediment dynamic in the Mekong River Estuary Vietnam. *Estuarine, Coastal and Shelf Science* 43: 565–582.

You, L. 1999. A study on temporal changes of river sedimentation in Lancang River basin. *Scientia Geographica Sinica* 54 (Suppl): 93–100 (in Chinese).

———. 2001. Scouring and silting changes of Lancang River (Mekong River) and its development tendency. *Geographical Research* 20(2): 178–183 (in Chinese).

Yunnan Provincial Development Planning Commission. 2004. Compiled by the Yunnan Provincial Department of Land and Resources, Yunnan Remote Sensing Integrated Survey of Land and Resources, 285. Kunming: Yunnan Science and Technology Press, May.

Chapter 7
Transboundary Environmental Effects of Hydropower: Fish Community

7.1 Overview

A growing number of studies are providing evidence that dam construction and operation of hydropower projects can lead to undesirable ecological responses in rivers (Nadon et al. 2015), and the influence of dams on fish diversity has become a major environmental problem (Ziv et al. 2012). Hydropower dams can lead to fish habitat fragmentation, obstruction of dispersal and migration movements, and loss of habitat, including spawning habitat, which has been directly linked to the loss of populations and species of freshwater fishes (Nilsson et al. 2005; Petesse and Petrere 2012; Gao et al. 2013).

Given general concern over the effects of hydropower projects on fish, as well as specific local evidence, concerns exist over fish communities in the Lancang-Mekong River. As one of the richest rivers in the world in terms of its fish diversity, the Lancang River supports a high level of endemism (Baran et al. 2012). However, in recent years, its fish fauna has experienced serious biodiversity loss due to increasing human impacts (e.g., construction of hydropower projects, facilitation of nonnative invasions, overfishing, land use changes, and introduction of pollution and contaminants) (Kang et al. 2009; Fan et al. 2015). Eighteen hydropower projects have been planned for the mainstem of the Lancang River, and 6 of them have already become operational (Sect. 1.2.2). In addition, more than 20 exotic species have been found in the Lancang River in recent decades (Liu et al. 2011). Given high levels of fish diversity and endemism within the Lancang River that are at risk from human impacts, as well as the potential for effects to extend downstream into the Mekong River, human disturbance to the Lancang River is of great public concern. As such, assessments of the impacts of anthropogenic disturbances to fish species and phylogenetic diversity were urgently needed.

The study of fish biodiversity has been published in Ecological Indicators (Zhang et al. 2018), with permission from Elsevier.

X. Yu et al., *Balancing River Health and Hydropower Requirements in the Lancang River Basin*, https://doi.org/10.1007/978-981-13-1565-7_7

Our approach to investigating transboundary effects on fish communities was to subdivide the Lancang River study area into regions in accordance with the locations of planned or operational dams and to compare fish biodiversity between historical (pre-project, prior to 1990) and current (post-project, between 2006 and 2015) periods. We selected an extensive study area that encompasses great variation in altitude and latitude, extending south for over 1000 km, and thereby represents many different kinds of fish habitat. We compiled fish community data from the literature and from surveys conducted in the Lancang River from 2008 to 2015. Our overall objectives were to (1) examine the temporal changes in species and phylogenetic diversity in the Lancang River; (2) investigate the relationship between fish biodiversity variabilities and dam presence in the Lancang River; and (3) investigate the fishery protection measures taken by China for the Lancang River in light of these concerns. The sections below introduce key concepts relevant to our investigation of potential transboundary impacts of Lancang hydropower projects on fish communities, which include fish biodiversity, migratory fish abundance, the importance of floodplain habitat, and conservation measures.

7.1.1 Fish Biodiversity

Potential impacts of hydropower projects related to fish biodiversity can include losses of biodiversity (e.g., extirpations) as well as introductions of exotic species. Most previous studies examining the effects of human-mediated biodiversity-related impacts on fish due to dam construction have focused on the level of species diversity (i.e., species richness and species composition among assemblages) (Agostinho et al. 2004, 2008; Limburg and Waldman 2009; Cheng et al. 2016). In contrast, the effects of human disturbances on the phylogenetic diversity of fish assemblages have been relatively seldomly documented (Jiang et al. 2015). Phylogenetic diversity, which can be defined as the average phylogenetic distance of species assemblages, has been acknowledged to represent the evolutionary history of species within an assemblage and can reflect the ecological differences among species (Webb et al. 2002). The invasions and extirpations of species are not random processes that occur within a hierarchical taxonomic tree but rather a reflection of species life history responses to environmental change (Freville et al. 2007; Winter et al. 2009). Extinctions usually befall specialized endemic or rare species, often from species-poor families, which form distinct parts of biotas (Gaston 1998; Vamosi and Wilson 2008). In contrast, invasive species are often ecological generalists with wide distributional ranges, belonging to species-rich families (Pysek et al. 2009). Hence, the loss of native species and the gain of exotic species should result in an alteration of phylogenetic diversity. Therefore, examining the effects of extirpations and invasions on phylogenetic diversity can significantly enhance the understanding of how evolutionary and ecological processes drive biodiversity patterns (Graham and Fine 2008; Winter et al. 2009).

Taxonomic distinctness (TD) indices developed by Clarke and Warwick (1998), which quantify the relatedness of the species within a sample based on Linnaean classification tree distances between species, have been widely used as a proxy for phylogenetic diversity in the absence of evolutionary data based on molecular phylogenies (Heino et al. 2005; Srivastava et al. 2012; Griffin et al. 2013; Jiang et al. 2015). The measure of the TD index is based on the presence/absence data and is insensitive to sampling methods and efforts (Clarke and Warwick 1998, 2001a). This approach has been applied in terrestrial, marine, and freshwater environmental assessments and biological conservation (Jiang et al. 2014; Hu and Zhang 2016) during the past 20 years. The use of the TD index, in comparison to a measure of species richness, was the approach taken in this study to investigate effects of the Lancang hydropower projects on fish biodiversity.

7.1.2 Abundance of Migratory Fish

Fish migration refers to the movement of fish among water bodies, or from one part of a water body to another, on a regular basis in order to meet the living conditions of different life periods (e.g., feeding, spawning, over-wintering). Fish can travel between the ocean and river, between different river sections, and between the mainstem of a river and its tributaries, with migration distances ranging from several kilometers to several thousand kilometers. According to the length of the migration route, the migratory fishes in the Lancang River can be divided into long-distance and short-distance migrants. Given their different life-history strategies, these two groups may be subject to different impacts and pressures. Since migratory fishes depend on movements to complete their life history cycles, a blockage, such as one that could be caused by a dam, may have serious impacts on populations. There are also other potential impacts in the Lancang River and downstream that may not be related to migratory distances or routes that may affect the abundance of fish species.

This study used information from the literature to (1) evaluate whether there is evidence for negative impacts of the Lancang hydropower projects on migratory fish within the Lancang-Mekong system due to disruption of migration or other causes, (2) investigate whether there is evidence for decreases in abundance of long-distance and short-distance migratory fishes, and (3) identify causes of any such decreases in abundance.

7.1.3 Floodplain Habitat

Flooding during the rainy season is a critical factor in maintaining fish productivity in the floodplain complexes of the Mekong River Basin (Junk 1997; Kummu and Sarkkula 2008). The food sources and cover provided for fish within this floodplain habitat make it the main site for fish fattening and breeding in the Mekong River, and most fishes in the Mekong River Basin depend on these resources early in their life cycle (Sverdrup-Jensen 2002). Thus, water level and duration of flood pulses are important environmental factors in maintaining productive fish habitat. The most important floodplain complexes in the Mekong River Basin are the Tonle Sap River and Tonle Sap Lake system in Cambodia and the Mekong Delta in Vietnam. The seasonally flooded area is much smaller in the upper part of the Mekong River Basin, which is in Thailand and Laos, and it is mainly associated with the tributaries of the Mekong River. Further upriver of Vientiane, floodplain habitat is further reduced in area due to the increased river gradient.

As investigated in Chap. 4, the Lancang hydropower projects have the potential to change water levels downstream through water regulation. This therefore has the potential to affect critical seasonal habitat of fish in the floodplain complexes of the Mekong River Basin. The potential for such impacts were considered at a high level in light of the hydrology analyses conducted on water level changes within the Lancang-Mekong River owing to hydropower development (Chap. 4).

7.1.4 *Fish Conservation Measures*

In order to reduce the impact of the cascade of hydropower development and environmental changes of the Lancang River on fish diversity and fishery resources of the Lancang-Mekong River, China fisheries administration, hydropower developers, research institutes, and other agencies are cooperating in developing protection measures for fish, which are dependent on determining the main threatening factors of the decline in fish in the Lancang-Mekong River. In this study, the potential effects of the Lancang cascade of hydropower projects have been investigated and the conservation measures that have been developed and implemented are summarized.

7.2 Methods

7.2.1 *Study Area and Regions*

The study area we selected for assessment of impacts of the Lancang hydropower projects on fish community extended from Guxue in the north to Guanlei in the south (Fig. 7.1). Within this area, 12 regions were delineated. Boundaries of the regions were based on locations of planned or operational dam sites and, in one case, at the southern boundary of the study area, by a hydrological station (Guanlei Portas). These 12 regions, which varied greatly in altitude and extended over many latitudes, represented different kinds of fish habitats. From north to south, as shown

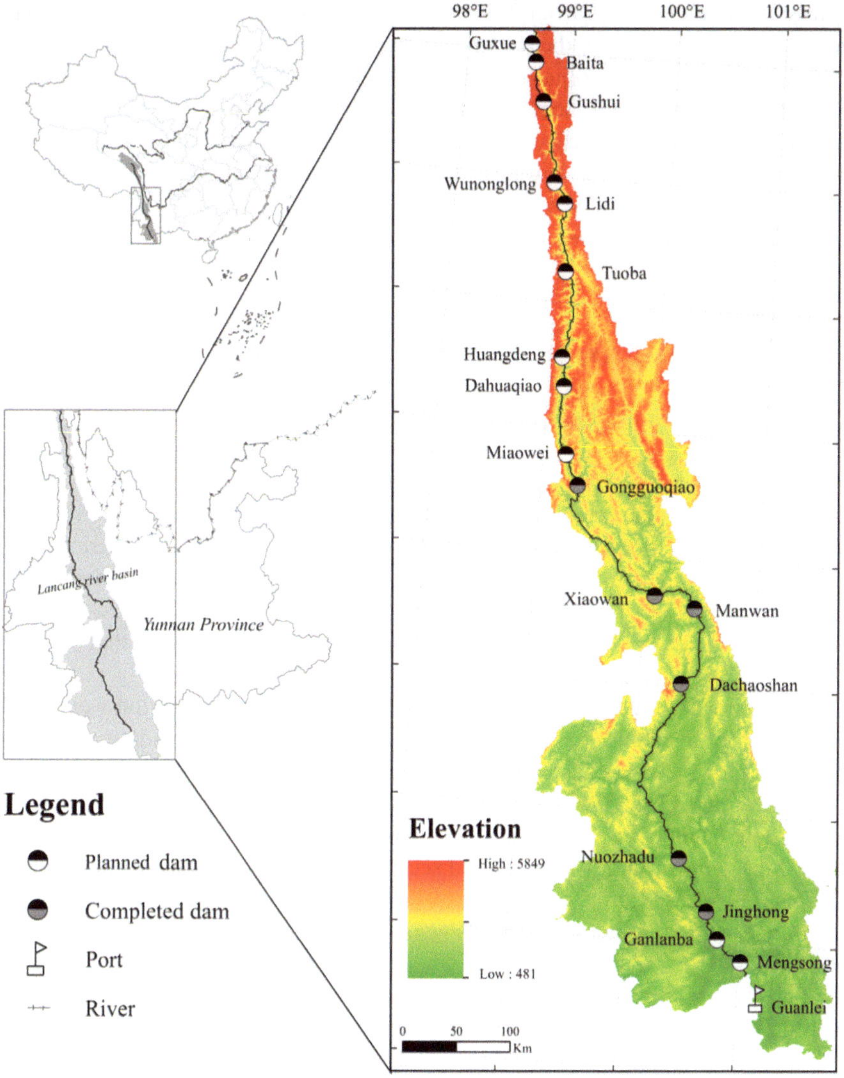

Fig. 7.1 Study area used for the analysis of transboundary effects on fish communities in the Lancang River

in Fig. 7.1, these 12 regions along with their abbreviations are Guxue-Gushui (GX-GS), Gushui-Wunonglong (GS-WNL), Wunonglong-Tuoba (WNL-TB), Tuoba-Huangdeng (TB-HD), Huangdeng-Gongguoqiao (HD-GGQ), Gongguoqiao-Xiaowan (GGQ-XW), Xiaowan-Manwan (XW-MW), Manwan-Dachaoshan (MW-DCS), Dachaoshan-Nuozhadu (DCS-NZD), Nuozhadu-Jinghong (NZD-JH), Jinghong-Ganlanba (JH-GLB), and Ganlanba-Guanlei (GLB-GL). From the northern to the southern end of the study area (GX-GS region to the GLB-GL region), the average altitude decreases from 2235 to 508.5 m. The 12 regions were also subdivided for some analyses by the presence of hydropower development. The Lancang River north of the Gongguoqiao dam has not been impacted by hydropower projects to date (Fig. 7.1), whereas the regions from the Gongguoqiao dam to Ganlanba dam have been under the influence of hydropower operations (see also Fig. 1.2 and Table 1.1). Thus, regions were also divided into two groups for some analyses: dammed regions, which were the regions from GGQ-XW to NZD-JH, and the undammed regions (all others).

7.2.2 Data Analysis, Methods, and Time Frame

7.2.2.1 Data Compilation

We compiled data on fish occurrences in the 12 regions of the mainstem of the Lancang River to quantify the temporal changes in species and phylogenetic diversity (taxonomic distinctness indices). We grouped these data into two periods corresponding to historical (pre-project) and current (post-project) periods. The historical, pre-project period was defined as the period before the 1990s, when no dams had been constructed, and the post-project period was defined as that between 2006 and 2015.

The historical data and some of the data for the post-project period were compiled from the available literature that provided species lists, including published literature (Zhang 2001; Kang et al. 2010; Liu et al. 2011; Lei 2012; Guo et al. 2014), books (Chu and Chen 1989, 1990; Chen 2013), scientific reports (environmental impact statements before and after the construction of each dam), and online databases. All scientific names were revised according to "Fishes of the World" (Nelson 2006) and FishBase (http://www.fishbase.org). In addition to compiling data on fish occurrences from the literature, we also carried out a series of sampling surveys in the Lancang River from 2008 to 2015, when the hydropower dams began operation. The objective of this sampling program was to conduct equal and sufficient fish sampling in each region to provide representative and unbiased results for the post-project period. This was conducted by setting up multiple sampling points in different areas of each region, such as within reservoirs and in the areas downstream of dams. Due to the seasonal dynamics of fish diversity, sampling was conducted at quarterly intervals: in January, April, July, and October. The fishing gear used for sampling included cast nets (mesh size 1 cm), drift gillnets (stretched mesh size 2.5 cm), and ground cages, with the method selected dependent on the ability of the different vessels to deploy them as well as the varying water depths and flow velocities in different parts of the regions. The duration of each sampling effort was about 20 min. Captured species were identified in accordance with the reference "The Fishes of Yunnan, China" (Chu and Chen 1989, 1990) and were then confirmed according to "Fishes of the World" (Nelson 2006). All specimens were classified and photographed in the field, and a few were taken back to the laboratory for further research. In order to minimize effects of sampling on fish community structure, all other fish were returned to the river.

Literature review was used to compile information and draw conclusions regarding the effects of the Lancang hydropower projects on migratory fish and to summarize key cooperative conservation measures. Effects of the Lancang hydropower projects on floodplain habitat were assessed by considering results of analyses of hydrology presented in Chap. 4. in light of potential impacts to critical fish habitat.

Fish Biodiversity Indices

Biodiversity indices were calculated and compared between historical and current periods. Based on fish species presence/absence data for both periods, three indices were calculated: species richness, average taxonomic distinctness (Δ^+), and variation in taxonomic distinctness (Λ^+). Species richness is simply the number of species documented which can be differentiated by total species, native species, and introduced species. Δ^+ measures the mean taxonomic (or phylogenetic) distance between all pairs of species in one assemblage (Clarke and Warwick 1998). Thus, a temporal decrease in Δ^+ indicates that the phylogenetic diversity has decreased whereas an increase indicates that phylogenetic diversity has increased. Λ^+ is defined as the unevenness of the taxonomic tree for a given sample; hence, a decrease in Λ^+ means the assemblages are becoming less phylogenetically variable (Clarke and Warwick 2001b). Δ^+ and Λ^+ were calculated as follows:

$$\Delta^+ = \left[\Sigma \sum_{i<j} \omega_{ij} \right] / \left[S(S-1)/2 \right]$$

$$\Lambda^+ = \left[\Sigma \sum_{i<j} \left(\omega_{ij-} \Delta^+ \right)^2 \right] / \left[S(S-1)/2 \right]$$

where ω is the branch length between species pairs, and S is the number of observed species in the sample.

We used a standard Linnaean classification (Nelson 2006) with 11 taxonomic levels for total and native fish species (Table 7.1). We used simple linear scaling, where the maximum distance throughout the tree is set at $\omega = 100$. When branch lengths are unweighted, the step between each taxonomic level in the tree is considered to be equal. However, all taxa were not defined to the same level of detail. For instance, some groups had defined subfamilies, while others did not. Therefore, we weighted the branch lengths by the number of species that were defined at a particular level following Tolimieri and Anderson (2010) and Jiang et al. (2015). Table 7.1 indicates the approximate weights and branch lengths used for all calculations of average taxonomic distinctness and variation in taxonomic distinctness. The weight is the proportion of species having a definition at the indicated taxonomic level.

Table 7.1 Weights and branch lengths for the calculation of average taxonomic distinctness (Δ^+) for fish in the Lancang River

	All species		Native species	
Taxon	Weight	Branch length	Weight	Branch length
Species	1	10.2	1	10.2
Genus	1	20.4	1	20.4
Subfamily	0.83	28.8	0.85	29.1
Family	1	39	1	39.2
Superfamily	0.89	48.1	0.92	48.6
Suborder	0.11	49.3	0.07	49.3
Order	1	59.5	1	59.5
Series	0.89	69.5	0.99	69.6
Superorder	0.99	79.6	0.99	79.6
Subdivision	1	89.8	1	89.8
Division	1	100	1	100

Branch length is the resulting branch length within the taxonomic tree to that level after weighting.

7.2.2.2 Statistical Comparisons

Tests of equal variance and normality for initial data of species richness and the TD indices provided confidence that the assumptions of the parametric statistical tests were satisfied.

We conducted t-tests to compare differences in species richness, Δ^+, Λ^+ between historical and current time periods for all regions, undammed regions, and dammed regions. By testing differences between historical and post-project time periods, we determined whether there was evidence for temporal changes in biodiversity indices.

We used multiple regression analysis to examine the relationships between temporal changes in Δ^+, Λ^+, and species introductions and disappearances. Due to the varied historical native species richness among regions (from cold-water to warm-water species), we used the total percentage of exotic species (i.e., ratio of introduced exotic/total species) and the total percentage of species that had disappeared (disappeared/historically native species) in each region as explanatory variables.

We also used linear regression analysis to determine the relationship between: (1) average altitude (explanatory variable) and species richness and Δ^+ and Λ^+(response variables) for historical and current periods (with t-tests used to test for differences between the regression lines for the two periods) and (2) reach length (explanatory variable; defined as the distance from one dam to the other) and the rate of disappearance of native species, the rate of invasion by nonnative species, and changes in Δ^+ and Λ^+ (response variables) in the five dammed regions. The latter analysis tested the influence of dam presence because reach length was inversely related to the frequency of hydropower projects, given that small reach lengths are associated with frequent projects.

We calculated species richness and TD indices in PRIMER version 6 (Clarke and Gorley 2006) and ran the paired *t* test and regression analyses in R version 3.4.

7.3 Results and Discussion

7.3.1 Fish Biodiversity

7.3.1.1 Temporal Changes in Species Diversity

The phylogenetic diversity of fish assemblages over all regions combined provided an overview of changes to fish biodiversity in the study area over time. In the historical period, the 12 regions combined contained 162 fish species belonging to 2 divisions, 8 orders, 21 families, and 89 genera. In the current (post-project) period, 49 native species previously present had disappeared, 22 nonnative species had become established, and the current fauna belonged to 1 division, 9 orders, 24 families, and 88 genera (Zhang et al. 2018). Thus, the total number of species had decreased by ~30% if only native species loss was considered. Based on these changes in phylogenetic diversity between historical and current periods, statistical comparison of biodiversity indices between periods indicated that, in all regions combined, the average species richness was significantly lower in the post-project period than in the historical period (paired *t* test: $t = 3.013$, $p = 0.012$), and this was due to the large decrease in native species richness ($t = 3.691$, $p = 0.004$) (Fig. 7.2a and Table 7.2).

Changes to species richness were also compared between upstream (undammed) and downstream (dammed) regions (Fig. 7.2a). In the five undammed regions upstream of the dammed area (GX-GS to HD-GGQ), three native species disappeared (equivalent to the disappearance of 23.8% of the native species in these regions), and six exotic species became established. All six exotic species that were introduced were found in the HD-GGQ region, which is the region furthest downstream. In the seven downstream regions combined (GGQ-XW to GLB-GL), a total of 49 native species disappeared, and 21 exotic species were introduced. Averages of 36.8 native species disappeared (47.2% of the historical native species), and 9.7 exotic species became established. Figure 7.2a also shows that, historically, species richness increased strongly with distance downstream; however, in the current period, this general relationship has weakened.

Species richness was also compared between periods, for total species and native-only species, for the region in our study area that was furthest downstream (GLB-GL; see Sect. 7.2.1). During the historical period, there were 121 native fish species in the GLB-GL region. As shown in Fig. 7.3, the majority of these were distributed in three orders: Cypriniformes (in which there were 4 families, 53 genera, and 88 species and which accounted for 72.7% of the total species), Siluriformes

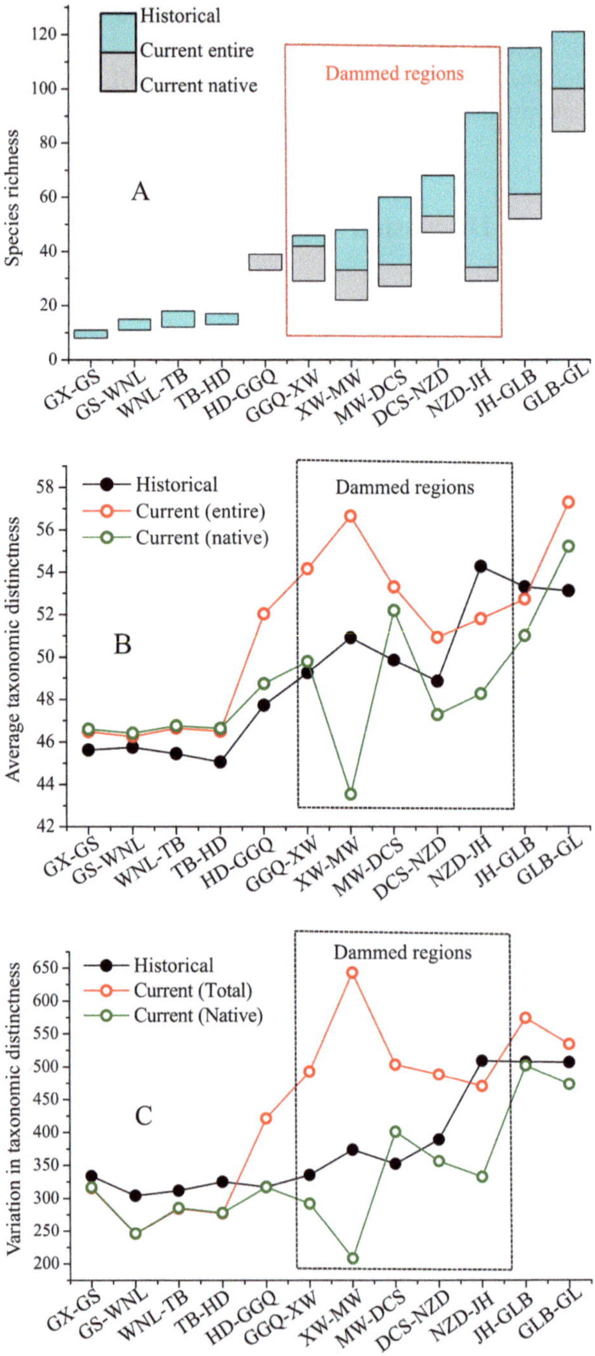

Fig. 7.2 Temporal changes in fish community between historical (pre-1990) and current (2006–2015) periods in 12 regions of the Lancang River for all species, including nonnative ones (entire), as well as for native species only: (**a**) temporal changes of species richness; (**b**) average taxonomic distinctness (Δ^{+}); and C) variation in taxonomic distinctness (Λ^{+})

Table 7.2 Results of paired t-tests of species richness (S), average taxonomic distinctness (Δ^+), and variation in taxonomic distinctness (Λ^+) between historical ("h"; pre-1990) and current ("c"; 2006–2015) periods in dammed and undammed regions of the Lancang River

Region	Comparisons	Entire		Native	
		t	*p*	*t*	*p*
All	Species richness (historical vs. current)	3.013	0.012	3.691	0.004
	Δ^+ (historical vs. current)	−3.032	0.011	0.766	0.454
	Λ^+ (historical vs. current)	−1.934	0.079	2.462	0.032
Undammed	Species richness (historical vs. current)	1.83	0.141	7.303	0.002
	Δ^+ (historical vs. current)	−2.433	0.072	−6.175	0.003
	Λ^+ (historical vs. current)	0.321	0.764	2.912	0.044
Dammed	Species richness (historical vs. current)	3.142	0.026	4.429	0.007
	Δ^+ (historical vs. current)	−1.682	0.153	1.694	0.151
	Λ^+ (historical vs. current)	−2.794	0.038	1.651	0.16

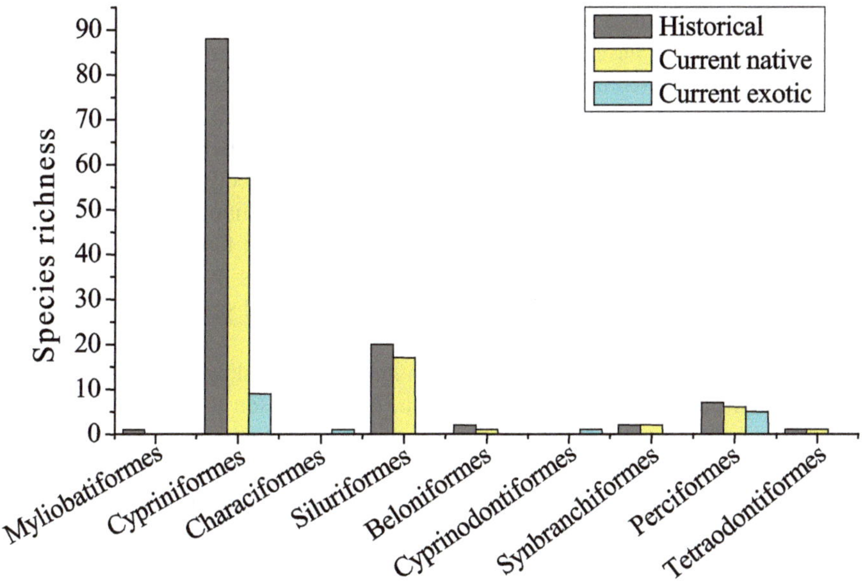

Fig. 7.3 Numbers of species within fish orders in the region furthest downstream in our study area (GLB-GL) during the historical and current periods and for native and exotic species during the current period

(in which there were 7 families, 13 genera, and 20 species and which accounted for 16.5% of the total species), and Perciformes (in which there are 4 families, 5 genera, and 7 species and which accounted for less than 10% of the total amount). In the post-project period, the number of native fish species in the region has decreased from 121 to 84 species; however, 16 exotic species have been introduced resulting in 100 total species. The species that have been lost are mainly Balitoridae (11 species lost) and Cyprinidae (20 species lost) that rely on fast-flowing water. In addition, several migratory fishes such as *Pangasius sanitwongsei* are also declining. Among the new exotic species are nine species belonging to Cypriniformes, five species belonging to Perciformes, and one species in each of Characiformes and Cyprinodontiformes.

Results of our analyses indicate that, consistent with conclusions of previous studies (Kang et al. 2009; Zheng et al. 2013), species richness of native fishes in the Lancang River, especially in its downstream regions, has been substantially reduced during the past 30 years. Native fish species decreased substantially, and nonnative species increased in the downstream regions; however, relatively few native species disappeared in the upstream regions.

A number of factors may explain the difference in fish biodiversity loss and exotic species introductions in the upstream and downstream regions of the Lancang River, including differences in human disturbance levels and environmental and climatic conditions (Kang and He 2007; Kang et al. 2009). Human disturbance resulting from multiple causes (e.g., dams, tourism, pollution), which can contribute to species disappearance and lead to other negative effects to the native fish fauna (Kang et al. 2009; Fan et al. 2015), is generally more severe in the downstream region than the upstream region. The high altitude and harsh environment in the upstream regions have prevented many human activities, such as tourism and fishing (Kang et al. 2009; Zefferman et al. 2015). Moreover, there are no completed dams in the upstream regions, allowing high habitat connectivity in the upstream river networks. However, it must also be recognized that although there has been no hydropower development in the upstream regions, the fish fauna has nevertheless been threatened by increasing anthropogenic disturbances over the past decades (e.g., industrial sewage from mineral exploitation and metal smelting) (Zhao and Li 2013).

In addition to the greater levels of human impacts in the downstream regions, the harsher environment in the upstream region, in comparison to the more moderate conditions in the downstream region, may directly affect the likelihood of exotic species introductions. Given that fish species have specific habitat requirements (e.g., water temperature, depth, current velocity), their establishment if introduced is dependent on particular conditions, and many nonnative species may not be able to adapt to the low temperatures or rapid currents in the upstream regions. For example, the optimal temperature for growth and reproduction for the widespread, invasive *Oreochromis* sp. in the downstream region of the Lancang River was reported to be 22–32 °C, and temperatures below 10 °C were found to be lethal (Azaza et al. 2008; Xie et al. 2011). Thus, this species cannot become established in the colder waters of higher elevations. Another invasive species, *Gambusia affinis*, that is found in the downstream regions of the Lancang River is also a warm-water fish and generally distributed in the warmer parts of the world (Pyke 2005). An example of an exotic species with specific flow requirements is the invasive genus *Rhodeus*, which occurs in downstream regions because it prefers lentic environments. Thus, the low altitude and moderate environment in the downstream regions of the Lancang River has promoted the successful establishment of many nonnative fishes. In addition, changes to the aquatic environment caused by damming have increased suitability for the dispersal and reproduction of nonnative fish species. For example, *Neosalanx taihuensis* is able to rapidly reproduce in reservoirs (Havel et al. 2005; Li et al. 2013). These exotic introductions and changes to critical habitat features therefore resulted in more native fish species disappearing in the downstream regions than further upstream.

7.3.1.2 Temporal Changes in Phylogenetic Diversity

When compared across regions and groups of regions, Δ^+ and Λ^+ did not show completely consistent temporal trends, and trends differed from that observed for species richness (Fig. 7.2b, c). For the entire fish fauna over all regions combined, Δ^+ was significantly greater in the current than the historical period (t = −3.032, p = 0.011; Table 7.2) (Fig. 7.2b), which indicates that phylogenetic diversity, overall, has increased over time. Comparison between upstream (undammed) and downstream (dammed) region groups indicated that the positive difference in Δ^+ between historical and current time periods was generally small but consistent for the upstream regions and that there was little difference between entire (native and introduced) and native values. In contrast, for the downstream regions, there was substantial among region variability in the difference in Δ^+ between time periods, the difference between time periods was generally large, and the difference between entire (native and introduced) and native values was great. Thus, for the native fish fauna, Δ^+ was significantly greater in undammed regions in the current relative to the historical time period (t = −6.175, p = 0.003; Table 7.2), indicating that phylogenetic diversity has increased overall. In dammed regions, average Δ^+ was lower for native species in the current relative to the historical period, but the difference between periods was not significant (t = 1.694, p = 1.51).

Analysis of Λ^+ indicated that, for the entire fish fauna over all regions combined, Λ^+ was significantly lower in the current than the historical period (t = 2.462, p = 0.032; Table 7.2), which indicates that phylogenetic variability in assemblages has generally decreased over time. Similar to results for Δ^+, differences in trends of Λ^+ were apparent between upstream (undammed) and downstream (dammed) areas. Phylogenetic variability in assemblages (Λ^+) in the upstream undammed regions was lower in the current relative to the historical period, and this reduction was relatively consistent among regions (Fig. 7.2c). Thus, for the undammed regions, the change in Λ^+ between periods was significantly different for the native species (t = 2.912, p = 0.044; Table 7.2). In the downstream (dammed) regions, differences in Λ^+ among time periods were larger, and there was greater variability among regions in these differences. In addition, current total and current native trends were different, such that for all fauna, phylogenetic variability was generally greater currently than during the historical period, whereas for native fauna only, variability was less during the current period. For the dammed regions, the change in Λ^+

Table 7.3 Results of multiple regression models for change in average taxonomic distinctness (Δ^+) and variation in taxonomic distinctness (Λ^+)

Assemblage	Comparisons	Model terms	Std b[a]	*t*	*p*	Model statistics
Entire	Change in Δ^+	Historical value	−0.285	−1.14	0.287	
		Introduction	0.92	5.12	0.001	F = 10.52, P = 0.004
		Disappearance	−0.23	−1	0.345	Adj.R2 = 0.722
	Change in Λ^+	Historical value	−0.232	−1.22	0.256	
		Introduction	0.851	5.5	0.001	F = 11.52, P = 0.003
		Disappearance	0.265	1.39	0.203	Adj.R2 = 0.742
Native	Change in Δ^+	Historical value	−0.168	−0.47	0.648	F = 3.32, P = 0.083
		Disappearance	−0.523	−1.47	0.175	Adj.R2 = 0.297
	Change in Λ^+	Historical value	0.059	0.16	0.875	F = 1.21, P = 0.343
		Disappearance	0.423	1.15	0.278	Adj.R2 = 0.036

[a]Std b: standardized regression coefficient

between periods was significantly different for all species only ($t = -2.794$, $p = 0.038$; Table 7.2).

Results of multiple regression analyses demonstrate that changes in Δ^+ and Λ^+ for entire assemblages were significantly predicted by the historical value, introduction rate, and disappearance rate (Table 7.3). The introduction rate had a significant positive effect, indicating that regions that experienced high pressure from introductions tended to become more phylogenetically diverse and variable (Table 7.3). The disappearance rate did not have a significant effect on changes in Δ^+ and Λ^+. In contrast to the results for the entire assemblages, for native assemblages, the changes in Δ^+ and Λ^+ were not significantly predicted by the historical value and disappearance rate (Table 7.3).

Our results indicate that temporal changes in the phylogenetic diversity indices (or TD indices) differed from those in species richness, indicating that the TD indices provide additional information about biodiversity. Inconsistency between TD indices and species richness has been documented in many previous studies (Ellingsen et al. 2005; Heino et al. 2005; Jiang et al. 2015), which suggests that the two kinds of indices reflect different facets of biodiversity. In our study, in contrast to species richness, the Δ^+ of entire assemblages increased with the introduction of exotic species and disappearance of native species, implying an increase in phylogenetic diversity as species richness decreased.

The reason for the difference in temporal changes between species richness and phylogenetic diversity indices is related to the high relatedness within native fish communities in contrast to the greater taxonomic distance of introduced exotic species relative to native species. In southwestern China, under evolution through adaptive radiation, rivers and lakes usually support highly endemic fish faunas composed of many taxonomically closely related species (Li 1982; Chu and Chen 1989). Such highly diverse assemblages of congeneric species commonly occur in the Lancang River (e.g., historically, there were 17 species from *Schistura*, 6 species each from *Tor* and *Glyptothorax*, and 5 species each from *Homatula* and *Garra*). The clusters of these congeneric species therefore contributed to the short taxonomic path lengths and induced a relatively low level of Δ^+ in the historical period (Jiang et al. 2015). However, in the current period, the loss of some congeneric native species from *Triplophysa*, *Glyptothorax*, and *Balitoropsis* in the upstream regions resulted in an increase in Δ^+. In the downstream regions, however, the introductions of distantly related exotic species from new orders (e.g., *Prochilodus lineatus*, *Gambusia affinis*, and *Neosalanx taihuensis* from the orders Characiformes, Cyprinodontiformes, and Osmeriformes) and a species-poor order (seven exotic species and one translocated species belonging to Perciformes, which historically contained only seven sporadically occurring species) contributed to the largest taxonomic path length. The introduction of phylogenetically distant species hence drove the increase in Δ^+ in downstream regions.

For Λ^+, the temporal changes were quite different between the entire and native-only assemblages. In downstream regions, the Λ^+ of entire assemblages was significantly higher in the current relative to historical period, but the Λ^+ of native-only assemblages was lower in the current relative to the historical period. The reason for these results is that nonnative introductions have made the taxonomic tree become more uneven (i.e., phylogenetic variability has increased), but the native disappearances have reduced the phylogenetic variability when native-only assemblages are considered. In the Lancang River, most nonnative introductions were from the families Cyprinidae (12 species contributing 60.8% of invasions) and Gobiidae (4 species contributing 17.6% of invasions), whereas Balitoridae experienced the most severe native disappearances (17 species disappeared from the species pool, and an average of 2.8 species disappeared per region). Such introductions and disappearances significantly increased the proportional richness of Cyprinidae (from 43.5% to 52.1%) and decreased that of Balitoridae (29.2% to 19.0%) and Sisoridae (11.7% to 9.4%) in the downstream regions, inducing greater unevenness of the taxonomic (or phylogenetic) structure. However, when considering native-only data, because the family Cyprinidae also experienced a loss of native species, the taxonomic structure only changed slightly. The decrease in Λ^+ in the native assemblages can also be attributed to the loss of congeneric native species, as high Λ^+ values are associated with species assemblages that contain clusters of species coming from few genera (some genera have more species than others), whereas low Λ^+ values are associated with assemblages belonging evenly to different higher taxa (different higher taxa that have nearly the same number of species) (Zintzen et al. 2011; Jiang et al. 2014).

Generally, an increase in Δ^+ is usually considered to be a positive change in biotic communities (Warwick et al. 2002; Milosevic et al. 2012). However, when Δ^+ increases as a result of introductions of distantly related exotic species and the loss of native endemism, the increase is of concern (Jiang et al. 2015). Meanwhile, high unevenness in taxonomic structure is considered to be a result of habitat heterogeneity (Leira et al. 2009), and the reduction in Λ^+ in the native assemblages indicates a decline in habitat heterogeneity. In the Lancang River, the TD indices markedly responded to nonnative introductions and native disappearances, reflecting the variation in phylogenetic diversity, which was quite different from the pattern detected for species richness. In environments hosting a high level of endemics, such as the Lancang River, massive disappearances of endemic species induce a definitive loss of biodiversity not only in species richness but also in phylogenetic structure, which mirrors the roles of the historical (dispersal mode, environmental sorting) and biotic (limiting similarity) forces that shaped the local fish fauna (Webb et al. 2002; Cavender-Bares et al. 2009). Therefore, changes in TD indices reflect the extent to which anthropogenic disturbances have modified the evolutionary patterns of biotic fauna (Abellan et al. 2006; Helmus et al. 2010).

7.3.1.3 Temporal Changes in Diversity-Altitude Relationships

In both historical and current periods, species richness (both total species and native-only species) was significantly negatively related to average altitude (Fig. 7.4a and Table 7.4). The intercept was significantly higher in the historical than in the current period ($p < 0.001$ for total and native-only species), and the slope was significantly flatter from current relative to the historical period ($p = 0.007$ and $p = 0.01$ for total and native-only species, respectively). This indicated that the negative relationship between species richness and altitude was stronger historically than it is now, such that a decrease in altitude used to be associated with a greater increase in species richness than it is now.

Δ^+ and Λ^+ were also significantly negatively related to the average altitude, except for Δ^+ for native species in the current period, which was marginally nonsignificant ($p = 0.062$, Fig. 7.4b and c; Table 7.4). For both Δ^+ and Λ^+, the intercept was greater, and the slopes were steeper for all species than for native-only species, but none of the intercepts or slopes differed significantly between time periods ($p > 0.05$).

7.3.1.4 Influence of Dams on Biodiversity

The rate of disappearance of native species, the rate of invasion by exotic species, and changes in Δ^+ and Λ^+ from the historical to current period differed by the reach length that was used to represent dam frequency (and, conversely, distance between dams). Within the five dammed regions, the rate of extirpation of native species exhibited a marginally nonsignificant negative correlation with reach length (i.e., distance between dams) (Fig. 7.5a, Adj.$R^2 = 0.552$, $p = 0.093$), while the rate of invasion by exotic species showed no correlation with reach length (Fig. 7.5b, Adj. $R^2 < 0$). Furthermore, the rate of change of the TD indices showed weaker patterns than those of the disappearance rate, with marginally nonsignificant negative correlations with reach length for both the Δ^+ (Fig. 7.5c, Adj.$R^2 = 0.468$, $p = 0.123$) and Λ^+ (Fig. 7.5d, Adj.$R^2 = 0.473$, $p = 0.122$) of native assemblages but no correlations for the entire assemblages (Fig. 7.5c, d, Adj. $R^2 < 0$). These results indicate that disappearance rate increased strongly with decreasing distance between dams (i.e., with increasing dam frequency), that the distance between dams has little effect on

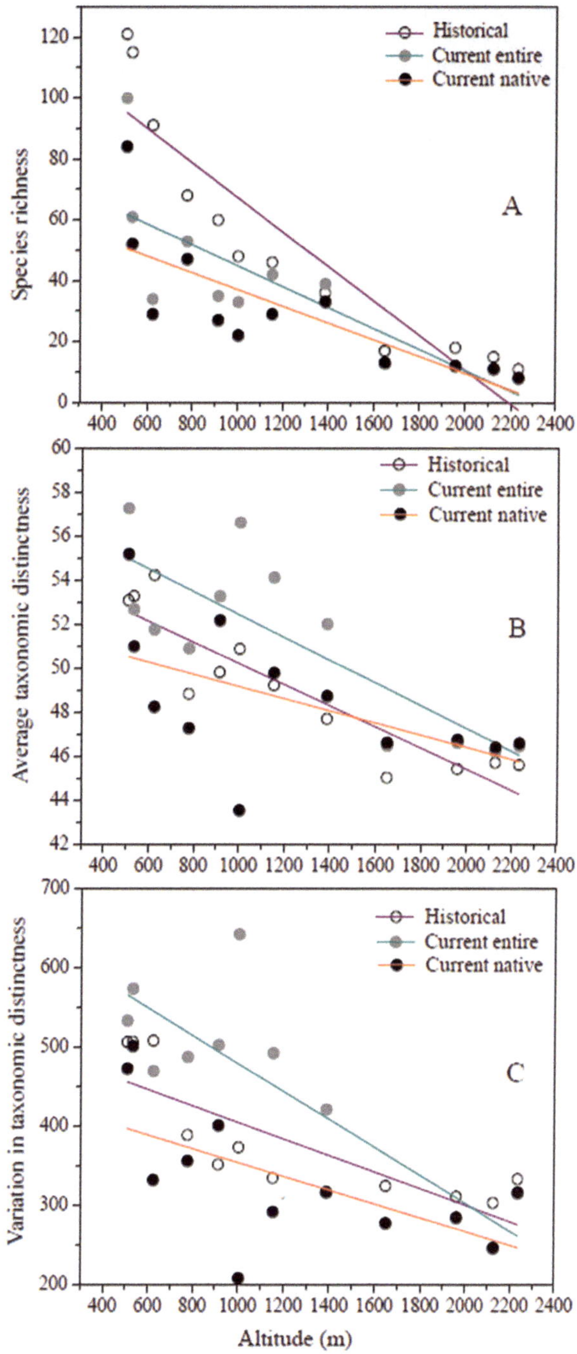

Fig. 7.4 Linear regression relationships between (**a**) species richness; (**b**) average taxonomic distinctness (Δ^+); and (**c**) variation in taxonomic distinctness (Λ^+) and average altitude in the historical and current periods for total (entire) and native-only assemblages (correlation coefficients, *p*-values, and differences in the intercept and slope among the three regression lines are given in Table 7.4)

Table 7.4 Results of linear regression analysis of species richness, average taxonomic distinctness (Δ^+), and variation in taxonomic distinctness (Λ^+) and average altitude in the historical and current periods for total (entire) and native-only assemblages

Biodiversity index	Model terms	Historical	Current	
			Total	Native
Species	Adj. R2	0.827	0.631	0.579
Richness	p	<0.001	0.001	0.002
	Intercept	123.9	79.2	64.6
	Slope	−0.057	−0.034	−0.027
Δ^+	Adj. R2	0.812	0.63	0.235
	p	< 0.001	0.001	0.062
	Intercept	55	57.6	52
	Slope	−0.0048	−0.0051	−0.0028
Λ^+	Adj. R2	0.618	0.703	0.328
	p	0.001	<0.001	0.03
	Intercept	509.5	655.8	442
	Slope	−0.104	−0.176	−0.087

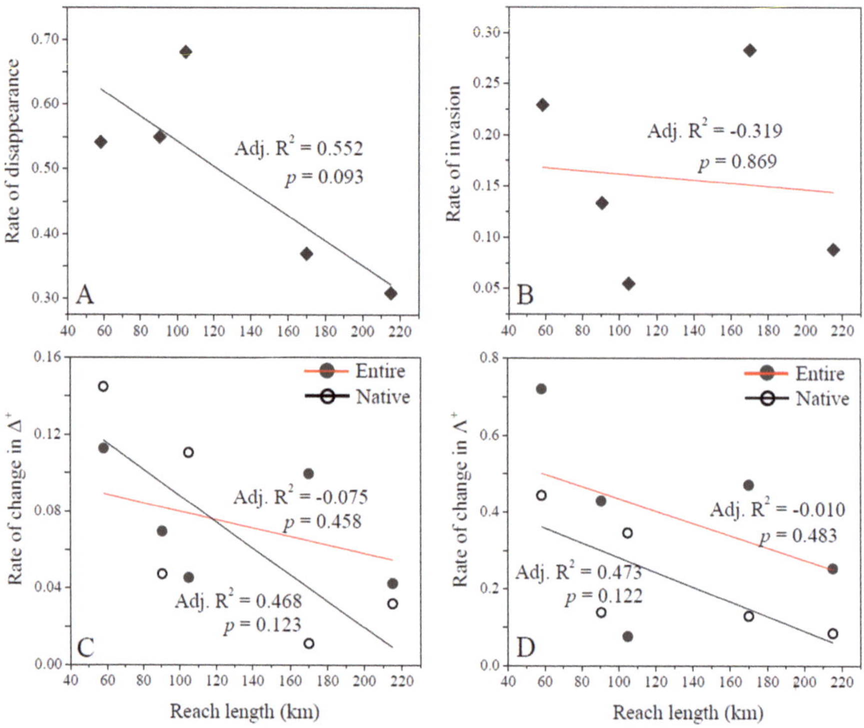

Fig. 7.5 Rate of change in (**a**) disappearance rate, (**b**) invasion rate, (**c**) average taxonomic distinctness (Δ^+), and (**d**) variation in taxonomic distinctness (Λ^+) by reach length in the 5 dammed regions (GGQ to JH). Rate of change in Δ^+ and Λ^+ are given for all species (entire) and native-only species

invasion rate, and that average taxonomic distinctness (Δ^+) and variation in taxonomic distinctness (Λ^+) were affected by distance between dams, but this relationship was stronger for native species.

The relationships between disappearance rate and TD indices with distance between dams suggest that there is an effect of dam frequency (which is inversely related to the area of habitat between dams) on fish biodiversity. Two likely impacts of dam frequency on fish biodiversity are habitat fragmentation and habitat alteration. Habitat fragmentation is considered to be one of the greatest causes of biodiversity loss (Newman et al. 2013). The disappearance of species increases when habitats are fragmented owing to increased risk of random factors (e.g., environmental perturbations) affecting small and isolated populations (Matthies et al. 2004). Further, the size of the unfragmented habitat patch (habitat area) has been positively associated with species richness, particularly for rare species, due to effects of habitat loss and edge effects (Cagnolo et al. 2006). Habitat alteration, which results when flows and other hydrological characteristics are affected by the construction of dams and creation of reservoirs, are likely to reduce habitat suitability for native fish species that have evolved under natural river conditions. When dams are close together, the proportion of altered habitat is greater than when dams are far apart. These ecological theories are consistent with the results of our study. Both the rate of species disappearance and the rate of change in TD indices in native assemblages showed negative correlations with the distance between dams (as a proxy for habitat area) (Fig. 7.5a, c, d) which indicates that dams have less impact on fish diversity when they are spaced farther apart. The results of our study indicate that more comprehensive impact assessments are important in the planning and operation of hydropower development programs. For example, extending the distance between dams in a cascade may be necessary to reduce the impact of hydropower development on native species.

Extrapolation Downstream

Evaluation of transboundary effects on fish biodiversity is complicated because there are many factors other than hydropower development that affect fish populations in the Lancang-Mekong River that change in space and time. For example, climate, hydrological conditions, land use, and fishing pressure may all change with distance downstream. Thus, it is difficult to separate out only those impacts to fish biodiversity that are caused by the Lancang hydropower projects. Our results above have shown that there is a general relationship between biodiversity and the frequency of dams within an area; however, these results do not directly address the potential for transboundary effects.

We investigated the potential for transboundary effects of the Lancang hydropower dams on fish biodiversity by comparing the change in fish biodiversity between historical and current periods between the region furthest downstream in our study area (the furthest section downstream in the Lancang River), where no dams currently exist, to regions within our study where dams currently exist (Fig. 7.2a; GLB-GL region in comparison to dammed regions). In general, Fig. 7.2a shows that the proportion of the fish species lost between the historical and current periods from the GLB-GL region of the Lancang River was much lower than that of the dammed region further upstream. Figure 7.2a also shows that current species richness is substantially greater downstream of the Jinghong dam and that species richness continues to increase as distance from the Jinghong dam increases. Further, the increase in species richness south of the Jinghong dam is quite abrupt, which is consistent with the idea that the existence of hydropower projects in the dammed region is depressing species richness but that the effect does not extend far downstream. Thus, these results suggest that the impacts of the cascade of hydropower projects on fish biodiversity decline with distance downstream. However, there are other potential factors that could be related to the increase in species richness in the GLB-GL region and that could potentially confound the relationship between hydropower development and species richness. Additional studies are required that specifically investigate the role of hydropower development in the loss of fish biodiversity and in which robust study design and associated data analysis allow partitioning of effects among key factors.

7.3.2 Abundance of Migratory Fishes

The best example of a long-distance migratory fish in the Lancang River is the pangasiid catfish, which is the largest fish in the Lancang-Mekong River. These fishes spend most of the annual cycle in the lower reach of the Mekong River where they grow to be, on average, 60–80 kg in weight. When the water rises in May each year, individuals migrate from the lower Mekong River up into Mengsong and then into the Nanban and Buyuan tributaries of the Lancang River for spawning (Yang et al. 2007). This long upstream migration involves moving through many barriers, such as torrents and shoals, and requires that the fish have strong swimming capabilities. After breeding, the fishes gradually come back to the open waters of the Mekong River, where they recover from breeding through massive food intake and remain until the following spring when they repeat their breeding migration.

Pangasiid catfish mainly migrate from the lower Mekong River to the lower Lancang River downstream of Jinghong (i.e., Buyuan tributary area), and this section of the Lancang Mekong River is currently undammed; thus, the migratory route for spawning is not blocked, and the current cascade of hydropower projects of the Lancang River has not affected the Pangasius breeding migration. In addition, the lower Lancang River is only a small part of the range of the pangasiid catfish during migration (Coates et al. 2000). Nevertheless, capture histories suggest that populations are in decline. In the Lancang River, sampling records exist for *Pangasius sanitwongsei* Smith, *Pangasius djambal* Bleeker, and *Pangasius micronemus* Bleeker, mostly from the 1960s and 1970s (four were collected in 1960; three were collected in 1978 weighing 62, 72, and 84 kg, respectively; one was collected in

1993 (Yang et al. 2007). All captures were in the lower reaches of the Lancang River in April or May, which is the spawning season of Pangasius fish and therefore provides evidence that movement into this area is for spawning rather than feeding. There were few captures of these species in the Lancang River after the 1980s, which is suggestive of a decline in numbers. Similar evidence for a decline also exists for the lower reach of the Mekong River: at the border between Thailand and Laos, 62 *Pangasianodon gigas* (Mekong Giant Catfish) were captured from 1986 to 1993, whereas only 18 and 16 were captured in 1994 and 1995, respectively. This apparent population decline has become a matter of common concern among the countries of the Mekong River, considering that pangasiid catfish are important for the Mekong River fishery.

Overfishing in the Mekong River Basin appears to be the cause of the observed population decline of pangasiid catfish in the last two decades. Over the past 50 years, the fish catch in Mekong River has doubled, and the human population involved in the fishery has increased threefold. For example, the total catch in Tonle Sap Lake increased from 125,000 tons in 1940 to 235,000 tons in 1995, and during this period, the number of fishermen increased from 360,000 to 1,200,000 (resulting in an overall decrease in catch per fisherman from 347 to 196 kg) (Van Zalinge et al. 2003). These catch results suggest that high-intensity fishing has caused a decline in pangasiid catfish numbers. Fishing has also indirectly affected populations through the bait fishery, which further exacerbates catfish population decline because bait fish are a natural food source for the catfish. A similar population decline as that observed for pangasiid catfish populations has also been reported for *Anguilla nebulosa* in the Nujiang-Saarwin River, which is a species that migrates between the mainstem and tributaries of the Nujiang River and the Indian Ocean, where it spawns. Although this species was previously common, it has not been captured for a decade (unpublished field survey data). This population decline is also not linked to hydropower development because China has not yet built dams in the Nujiang River.

Many short-distance migratory fish species occur in the middle and lower reaches of the Lancang River where hydropower development has occurred and is planned. These include species in the Cyprinidae, Cobitidae, Balitoridae, Siluriformes, Bagridae, and Sisoridae families, and some direct impacts from hydropower developments have been observed. Short-distance migrants generally move short distances for spawning, such as a short distance up the mainstem or into a nearby tributary. After the larvae hatch, they feed in the vicinity of the spawning ground and then slowly drift downstream, also for relatively short distances. At present, the short-range migratory fish, such as *Bagarius yarrelli*, *Platytropius sinensis*, and *Tor sinensis*, are still the main catch species in the lower Lancang River, particularly in the Buyuan River where the biomass of these species is very large. The main risk to these types of fish associated with Lancang hydropower development that has been identified to date is mortality risk associated with infrastructure. For example, in 2014, a large number of dead *Bagarius* catfish were found at the discharge opening of the Jinghong dam. A possible explanation for the deaths in this situation was low oxygen concentration in the discharge water (China News 2014).

7.3.3 Floodplain Habitat

Impacts of the Lancang cascade of hydropower projects on water levels downstream were analyzed in Chap. 4. Results of the analyses indicated that monthly average water levels at key hydrological stations in the lower Mekong River were higher post-project during the dry season (November–May) and lower during the flood season (June–October) than during the baseline period. Thus, reduced water levels during the flood season have the potential to impact critical seasonal habitat of fish in the floodplain complexes of the Mekong River Basin.

Although reductions in water level due to water regulation of the hydropower projects can cause substantial changes in water levels downstream relative to baseline, results of the analyses (Chap. 4) also demonstrated that the effect of the hydropower projects dissipates with distance downstream from the dams and that the speed of this dissipation differs by season. In particular, although higher water levels post-project can be detected as far downstream as Kratie, lower water levels during the flood season were not detectable south of Mukdahan (northeast Thailand). For example, at Chiang Saen (northern Thailand), monthly average water levels during the flood season were between 0.60 and 2.26 m lower post-project (2013 and 2014) relative to the baseline period (with the most pronounced effect observed in July and August). However, such differences were not apparent at Kratie, which is well north of the main floodplain complexes in the system (Tonle Sap River and Tonle Sap Lake system in Cambodia and the Mekong Delta in Vietnam). The reasons for the rapid dissipation downstream of the water level effects of the hydropower projects during the flood season were identified to be the influence of regional precipitation and inflow of major tributaries, which outweigh the stabilizing impacts of water regulation.

The flood areas in the upper part of the basin, in Thailand and Laos, that are mainly associated with the tributaries of the Mekong River, will, however, be affected by water regulation of the Lancang hydropower projects during the flood season. Effects of water regulation during the flood season, through reduced water level and flow relative to the baseline period, were observed to extend to Mukdahan and Vientiane for water level and flow, respectively (Chap. 4). Thus, the area of floodplain in Thailand and Laos will be affected to some extent, and the input of exogenous nutrients into the river due to flooding will be reduced, which is likely to eventually lead to the decrease of fish productivity in this area. However, the increase in water level during the dry season caused by water regulation may compensate somewhat by providing spawning grounds and habitat for settled (non-migratory) fishes such as *Anabas testudineus*, *Clarias batrachus*, and *Chana striata*.

7.3.4 *Fish Conservation Measures*

In the last three decades, species of medium-sized fish, such as those in the families Cyprinidae and Balitoridae, have had greatest losses in numbers below the Jinghong dam in the Lancang River (Fig. 7.3). There are three key reasons for this loss: (1) habitat alteration, (2) biological invasion, and (3) overfishing.

Transformation of habitat is related to the construction of the cascade of dams and changes in land use. Construction of dams transforms hydrological conditions and the physical and chemical characteristics of the water resulting in the loss of suitable habitat for some species. Land use changes may affect food sources for fish. The replacement of the original natural forest in the lower reaches of the Lancang River with rubber forests and banana trees has caused a reduction in exogenous food that previously entered the river from natural forests.

Biological invasion has resulted largely from aquaculture. In the past 30 years, aquaculture has become highly developed in the lower reaches of the Lancang River, and as a consequence, a large number of exotic fishes have escaped. This has resulted in a large-scale biological invasion in the lower reaches of the Lancang River. For example, invasive species (the number of which has reached 16) are found within the entire mainstem and side channels of the downstream portion of the Lancang River, and the biomass of the invasive Nile tilapia (*Oreochromis niloticus*), along with other invasive species, accounts for 80% of total biomass in some reaches. These invasive species prey on native fishes and compete with them for food and habitat, leading to the decline of populations or even extinction of some species.

Overfishing is the third key cause of biodiversity loss. In addition to general overfishing issues, over the last 10 years, fishes have sometimes been caught in the downstream portion of the Lancang River by electricity and poison, resulting in excessive mortality of wild fish communities.

Three key conservation measures are being taken by China to protect fish communities in the Lancang-Mekong River in light of the observed adverse changes. These conservation measures will provide compensation for potential impacts of hydropower development and will also address other reasons for loss of biodiversity and decreases in abundance. These conservation measures, which include the designation of a conservation area, enhancement and release of native fish species, and management and cross-border cooperation, are summarized below.

7.3.4.1 Designation of Buyuan Tributary as a Fish Conservation Area

An important contribution to the conservation of fish communities in the Lancang River has been the recent designation of important fish habitat as a conservation area. The Buyuan Tributary is an important and large tributary of the lower reaches of the Lancang River (tributary length, basin area, annual rainfall, and mean runoff are 319 km, 7747 km^2, 764 mm, and 57.89 × 108 m^3, respectively; Zhong et al.

2011), which, due to its large discharge and variable flow rates (i.e., contains areas of slow flow as well as rapids), is the destination for a variety of migratory fish in the Lancang-Mekong River. A total of 106 fish species are recorded for the Buyuan Tributary, which accounts for more than 70% of total fish species of the lower Lancang River (Kang et al. 2009). Given that Buyuan Tributary is an important habitat for many fish species, and to implement a significant fish conservation measure, the Buyuan Fish Conservation Zone was established by the government of Yunnan Xishuangbanna in 2007. In 2011, the fish conservation management office for the lower Lancang River was founded, and fishing was closed within the area year round.

7.3.4.2 Enhancement and Release of Native Fish Species

Enhancement of native fish populations in the Lancang River through captive (artificial) breeding and release is being implemented to enhance native fish populations. To date, captive breeding techniques have been developed by the Chinese fishery research department for seven native fish species of the Lancang River (*Wallago attu*, *Mystus wyckioides*, *Tor sinensis*, *Bagarius yarrelli*, *Sinilabeo yunnanensis*, *Schizothorax lissolabiatus*, and *Schizothorax lantsangensis*) (Fig. 7.6). In addition, research is being conducted to develop artificial breeding techniques for other species, such as *Platytropius sinensis* and *Percocypris pingi retrodorslis*. In total, since 2013, China has artificially bred more than ten million larval fish (e.g.,

Fig. 7.6 Artificial breeding and release of native fish in the lower Lancang River (Photos by Jiankuan He)

Wallago attu, *Mystus wyckioides*, *Tor sinensis*, *Bagarius yarrelli*, *Sinilabeo yunnanensis*, and *Schizothorax lissolabiatus*) that have been released to the lower reaches of the Lancang River. These fishes represent typical short-range migratory fish and are targets of the main fishery in the lower Lancang River. These releases supplement the fish populations in the Lancang-Mekong River and promote stability and restoration of fisheries resources, which in turn, through increased fish productivity, enhance the income of fishermen.

7.3.4.3 Fisheries Management and Cross-Border Cooperation

Effective fisheries management for the Lancang-Mekong River is dependent on a sound and cooperative approach that includes involvement at local and international scales, as well as education of the public. The fisheries administration in lower Lancang River uses an approach of government-led joint management by fishery departments and villagers. The government plays a decisive role in developing regulations (e.g., "Xishuangbanna the Lancang River Protection Ordinance") to provide legal protection for aquatic organisms. Promotional materials have also been developed and distributed to educate the public on these regulations as well as the purpose and importance of protecting aquatic organisms. In 2014, a total of 8500 pamphlets were handed out, 150 notices were put up, and 20 billboards were installed to inform the public on fishing closures.

A serious threat to fish management and conservation is the use of illegal fishing methods, such as electricity, poison, and dynamite, and enforcement efforts are required to effectively halt such practices. An integrated enforcement team involving the department of agriculture, public security, industry and commerce, and maritime affairs is conducting routine inspections on the fishery resources in important sections of the lower Lancang River and border waters, which effectively strengthen the supervision and management efforts (Fig. 7.7a, b, c). In 2014, the state

Fig. 7.7 Fishery management of the lower Lancang River (Photos by Wenlong Qi)

government dispatched 775 fishery law enforcement officers, conducted 203 vehicle trips, and used the law enforcement speedboat 52 times. During these enforcement efforts, 76 detonators, 131 electric fish catching machines, 1326 fishing nets, 11 kayaks, and 150 kg of illegal fish catches were confiscated. In addition, 62 shops were inspected and within these 324 pieces of banned fishing gear (e.g., electric fish catching machines), 158 chargers, and 66 batteries were seized. In addition, 15 workshops were conducted to inform people on prohibited fishing methods (e.g., by electricity, poison, and dynamite), which involved more than 800 people. Another potentially significant impact on fish stocks results from the construction of dams to catch migrating. In Dai communities, which are close to the Buyuan River, catching fish by building a dam in the fish migration channel remains a local tradition (Fig. 7.7d). Through the fishery promotion and economic compensation provided by fishery department, several dams built in the Buyuan River were removed and river connectivity was restored (Fig. 7.7e, f).

In view of the difficulties faced by fishery management given the sometimes difficult terrain (e.g., steep mountain streams), extensive areas (wide basin area of the lower Lancang River), and limited law enforcement officials, a joint management approach involving state, county, township, and village has been adopted by the fisheries management department. This approach addresses a number of management needs. Key components include:

- Dividing the Lancang River into a number of small sections, with each community responsible for a nearby section as the administrative responsibility through clearly defined obligations
- Promoting the development of industries such as fisheries, planting, aquaculture, and tourism in nearby villages
- Reducing the use of pesticide and chemical fertilizers by promoting the implementation of environment-friendly fisheries and agriculture
- Prohibiting and reducing illegal and unsustainable fishing through village autonomy
- Promoting the recovery of fish population in the Lancang River Basin
- Protecting aquatic habitats in watersheds and improving the sustainable income level of the villagers

To protect the Lancang-Mekong cross-border fish resources, joint fisheries management is conducted in cooperation with downstream countries. For example, in 2015, China and Laos signed a "Cooperation Agreement on Fishery Resources Protection" (Fig. 7.8). The signing of the agreement escalated fishery resources protection efforts from a domestic regional management level to an international cooperative management level. The Cooperation Agreement delineated the 40 km river border between China and Laos as a common protected area and developed a framework for future cooperation between the two countries that clearly defined responsibilities and obligations. The two countries also cooperated in carrying out watershed fisheries protection through the enhancement and releasing of fish and joint fishery enforcement (Fig. 7.8).

Fig. 7.8 China and Laos signed the "Cooperation Agreement on Fishery Resources Protection" and carried out joint fishery law enforcement (Photos by Jiankuan He)

7.4 Conclusions

Results of our analyses indicate that species richness of native fishes in the Lancang River, especially in its downstream (dammed) regions, has been substantially reduced during the past 30 years. Native fish species decreased substantially in number, and nonnative species increased in the downstream regions. In contrast, only relatively few native species disappeared in the upstream regions (undammed regions). Our results also indicated that temporal changes in the phylogenetic diversity indices differed from changes in species richness. In contrast to species richness, the Δ^{+} of entire assemblages increased with the introduction of exotic species and disappearance of native species, implying an increase in phylogenetic diversity as species richness decreased. The reason for the difference in temporal changes between species richness and phylogenetic diversity indices is related to the high relatedness within native fish communities of which species have been lost, in contrast to the greater taxonomic distance of introduced exotic species.

The potential for abundance of long- and short-range migratory fishes to be affected by hydropower development was investigated to (1) determine whether there is evidence for negative impacts within the Lancang-Mekong system due to disruption of migration or other causes, (2) determine whether there is evidence for decreases in abundance, and (3) identify causes of any such decreases in abundance. Although decreases in abundance in long-range migrating pangasiid catfish are apparent from declining capture rates over time, hydropower development was not identified as a cause, given the lack of such developments within the annual range of these fish. Overfishing was identified as the key issue of concern. Direct mortality associated with infrastructure was identified as a risk for short-range migrant species that occur in the middle and lower reaches of the Lancang River where hydropower development has occurred and is planned.

Potential impacts of hydropower development on critical fish habitat in the floodplain complexes of the Mekong River Basin during the flood season were investigated by assessing effects of water level regulation on downstream areas as analyzed in Chap. 4. Results suggested that although reduction of water level due to water regulation of the hydropower projects can cause substantial changes in water levels downstream relative to baseline, during the flood season when flooding of the floodplain complexes provides critical seasonal fish habitat, these effects do not extend downstream to the main floodplain complexes in the system (Tonle Sap River and Tonle Sap Lake system in Cambodia and the Mekong Delta in Vietnam). The flood areas in the upper part of the basin, in Thailand and Laos, that are mainly associated with the tributaries of the Mekong River, will, however, be affected by water regulation of the Lancang hydropower projects, and there may eventually be a decrease in fish productivity in this area, although there may be some compensatory effects due to the increase in water level during the dry season.

Evaluation of transboundary effects on fish communities is complicated because there are many factors other than hydropower development that affect fish communities in the Lancang-Mekong River and that change in space and time. We investigated the relationship between the distance between dams and biodiversity measures and found that extirpation rate increases strongly with decreasing distance between dams (i.e., with increasing dam frequency). These results are suggestive of hydropower project impacts and likely causes were identified to be habitat alteration and fragmentation associated with dam and reservoir creation. To directly address the potential for transboundary effects, biodiversity changes in the region furthest downstream of the Lancang River (where no dams currently exist) were compared to those in the dammed portion of our study area. These results indicated that current species richness is substantially greater downstream of the Jinghong dam and that species richness continues to increase as distance from the Jinghong dam increases. This comparison suggests that the impacts of the cascade of hydropower developments on fish biodiversity decline with distance downstream. However, there are potentially confounding factors that may affect this relationship that have not been quantified; as such it is recommended that additional studies are conducted that specifically investigate the role of hydropower development in loss of fish biodiversity through a robust study design and associated data analysis.

Three key conservation measures taken by China to protect fish communities in the Lancang-Mekong River were identified. These include the designation of the Buyuan Tributary as a fish conservation area, enhancement and release of native fish species, and management and cross-border cooperation.

Four key potential effects (hydrology, water temperature, sediment transport/geomorphology, and fish community) of the Lancang cascade of hydropower projects were documented to extend downstream in the Lancang-Mekong River beyond the Chinese border. Stabilizing effects due to water regulation and storage were documented for both hydrological variables investigated (water level and discharge) and water temperature, such that extremes and ranges of values had generally been dampened (e.g., water levels and flows increased during the dry season and decreased during the flood season; water temperature increased in winter and decreased in summer) and in some cases the timing of seasonal patterns had shifted. Effects of the hydropower projects caused by water regulation on hydrology metrics were most pronounced and extended furthest downstream during the dry season, whereas during the flood season, the high levels of precipitation and the inflow of large tributaries in countries to the south limited the potential for project effects to extend south. Although all documented effects of hydrology and water temperature decreased with distance downstream, effects could be detected south of the Chinese border, and, in some cases (e.g., hydrological effects during the dry season) extended to the entire lower Mekong River.

Effects of the hydropower projects on sediment transport/geomorphology and fish community had less potential to extend for great distances downstream, at least in the immediate future. Changes to sediment transport downstream of the Lancang hydropower projects were evident in a steep decrease in sediment discharge coincident with the timing of dam completion (owing to sediment trapping within the reservoir), although effects did not appear to persist into the southern Mekong River. Further, erosional effects were anticipated to gradually create channel changes that would eventually progress downstream through a slow and complicated process. Identified effects of hydropower projects on fish community included habitat alteration and fragmentation associated with dam and reservoir creation (evident from the relationship between the distance between dams and biodiversity measures), direct mortality associated with infrastructure, and effects to fish habitat in the upper part of the basin (Thailand and Laos) during the flood season due to water regulation. These effects were generally anticipated to dissipate relatively quickly with distance downstream.

Our results on transboundary effects on hydrology, water temperature, sediment transport/geomorphology, and fish community indicate potential for negative effects to natural ecosystems that extend for differing distances downstream of the Chinese border. However, positive social transboundary effects have also been documented. Specifically, the hydropower reservoirs have high regulatory capacity and therefore have the potential to mitigate droughts and floods downstream. For example, the release of water from the Lancang dams supplemented the low discharge and eased the regional drought of 2016 in the lower Mekong River Basin and alleviated salinity intrusion in the Mekong Delta (MRC 2017). The ability of the Lancang dams to regulate water also has implications for potential water issues during the dry season related to climate change. Although water regulation potential is less in the flood season, the regulation and storage of floodwaters has the potential to lower the peak flow in downstream areas and thereby provide flood control.

Our analyses were conducted with data available to date. However, owing to the short period of time that the Lancang hydropower projects have been operational, confidence in our conclusions is limited. It is therefore recommended that further analyses incorporating additional data from future years are conducted to fully investigate the effect of hydropower project-related effects over large distances downstream and to document patterns and identify causes which may be complex and variable. Further, evaluation of the role of the hydropower projects on the effects investigated can be difficult because there can be many factors other than hydropower development that can affect the key variables investigated. For this study, this was especially evident for analyses related to effects of sediment transport and channel geomorphology and impacts to fish communities, owing to a variety of spatial and temporal changes that may coincide with hydropower development. For example, changes in land use and human population size have occurred during the period of hydropower development and therefore complicate interpretation of spatial and temporal changes in sediment yield/transport and fish community. Moreover, changes in climate and topography that are correlated with distance downstream confound interpretation of the potential for effects of water temperature and hydrology to extend downstream. Such considerations were addressed with our analyses by using multiple approaches or by explicitly choosing appropriate comparisons. It is nevertheless recommended for future studies that potential confounding effects are identified and incorporated into robust study design and statistical analyses that allow partitioning of effects among key factors.

Identification of key transboundary effects highlights the need for strong transboundary cooperation for managing the downstream impacts (Räsänen et al. 2017). A key recommendation is therefore the continued and augmented cross-border communication and collaboration that is a component of fisheries management (Sect. 7.3.4) and that is documented through a case study of experience sharing in Laos (Chap. 9).

References

Abellan, P., D.T. Bilton, A. Millan, D. Sanchez-Fernandez, and P.M. Ramsay. 2006. Can taxonomic distinctness assess anthropogenic impacts in inland waters? A case study from a Mediterranean river basin. *Freshwater Biology* 51: 1744–1756.

Agostinho, A.A., L.C. Gomes, S. Verissimo, and E.K. Okada. 2004. Flood regime, dam regulation and fish in the upper Parana' river: Effects on assemblage attributes, reproduction and recruitment. *Reviews in Fish Biology and Fisheries* 14: 11–19.

Agostinho, A.A., F.M. Pelicice, and L.C. Gomes. 2008. Dams and the fish fauna of the Neotropical region: Impacts and management related to diversity and fisheries. *Brazilian Journal of Biology* 68 (11): 1119–1132.

Azaza, M.S., M.N. Dhraief, and M.M. Kraiem. 2008. Effects of water temperature on growth and sex ratio of juvenile Nile tilapia *Oreochromis niloticus* (Linnaeus) reared in geothermal waters in southern Tunisia. *Journal of Thermal Biology* 33: 98–105.

Baran, E., M. Chum, T. Fukushima, K. Hand, T. Hortle, and B. Kang. 2012. Fish biodiversity research in the Mekong Basin. In *The biodiversity observation nsetwork in the Asia-Pacific region*, Ecological research, ed. S. Nakano, T. Yahara, and T. Nakashizuka, 149–164. Tokyo: Springer.

Cagnolo, L., M. Cabido, and G. Valladares. 2006. Plant species richness in the Chaco Serrano Woodland from Central Argentina: Ecological traits and habitat fragmentation effects. *Biological Conservation* 132: 510–519.

Cavender-Bares, J., H. Kozak, P. Fine, and S. Kembel. 2009. The merging of community ecology and phylogenetic biology. *Ecology Letters* 12: 693–715.

Chen, X. 2013. Checklist of fishes of Yunnan. *Zoological Research* 34: 281–343 (in Chinese).

Cheng, S., E. Herricks, W. Tsai, and F. Chang. 2016. Assessing the natural and anthropogenic influences on basin-wide fish species richness. *Science of the Total Environment* 572: 825–836.

China News. 2014. http://news.china.com/domestic/945/20140701/18602228.html.

Chu, X., and Y. Chen. 1989. *The fishes of Yunnan, China: Part I Cyprinidae*. Beijing: Science Press (in Chinese).

———. 1990. *The fishes of Yunnan, China: Part II Cyprinidae*. Beijing: Science Press (in Chinese).

Clarke, K.R., and R. Gorley. 2006. *PRIMER v6: User manual/tutorial*. Plymouth: PRIMER-E.

Clarke, K.R., and R.M. Warwick. 1998. A taxonomic distinctness index and its statistical properties. *Applied Ecology* 35: 523–531.

———. 2001a. A further biodiversity index applicable to species lists: Variation in taxonomic distinctness. *Marine Ecology Progress Series* 216: 265–278.

———. 2001b. *Changes in marine communities: An approach to statistical analysis and interpretation*. Plymouth: PRIMER-E.

Coates, D., Poulsen, A. F., and Viravong, S. 2000. *Governance and transboundary migratory fish stocks in the Mekong River Basin*. Paper presented at the MRC Third Fisheries Technical Symposium. Mekong River Commission, Phnom Penh, Cambodia, 8–9th December 2000.

Ellingsen, K.E., K.R. Clarke, P.J. Somerfield, and R.M. Warwick. 2005. Taxonomic distinctness as a measure of diversity applied over a large scale: The benthos of the Norwegian continental shelf. *Journal of Animal Ecology* 74: 1069–1079.

Fan, H., D. He, and H. Wang. 2015. Environmental consequences of damming the mainstream Lancang-Mekong River: A review. *Earth-Science Reviews* 146: 77–91.

Frevill, H., K. McConway, M. Dodd, and J. Silvertown. 2007. Prediction of extinction in plants: Interaction of extrinsic threats and life history traits. *Ecology* 88: 2662–2672.

Gao, X., M. Li, P. Lin, Z. Duan, and H. Liu. 2013. Environmental cues for natural reproduction of the Chinese sturgeon, *Acipenser sinensis* Gray, 1835, in the Yangtze River, China. *Journal of Applied Ichthyology* 29: 1389–1394.

Gaston, K.J. 1998. Species-range size distributions: Products of speciation, extinction and transformation. *Philosophical Transactions of the Royal Society of London. Series B, Biological Sciences* 353: 219–230.

Graham, C.H., and P.V.A. Fine. 2008. Phylogenetic beta diversity: Linking ecological and evolutionary processes across space in time. *Ecology Letters* 11: 1265–1277.

Griffin, J.N., J.E.K. Byrnes, and B.J. Cardinale. 2013. Effects of predator richness on prey suppression: A meta-analysis. *Ecology* 94: 2180–2187.

Guo, Z., L. Li, W. He, and H. Li. 2014. Research on fish resource in the upstream of Lancang River. *Modern Agricultural Science and Technology 14*(228): 337 (in Chinese).

Havel, J.E., C.E. Lee, and M.J. Vander Zanden. 2005. Do reservoirs facilitate invasions into landscapes? *Bioscience* 55: 518–525.

Heino, J., J. Soininen, J. Lappalainen, and R. Virtanen. 2005. The relationship between species richness and taxonomic distinctness in freshwater organisms. *Limnology and Oceanography* 50: 978–986.

Helmus, M.R., W. Keller, M.J. Paterson, N. Yan, C.H. Cannon, and J.A. Rusak. 2010. Communities contain closely related species during ecosystem disturbance. *Ecology Letters* 13: 162–174.

Hu, G., and Q. Zhang. 2016. Seasonal variations in macrobenthic taxonomic diversity and the application of taxonomic distinctness indices in Bohai Bay, northern China. *Ecological Indicators* 71: 181–190.

Jiang, X., Y. Song, J. Xiong, and Z. Xie. 2014. Can excluding non-insect taxa from stream macroinvertebrate surveys enhance the sensitivity of taxonomic distinctness indices to human disturbance? *Ecological Indicators* 41: 175–182.

Jiang, Z., S. Brosse, X. Jiang, and Z. Zhang. 2015. Measuring ecosystem degradation through half a century of fish species introductions and extirpations in a large isolated lake. *Ecological Indicators* 58: 104–112.

Junk, W.J. 1997. The central amazon floodplain: Ecology of a pulsing system. In *Ecological Studies Volume 126*, 525 pp. Berlin: Springer.

Kang, B. and D. He. 2007. Research progress of biodiversity of fish species in the Lancang Jiang River. *Resources Science 29*(5): 195–200 (in Chinese).

Kang, B., D. He, L. Perrett, H. Wang, W. Hu, W. Deng, and Y. Wu. 2009. Fish and fisheries in the upper Mekong: Current assessment of the fish community, threats and conservation. *Reviews in Fish Biology and Fisheries* 19: 465–480.

Kang, B., W. Hu, W. Qi, C. Yang, and J. Li. 2010. Research on fish species diversity in the Buyuan River. *Progress in Fishery Sciences, 31*(3): 6–14 (in Chinese).

Kummu, M., and J. Sarkkula. 2008. Impact of the Mekong River flow alteration on the Tonle Sap flood pulse. *Royal Swedish Academy of Sciences* 37 (3): 185–192.

Lei, C. 2012. *The influence and protection of hydropower construction on fish in the mainstream of Lancang River: A case study of Jinghong, Xiaowan and Gongguoqiao Dams*. Yunnan University (in Chinese).

Leira, M., G. Chen, C. Dalton, K. Irvine, and D. Taylor. 2009. Patterns in freshwater diatom taxonomic distinctness along an eutrophication gradient. *Freshwater Biology* 54: 1–14.

Li, S. 1982. Fish fauna and its differentiation in the upland lakes of Yunnan. *Acta Zoologica Sinica 28*: 169–176 (in Chinese).

Li, J., S. Dong, M. Peng, Z. Yang, S. Liu, X. Li, and C. Zhao. 2013. Effects of damming on the biological integrity of fish assemblages in the middle Lancang-Mekong River basin. *Ecological Indicators* 34: 94–102.

Limburg, K.E., and J.R. Waldman. 2009. Dramatic declines in North Atlantic diadromous fishes. *Bioscience* 59: 955–965.

Liu, M., D. Chen, X. Duan, X. Wang, and S. Liu. 2011. Ichthyofauna composition and distribution of fishes in Yunnan section of Lancang River. *Journal of Fishery Sciences of China 18*(1): 156–170 (in Chinese).

Matthies, D., I. Brauer, W. Maibom, and T. Tscharntke. 2004. Population size and the risk of local extinction: Empirical evidence from rare plants. *Oikos* 105 (3): 481–488.

Milosevic, D., V. Simic, M. Stojkovic, and I. Zivic. 2012. Chironomid faunal composition represented by taxonomic distinctness index reveals environmental change in a lotic system over three decades. *Hydrobiologia* 683: 69–82.

MRC (Mekong River Commission). 2017. *The effects of Chinese dams on water flows in the Lower Mekong Basin*. http://www.mrcmekong.org/news-and-events/news/the-effects-of-chinese-dams-on-water-flows-in-the-lower-mekong-basin/. Accessed 12 June 2017.

Nadon, M.J., R.A. Metcalfe, C.J. Williams, K.M. Somers, and M.A. Xenopoulos. 2015. Assessing the effects of dams and waterpower facilities on riverine dissolved organic matter composition. *Hydrobiologia* 744: 145–164.

Nelson, J.S. 2006. *Fishes of the world*. New York: Wiley.

Newman, B.J., P. Ladd, M. Brundrett, and W.D. Kingsley. 2013. Effects of habitat fragmentation on plant reproductive success and population viability at the landscape and habitat scale. *Biological Conservation* 159: 16–23.

Nilsson, C., C.A. Reidy, M. Dynesius, and C. Revenga. 2005. Fragmentation and flow regulation of the world's large river systems. *Science* 308: 405–408.

Petesse, M.L., and M. Petrere. 2012. Tendency towards homogenization in fish assemblages in the cascade reservoir system of the Tiete river basin, Brazil. *Ecological Engineering* 48: 109–116.

Pyke, G.H. 2005. A review of the biology of Gambusia affinis and G-holbrooki. *Reviews in Fish Biology and Fisheries* 15: 339–365.

Pysek, P., V. Jarosik, J. Pergl, R. Randall, M. Chytry, I. Kuhn, L. Tichy, J. Danihelka, J. Chrtek, and J. Sadlo. 2009. The global invasion success of Central European plants is related to distribution characteristics in their native range and species traits. *Diversity and Distributions* 15: 891–903.

Räsänen, T.A., P. Someth, H. Lauri, J. Koponen, J. Sarkkula, and M. Kummu. 2017. Observed river discharge changes due to hydropower operations in the upper Mekong Basin. *Journal of Hydrology* 545: 28–41.

Srivastava, D.S., M.W. Cadotte, A.A.M. MacDonald, R.G. Marushia, and N. Mirotchnick. 2012. Phylogenetic diversity and the functioning of ecosystems. *Ecology Letters* 15: 637–648.

Sverdrup-Jensen, S. 2002. *Fisheries in the Lower Mekong Basin: status and perspectives*, MRC Technical paper, no. 6, 103 pp. Phnom Penh: Mekong River Commission.

Tolimieri, N., and M.J. Anderson. 2010. Taxonomic distinctness of demersal fishes of the California current: Moving beyond simple measures of diversity for marine ecosystem-based management. *PLoS One* 5: e10653.

Vamosi, J.C., and J.R.U. Wilson. 2008. Nonrandom extinction leads to elevated loss of angiosperm evolutionary history. *Ecology Letters* 11: 1047–1053.

Van Zalinge, N.P., L. Deap, P. Ngor, S. Juha, K. Jorma, and R. Huon. 2003. *Mekong flood levels and Tonle Sap fish catches*. Abstract submitted to the Second International Symposium on the Management of Large Rivers for Fisheries. Phnom Penh: Mekong River Commission.

Warwick, R.M., C.M. Ashman, A.R. Brown, K.R. Clarke, B. Dowel, B. Hart, R.E. Lewis, N. Shillabeer, P.J. Somerfield, and J.F. Tapp. 2002. Inter-annual changes in thebiodiversity and community structure of the macrobenthos in Tees Bay and the Tees estuary, UK, associated with local and regional environmental events. *Marine Ecology Progress Series* 234: 1–13.

Webb, C.O., D.D. Ackerly, M.A. McPeek, and M.J. Donoghue. 2002. Phylogenies and community ecology. *Ecology and Systematics* 33: 475–505.

Winter, M., O. Schweiger, S. Klotz, W. Nentwig, P. Andriopoulos, M. Arianoutsou, C. Basnou, P. Delipetrou, V. Didziulis, M. Hejda, P.E. Hulme, P.W. Lambdon, J. Pergl, P. Pysek, D.B. Roy, and I. Kuhn. 2009. Plant extinctions and introductions lead to phylogenetic and taxonomic homogenization of the European flora. *Proceedings of the National Academy of Sciences of the United States of America* 106: 21721–21725.

Xie, S., K. Zheng, J. Chen, Z. Zhang, X. Zhu, and Y. Yang. 2011. Effect of water temperature on energy budget of Nile tilapia, *Oreochromis niloticus*. *Aquaculture Nutrition* 17: E683–E690.

Yang, J., X. Chen, and Y. Chen. 2007. On the population status and migration of pangasiid catfishes in Lancangjiang River basin, China. *Zoological Research 1*: 63–67 (in Chinese).

Zefferman, E., J.T. Stevens, G.K. Charles, M. Dunbar-Irwin, T. Emam, S. Fick, L.V. Morales, K.M. Wolf, D.J.N. Young, and T.P. Young. 2015. Plant communities in harsh sites are less invaded: A summary of observations and proposed explanations. *AoB Plants*: 7, plv056.

Zhang, R. 2001. Post-evaluation of ecological environment of Manwan Hydroelectric Station on the Lancang River. *Design of Hydroelectric Power Station 17*(4): 27–32 (in Chinese).

Zhang, C., L. Ding, C. Ding, L. Chen, J. Sun, and X. Jiang. 2018. Responses of species and phylogenetic diversity of fish communities in the Lancang River to hydropower development and exotic invasions. *Ecological Indicators* 90: 261–279.

Zhao, Z., and Y. Li. 2013. Current pollution situation and pollution control in Yunnan section of upper Lancang River. *Environmental Science Survey 32*(5): 68–70 (in Chinese).

Zheng, L., X. Chen, and J. Yang. 2013. Status and conservation of fishes in the middle and lower Lancangjiang River. *Zoological Research*, 34(6): 680–686 (in Chinese).

Zhong, R., C. Yang, K. Fu, and D. He. 2011. Analysis of characteristics of water and sediment and their variability in Buyuan River basin. *Yangtze River 42*(24): 29–33 (in Chinese).

Zintzen, V., M.J. Anderson, C.D. Roberts, and C.E. Diebel. 2011. Increasing variation in taxonomic distinctness reveals clusters of specialists in the deep sea. *Ecography* 34: 306–317.

Ziv, G., E. Baran, S. Nam, I. Rodriguez-Iturbe, and S.A. Levin. 2012. Trading-off fish biodiversity, food security, and hydropower in the Mekong River basin. *Proceedings of the National Academy of Sciences of the United States of America* 109: 5609–5614.

Chapter 8
Changes in Women's Livelihood in Areas Affected by Hydropower Projects

8.1 Background

Beginning in the mid-1980s, China began to adhere to policies set out by the World Bank requiring the assessment of forced resettlements as an integral and mandatory component of infrastructure projects (Cernea 1998). Nevertheless, until recently, hydropower developers did not comprehensively consider the social impacts of their projects and consequences to livelihoods resulting from the displacement of people.

The resettlement caused by hydropower development can have significant adverse impacts by simultaneously affecting the economic, social, and cultural construct of resettlers (Koenig and Diarra 2002). However, due to rapid social and economic changes over more than two decades (1980–2004), the approach to mitigating adverse effects to displaced people has been advancing, including adopting long-term compensation (up to 20 years). For example, infrastructure and services are established through a gradual resettlement process to support displaced people.

In 2006, two significant policies were implemented in China to address effects to displaced people. These are (1) the *State Council Decree No. 471 on Large- and Medium-scale Hydraulic and Hydropower Projects' Land Acquisition and Resettlement Compensation Regulation* (No. 471 policy) and (2) the *Suggestions of the State Council No. 17 on the Improvement of Follow-on Support to the Large- and Medium-Scale Hydraulic and Hydropower Projects' Resettled Migrants* (No. 17 policy). With these policies, the concept of compensation was expanded in scope and scale, with compensation to resettled communities being combined with subsidization and follow-up support. For example, the No. 17 policy provides up to 20 years of financial compensation for qualified migrants at a rate of 600 yuan per person per year. The financial support focuses on the importance of vocational training and education to support migrants to effectively transition to other forms of livelihood. The new policies provide guidance on the procedures for migrant

X. Yu et al., *Balancing River Health and Hydropower Requirements in the Lancang River Basin*, https://doi.org/10.1007/978-981-13-1565-7_8

negotiations, including consulting with impacted individuals, holding public hearings, providing publicly available details about the compensation standards, etc.

When considering gender in relation to impacts from hydropower projects in developing countries, women are generally more adversely impacted than men (Simon 2013). Following forced displacement, women experience more adverse economic, social, and cultural impacts than men. Prior to recent advancements in policy and practice, women's needs were not given specific consideration during hydropower development, even though women have different societal roles and therefore different needs and priorities for water, sanitation, and livelihood than men. Historically, hydropower projects have exacerbated existing gender biases and adversely impacted women's roles and position within the home and community. Several studies have documented the negative outcomes of dam construction to women's livelihoods (Scudder 2005; WCD 2000). Impoverishment, health impacts, and trauma, as a result of displacement, have been documented to be more severe for women (Scudder 2005; WCD 2000). However, hydropower development, like other infrastructure projects, has the potential to play a positive role in gender relations and equality (WCD 2000).

As a whole, the gender issue in China has unique context due to the social and political environment and context. Scholars have researched the impact of hydropower development on women, including their livelihoods of displaced communities, their social security, their self-identity, and how resettlers rebuilt their social relations (Chen 2007, 2013a, b). The study by Chen (2007) on women resettlers from the Three Gorges Hydroplant found that there are gender distinctions in the reconstruction of migrants' self-identities following resettlement and that some women were further marginalized within the family and society following displacement. Chen (2013b) found that women resettlers faced more significant challenges than their male counterparts in terms of finding employment, obtaining skills training, and participating in collective affairs, which in turn increased women's poverty and lowered their economic status in the family.

These studies focused on the short-term impacts of resettlement, without fully considering the medium- and long-term changes to resettlers' livelihoods and social status. Typically, resettlers are displaced in the first few years of dam construction; however, replacing a community and livelihood usually takes 10–30 years. As such, it is important to understand the long-term effects and changes in order to continue to improve resettlement policy and practice. Moreover, since the No. 17 and No. 471 policies were implemented, government and hydropower developers have gained valuable insights on how to reduce resettlement impacts. However, no study has yet described the lessons learned from these larger institutional changes and the linkages to addressing impacts to women.

This study aims to examine the impacts to women from hydropower development using two case studies in the Lancang River Basin: the Manwan and Jinghong hydropower projects. The Manwan case study focuses on long-term social changes and effects. The Manwan hydropower project was built in a remote and isolated mountainous area in southwest China. The project's resettlement policy evolved as the project was developed. During the initial policy, Manwan resettlers faced challenges in sustaining their livelihoods. In response, the policy was revised to consider social development at the macro-level which resulted in improvements to resettlers' livelihoods. For this study, we explored the changes in livelihood of Manwan resettlers, with specific emphasis on women's livelihood and social status. We also considered the reasons for this change from the perspective of macro-social and environmental policy. Moreover, we conducted a comparison with the Jinghong case study discussed below.

The Jinghong case study explores how larger institutional changes addressed impacts to women and their ability to adapt during resettlement due to the development of the Jinghong hydropower project. The hydropower project was developed following the No. 471 and 17 policies and provides an opportunity to examine how the changes in these government policies and institutional arrangements addressed gender issues. Furthermore, we examined the different scales of institutional arrangements (macro, meso, and micro) set to protect the interests of women during hydropower development.

8.2 Manwan Case Study

The Manwan hydropower project is located in the middle of the Lancang River and was the first hydropower project to be developed on the river's mainstem. The project began construction in 1986 and began storing water within the reservoir in 1993. At the location of the project, the Lancang Valley is deep with high steep mountains. The reservoir is on average 337.1 m wide, encompassing an area of 23.6 km^2 (Li 2012). After the construction of the dam, the water level rose from 891 m to 994 m, and 70 km of the valley was inundated (He et al. 2001)

8.2.1 Background

Villagers in Yun County, Jingdong County, Fengqing County, and Nanjian County were affected by the project, with the Yun and Jingdong Counties most severely affected. Our research focused on resettlements in the Yun County. The climate of Yun County is a subtropical mountainous monsoon climate, with an average temperature of 19.4 °C and average rainfall between 920 and 1330 mm per year (Wang et al. 2000). Prior to and following development of the dam, agriculture was and continues to be the primary industry in this region.

In 1993, the resettlement of people began, with approximately 3410 households in Yun County requiring resettlement as their houses and/or farmlands became submerged by the reservoir. Three models of resettlement were adopted, two involving agricultural compensation and one non-agricultural:

1. *Out-migration.* Out-migration involved villagers resettling out of the Lancang River Basin, usually moving to towns far away from their original homes. Initially, 375 people from Yun County were resettled under this model. Subsequently, due to poverty and landslide risk, 1107 additional people were moved out of the reservoir area to new settlements close to towns or key transportation routes between 2004 and 2008. This process was referred to as *second-round out-migration.*

   Under this model, resettlers were responsible for building their own homes with compensation from the government. Compensation varied between villages. In the case of Hongyan, each resettler, men or women of all ages, was given 2.95 mu (1 mu = 1/15 ha) of farmland, including 1 mu of paddy fields. In addition, 800 mu of firewood forest and 50 mu of orange orchards were given to the village as a whole. Due to lack of irrigation water, the paddy fields were converted to dry farmland and resettlers mainly depended on sugarcane cultivation for income.
2. *Back-up resettlement.* Back-up resettlement involved people who moved but remained within the Lancang River Basin. Farmers relocated to higher elevation areas within or in the vicinity of their original hometown. New farmlands were allocated to the displaced farmers as compensation, either by clearing forest or by re-allocating existing farmland. For those who lost buildings, compensation was provided to rebuild. However, the amount of agricultural land available was less and of lower quality than what the resettlers had prior to resettlement. In addition, there were adverse environmental effects from converting forest lands to provide agricultural land.
3. *Non-agricultural resettlement.* Non-agricultural resettlement involved people receiving financial compensation only. This was the case for the resettlers in Tianba, located within the footprint of the dam. They received money to compensate for the loss of their farmland instead of receiving new agricultural land. In order to help them make a sustainable living, the government helped resettlers organize and develop a company that ran restaurants, hotels, and small shops, for which they became stockholders. The resettlers also changed their residential identity from farmer to citizen. This allowed them to benefit from the welfare services only available to citizens, regardless of age or gender, including highly subsidized grain and social insurance. However, a few years later, the economic system in China was reformed from a centrally planned economy to a free-market-oriented economy. Thus, the government gave up control over the price and allocation of grain and other goods. As a result, the resettlers had to pay market price, which was four times higher than under the previous economic model. With the decline in the temporary direct and indirect construction population after the dam was completed, the hotel and restaurant services company became bankrupt due to the loss of the customer base related to dam construction. The resettlers lost their income source as expenditures increased, leading to impoverishment.

On-site settlement was another form of compensation. While not specifically a model of resettlement, as people impacted did not lose their homes, they did nevertheless lose portions of their land. In summary, on-site resettlement involved people receiving compensation for the loss of their land.

Due to challenges in making and adjusting to a new living following displacement, thousands of resettlers launched a protest campaign in 2003 at the hydropower project to request further support for their livelihood. In 2006, the State Council issued the No. 17 policy to provide support and reduce the effects of resettlement caused by the project. Under this policy, the standard of subsidy was increased from 240 to 400 yuan per year to 600 yuan per year, and the subsidy period was extended from 10 to 20 years. In the study area, Yun County, the resettlers who lost buildings during resettlement could receive this subsidy. However, those who only lost farmlands could not receive the subsidy but were supported through infrastructure and production support projects. These changes led to improving resettlers' livelihood.

8.2.2 Methods

We compared our research on the Manwan case study with previous studies to identify changes to resettlers' livelihoods over the longer term, with a focus on women. The previous studies included those conducted by the Research Group of Impacts of Manwan Hydropower Station Lancang River (RGIMHSLR 2002; Li 2012; Zheng 2014).

The RGIMHSLR study involved an in-depth examination of the changes to Manwan resettlers' livelihoods, including from a gender perspective (RGIMHSLR 2002). Zheng (2014) conducted an anthological study in the administrative village of Manjiu, to which most of the studied villages in our study belong. Zheng (2014) carefully studied the historic livelihoods and changes in 2004, from a human ecology perspective. These two studies provided the baseline for our livelihood analysis. In her thesis *Ecological Change and their Impacts in Manwan Reservoir in Recent 20 Years*, Li (2012) recorded the status of resettlers' livelihood around 2011. For our study, we also conducted surveys in 2016 to obtain the current livelihood status of resettlers to compare with previous studies and to identify any status changes.

To facilitate data comparison, we selected the villages described in the RGIMHSLR study (2002). Our surveys were conducted in eight distinct villages representing four types of displacement including a control (Table 8.1). In each village, 8 to 16 households were randomly selected for interviews (Table 8.1). As the size of villages varied considerably, the sampling ratios varied between depending on village size. Nevertheless, between 15 to 42% of households present in the village were interviewed.

Face-to-face interviews were conducted within each household, to obtain information about basic demographics, access to resources, income and expenditures, gender roles in households, participation in collective activities, and degree of satisfaction about current living situations. In-depth interviews were conducted with village leaders to obtain comprehensive information about the villages.

Table 8.1 Information on villages interviewed as part for Manwan case study survey (2016)

Model of resettlement	Name of village	Time of resettlement	No. of interviewed households	Total households	Sampling ratio
Out-migration	Hongyan	1993	12	80	0.15
Second-round out-migration	Xiabatian	2009	15	65	0.23
	Yakou	2008	8	47	0.17
Back-up resettlement	Lannitang	1997	16	58	0.28
Non-agricultural settlement	Tianba	–	10	28	0.36
On-site settlement	Pingzhang	–	10	29	0.34
	Songshan	–	13	52	0.25
Control[a]	Luqing	–	11	26	0.42
Total	**8**		**95**	**385**	**0.25**

[a]The control village of Luqing was affected by the dam, but the effects were minimal. Luqing was determined to be the most effective control as the natural conditions in completely unaffected villages are quite different than the villages in the study

Table 8.2 Summary of age and gender of household interviewees

Gender	Number of interviewees by age class				
	≤20 years	21–40 years	41–60 years	≥61 years	Total
Male	2	17	23	5	47
Female	1	17	25	5	48
Total	**3**	**34**	**48**	**10**	**95**

The interviews were conducted with the adult members of the households. In total, 95 individuals were interviewed, including 47 men and 48 women (Table 8.2). Most (86%) of interviewees were between 20 and 60 years old. The number of women and men interviewed by age class was similar. Most of the interviewees were Yi ethnicity, followed by Han, and a few Bulang and Lisu (Fig. 8.1). The most common highest level of education reported was primary school or middle school. Male interviewees reported slightly higher education levels than the female interviewees, as most male interviewees attended primary school and middle school, while most female interviewees only attended primary school. Some of the female interviewees (13.7%) did not receive any formal education.

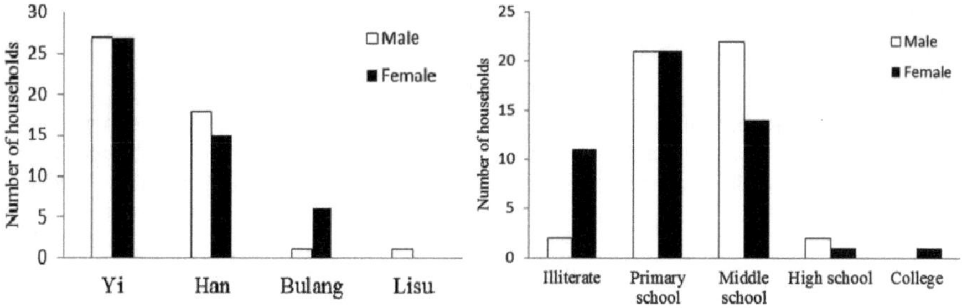

Fig. 8.1 Nationality (*left*) and level of education (*right*) of household interviewees

8.2.3 *Results and Discussions*

8.2.3.1 Changes in Resettlers' Livelihoods

Since the development of the Manwan hydropower project, the resettlers experienced a decline and subsequent recovery in their livelihoods. In the initial years of resettlement, the resettlers' access to natural resources and infrastructure declined, including access to farmland, forest, water, and transportation. Moreover, many of the resettled people experienced a decline in their livelihoods due to lack of employment opportunities, low compensation offered from the government and the hydropower company, and drastic social transformation.

In the last 10 years, the overall livelihoods of both women and men improved concurrent with the improvement of the resettlers' social and economic environment. This improvement was particularly significant for women given women's inferior status prior to resettlement due to physical and social reasons. Resettlers' livelihoods diversified with the main source of income derived from non-agricultural sources. With less intensive agricultural labor demands due to the decline of agricultural land area and crop changes, women's workload declined, while family income increased. Women's access to medical care, education, and financial services also improved. Access to public services also improved, including housing, energy, transportation, and drinking water. These infrastructural improvements made it easier for women to care for their families. The general changes in the livelihoods of resettled families before and after the dam are summarized in Table 8.3.

With improvements in the social and economic environment since 2000, the resettlers' livelihoods have changed and their living conditions have improved. On one hand, the amount of farmland in the resettled communities decreased, especially the proportion of paddy rice fields, due to the rise of water level, landslides, population growth, and underdeveloped irrigation systems. On the other hand, more job opportunities are now available to the resettlers due to improved transportation

Table 8.3 Changes in Manwan resettlers' livelihood: a profile over time

Metric of livelihood	Before the dam (pre-1986)	Around 2000	Around 2011	In 2016
Source of income	Rice, fishing, animal husbandry, shellac production	Corn, animal husbandry, migration workers, small businesses, orange and sugarcane production	Corn, sugar cane, nuts, migration worker, small businesses	Migration workers, small businesses, animal husbandry, sugarcane, nuts, coffee, medicinal plant production
Farmland	Main crops included corn, wheat, rice, beans, sugar cane, tea, tobacco, and walnuts	Decline in total area of farmland, especially paddy rice fields	Further decline of per capita farmland and paddy rice fields	Further decline of paddy rice fields
Transportation	Convenient at local level	Deteriorated	Improved but varied among villages	Overall improved from before the dam
Water	Abundant, from springs, wells, and creeks	Difficult to have enough safe drinking water. Lack of irrigation water	Lack of drinking water and irrigation water due to insufficient water source	Drinking water access and quality generally improved; however, irrigation conditions have not significantly improved
Energy	Wood	Wood	Wood, electricity, methane	Electricity, wood, solar energy
Electricity	From small hydropower stations. Electricity was not available in some villages	From Manwan and other hydropower stations. High price. Some households could not afford electricity	Price varied; people in some villages could not afford electricity	The price of electricity declined. All households could afford electricity
Housing condition	Earth homes	Improved in some villages, worsened in other villages	Improved for second-round out-migration households	Improved generally
Data source	RGIMHSLR 2002, Zheng 2014	RGIMHSLR 2002, Zheng 2014	Li 2012	This study (2016)

infrastructure and increased proximity to markets. Cultivation of high market value crops, such as coffee, nuts, walnuts, and herbs, has increased. Transformation of rice cultivation to perennial fruit and crop production reduced workload, especially for women who are instrumental in rice production. The resettlers have become increasingly engaged in non-agricultural activities. Migration labor, in particular for men, became the primary income source of the resettled households. Some households have become engaged in large-scale intensive animal husbandry. This evolution and change in income source has increased families' incomes.

The resettlers' living conditions generally improved through increased access to drinking water, energy, housing, and transportation. All villagers have access to tap water, and although some villages have water quality challenges, there has been a decline in periods when safe drinking water is not available. With regard to energy, the main sources for local families are electricity, wood, and solar energy. All households now have access to inexpensive electricity. The increased use of electricity for lights and cooking and solar energy for heating water, combined with the decline in animal husbandry, reduced resettlers' dependency on wood as a source of energy. The resettlers' housing conditions have also improved through government funding. Most resettlers, except for those of Pingzhang village, have brick houses. Moreover, in general, the transportation conditions have improved, and the previously isolated communities are increasingly connected to the external world. Paved roads connect all of the villages to markets and the economic and political centers. Approximately half of the resettled households own motorcycles or tractors. Public transportation is available in most villages.

As women are the primary care givers in the family, improvements in family income and living conditions have benefitted women. Increased access to drinking water and inexpensive and convenient energy has greatly reduced women's workloads. Although both men and women benefit from such improvements, women benefited more given that women conduct most of the work in the home. Although women's workload around the home has decreased, their role and workload in agriculture have increased as men began to work more as migration labor (i.e., far away from the community). Thus, women conducted most of the agriculture activities. However, this increased workload was alleviated to some extent through the diversification of livelihood, planting of perennial cash crops, change of cropping patterns, and the decline in farmland area. In addition, with men increasingly working as migrant labor, their absenteeism created more space for women to participate in public/collective activities.

The details of changes in diversification of livelihoods, infrastructure and access to public services, participation in collective activities, and degree of satisfaction with current living conditions are discussed below.

Diversification of Livelihood

Changes in Farmland Resources and Crop Pattern

Paddy rice fields are the most productive farmland in the study area. Before dam construction, in the study area, most paddy fields were located in the Lancang River valley and rice was the primary crop and a key income source for local people. Highly productive large paddy fields made it possible for most households to produce and sell extra rice. A reliance on rice cultivation resulted in high workloads for women, as they are the main workforce in rice cultivation, especially in seedlings, transplants, weeding, and harvesting.

When the reservoir began to store water, most paddy fields were inundated (RGIMHSLR 2002; Li 2012). Although the government tried to mitigate the impacts by developing new paddy fields on the mountains along the reservoir and subsidizing farmland in resettled areas, the total area and quality of farmland generally declined. Surrounding the reservoir, the newly claimed lands generally had insufficient infrastructure, irrigation, and soil nutrients for rice cultivation or were threatened by landslide. As such, some farmlands were abandoned, and some paddy fields were changed into rain-fed fields. In the out-migration communities, many resettled households did not purchase as much farmland as the government expected, partly due to lack of funds and partly due to increased land prices. The result was an overall decrease in farmland area, especially paddy fields, compared to before the dam was developed.

Since 2012, the amount of farmland has not increased in the resettled villages due to conservation and enhanced regulation of natural vegetation. With population growth, the area of farmland per capita declined further. Deterioration of irrigation conditions in some villages further reduced the quantity of paddy rice fields. Table 8.4 summarizes the farmland resources and crop patterns in the villages surveyed in our study in 2016. The per capita farmland and paddy rice fields were higher than average in Luqing village (control), which was not heavily impacted by the dam. Although per capita farmland in Hongyan village was highest among the eight villages, the per capita paddy fields were very low due to lack of irrigation water. Consequently, the per capita paddy fields are relatively higher in villages along the reservoir rather than those that are further away.

In response to the decline of paddy rice fields and increasing market opportunities through improved transportation conditions, the villages changed the types of crops produced (Table 8.4). They reduced rice cultivation, which requires intensive labor and irrigation, while increasing production of cash crops and fruits. Oranges became a specialty of Lannitang village, and sugarcane became a primary income source in Hongyan village. Fruit tree saplings were given to resettlers for free or heavily subsidized, and skill training was provided by the local government. Although some farmers were dissatisfied with the quality of the subsidized saplings, farmers began to plant more trees than those provided by government as the introduction of these cash crops led to the spontaneous emergence of new markets. The shift in crops increased resettlers' incomes and reduced their workloads as fruit and nut trees require less water and care than rice paddies. Women in particular benefitted from the agricultural efficiencies given they are now the primary agricultural workforce.

Crop production continues to evolve. In recent years, most resettlers gave up shellac cultivation due to poor market opportunities. Orange productivity declined, and many households stopped further investing in this crop due to pests, disease, and changes in climate. The resettlers are nevertheless actively adapting to adverse conditions by trying new crops and crop productions. New crops, such as nuts and coffee, emerged in the study area complimenting the traditional crops (rice, corn, sugarcane). The farmers are also shifting to commercial agriculture. We found that 71% of households now buy rice for domestic use.

Table 8.4 Summary of farmland resources and crop patterns in surveyed villages

Model of resettlement	Name of village	Area of farmland per capita (mu)	Area of paddy rice fields per capita (mu)	Proportion of paddy rice fields in farmland (%)	Main crops
Out-migration	Hongyan	3.19	0.22	7	Corn, sugarcane, coffee, nut
Second-round out-migration	Xiabatian	0.84	0.29	34	Rice, corn, vegetables, condiment, herbs, nut
	Yakou	1.93	0.48	25	Rice, corn, sugarcane
Back-up resettlement	Lannitang	0.98	0.48	34	Corn, walnut, orange, nut, coffee, tobacco
Non-agricultural settlement	Tianba	1.11	0	0	Corn, bean, orange, nut, coffee
On-site settlement	Pingzhang	1.16	0.63	54	Rice, corn, orange, walnut, bean, bamboo shoots, tea
	Songshan	1.62	0.23	14	Corn, rice, orange, walnut, bean, coffee
Control	Luqing	2.38	0.79	33	Rice, corn, walnut, nut, tea, herbs
Total		**1.46**	**0.25**	17	

Diversification of Income Source

Prior to the construction of the dam, the communities around the reservoir were isolated. They relied on self-sufficient economies with very limited connections to the outside world. The primary income source was cultivation, in particular rice production. Animal husbandry, usually pigs, cattle, and sheep, and fishing provided extra source of protein to local communities (Zheng 2014). With the construction of the dam, many outsiders interacted in households and community members in the study area, resulting in diversification of livelihoods. In addition, the communities were increasingly connected with the outside world. In addition to agriculture, some villagers began to engage in shipping or short-distance transportation, fishing, and sand collecting activities. Migration labor also increased (Zheng 2014). However, it should be noted that working as a migrant worker was only considered when there were no other alternatives to supplementing the loss of their livelihood (RGIMHSLR 2002).

Overall, resettlers' incomes increased and diversified between 2001 and 2011 (Li 2012). Farmers reduced their reliance on rice, and to an extent corn production, for income. The collection of non-timber forest products, fishing, and migration labor

Table 8.5 The number and proportion (%) of households surveyed by primary income type and source

Model of resettlement	Name of village	Income type					Total
		Migration worker	Small business	Animal husbandry	Cultivation	Subsidy	
Out-migration	Hongyan	4 (33.3%)	–	6 (50.0%)	2 (16.7%)	–	12 (100%)
Second-round out-migration	Xiabatian	13 (86.7%)	–	–	–	2 (13.3%)	15 (100%)
	Yakou	5 (62.5%)	2 (25.0%)	–	1 (12.5%)	–	8 (100%)
Back-up resettlement	Lannitang	11 (68.8%)	3 (18.8%)	1 (6.2%)	1 (6.2%)	–	16 (100%)
Non-agricultural settlement	Tianba	8 (80.0%)	2 (20.0%)	–	–	–	10 (100%)
On-site settlement	Pingzhang	5 (50.0%)	2 (20.0%)	2 (20%)	1 (10.0%)	–	10 (100%)
	Songshan	11 (84.6%)	1 (7.7%)	–	1 (7.7%)	–	13 (100%)
Control	Luqing	6 (54.5%)	3 (27.3%)	1 (9.1)	1 (9.1%)	–	11 (100%)
Total		**63 (66.3%)**	**13 (13.7%)**	**10 (10.5%)**	**7 (7.4%)**	**2 (2.1%)**	**95 (100%)**

became the main sources of income. As previously mentioned, the primary income sources are now non-agricultural (Table 8.5), with the majority of households (66.3%) depending on migration labor as their primary source of income. Small businesses, such as retail and transportation, form the primary income source in 13.7% of households. For those households who rely on agriculture for their primary income, 10.5% depend on animal husbandry (in particular pigs) and 7.4% on specialized cultivation (mainly sugarcane, herbs, and nuts).

Primary Income Sources Among the Communities

The source of livelihood varied between communities. More than 60% of households depend on migration labor for the primary income source in most villages except for Pingzhang, Luqing, and Hongyan villages (Table 8.5). In Xiabatian, Songshan, and Tianba, 80% or more households rely on migration labor. By comparison, only 33% of Hongyan households depend on migration labor. Hongyan is not near markets or the main transportation route. In addition, the amount of farmland per capita is higher than average (Table 8.4). Animal husbandry is the primary income source in half of households in Hongyan. In Pingzhang and Luqing, livelihoods were more diversified. Although approximately half of the households obtain their primary income from migration labor, there were still several households mainly dependent on small business, animal husbandry, and cultivation (Table 8.5).

Generally speaking, the livelihoods are more diversified in villages along the reservoirs than out-migrated villages.

In some households, livelihoods are becoming specialized, including with the types of agriculture and animal husbandry (e.g., pigs and sheep). For example, three households in Hongyan village specialized in sugarcane production, averaging 8 mu of land planted with the crop. In Lannitang and Xiabatian, some households specialized in plant herbs and Chinese prickly ash, planting 5 mu of land with these crops.

Increase of Non-agricultural Income

In recent years, the resettlers' views on migration labor as a livelihood have evolved from a last-resort occupation in 2000 (RGIMHSLR 2002) to what is now considered a better employment opportunity. Most migrant workers work in other provinces (50%), other counties within in Yunnan Province (23%), and towns in Yun County (23%). Very few migrant workers (4%) work in Laos and Burma. Compensation for migrant workers increases in correlation with distance from their homes.

There are obvious gender differences in migration labor. Men are the migrant workers in 90% of households that rely on migrant labor for income, with jobs primarily in major infrastructure construction (e.g., highway, railway, etc.), housing construction, and factories. In contrast, for households that rely on migrant labor income, only 37% of women are migrant workers. Women migrant workers work primarily in the service sector and factories. The factors that lead to this gender difference include that the work available is often heavy labor, men are typically higher compensated, and women are traditionally expected to be responsible for caring for the family. For example, a man working in highway or railway construction could be paid around 5000 yuan per month, while a woman working in factories could be paid no more than 3000 yuan per month. Those women who work in migration labor tend to be young (typically 16–22 years old). Women generally return to rural communities after they get married, where they usually work in agriculture or small businesses.

The gender differences in migration labor led to women becoming the main force in agriculture production in their communities. However, as previously discussed, the change in crop production reduced women's overall workload as compared to 20 years ago. With less time required in agricultural production and the absence of men, women are increasingly engaged in the public space, collective activities, and domestic decisions.

Improved Infrastructure and Access to Public Services

During construction of the dam, the local communities were increasingly connected to the external world. With the rapid economic development in China in recent decades, the government invested significant funds to greatly improve infrastructure in rural areas. In addition, the hydropower company invested directly in the impacted area. The reservoir fund, which was collected from the profits of the hydropower company, also financially contributed to infrastructure development. Improved infrastructure increased access to public services in rural areas. Although both men and women benefited from improved infrastructure, women have benefited more significantly given they had inferior status prior to dam construction.

Housing Conditions

Housing conditions in 2001 varied considerably among the resettlements (RGIMHSLR 2002). In some villages (e.g., Lannitang), housing conditions improved following resettlement. For resettlers in other villages, housing conditions declined. For example, in Hongyan, the resettlers' homes were built without sufficient local knowledge and the new houses were degraded by red ants, becoming unsafe in a few years. However, since 2001, residential conditions have greatly improved, due to increasing income and government-funded projects. The type of construction has improved over time with brick construction now being the primary type of housing, and the proportion of households that live in unsafe buildings has declined (Fig. 8.2 and Table 8.6).

Energy

Before the construction of the Manwan dam, most villages did not have access to electricity. A few communities had access to electricity from small hydropower plants. Firewood was the main source of energy year-round. Villagers cut tree branches on mountains and collected floating wood from the Lancang River for firewood. After the dam was built and the forests were submerged, the resettlers no longer had sufficient sources of firewood to use as an energy source. Conflicts arose among resettlers' communities and neighboring communities as resettlers would cut trees in other communities (RGIMHSLR 2002). Many resettlers could not afford electricity as infrastructure was lacking and those communities far from the dam had very high electricity fees (Li 2012).

Fig. 8.2 Changes in resettlers' housing conditions (wood house in the 1990s, brick house in the early 2000s, and two-story brick house in the 2010s) (Photos by Yanbo Li)

Table 8.6 Summary of housing conditions in surveyed villages

Model of resettlement	Name of village	Housing condition	Households in unsafe or very old houses (%)
Out-migration	Hongyan	Single-story brick house	3
Second-round out-migration	Xiabatian	Two-story brick house	0
	Yakou	Two-story brick house	0
Back-up resettlement	Lannitang	Single-/two-story brick house	10
Non-agricultural settlement	Tianba	Single-/two-story brick house	21
On-site settlement	Pingzhang[a]	Tile-roofed house	40
	Songshan	Tile-roofed house, single-/ two-story brick	19
Control	Luqing	Single-/two-story brick house	12

[a]Pingzhang village is threatened by landslides. The government required the villagers to move to Yakou village and stopped investing in housing. Most former Pingzhang residents have a house or are planning to build a house in Yakou

Table 8.7 Primary source of energy of households surveyed, including number of households and proportion (%)

Model of resettlement	Name of village	Energy type: Electricity	Firewood	Solar energy	Methane
Out-migration	Hongyan	12 (100%)	12 (100%)	10 (83.3%)	3 (25%)
Second-round out-migration	Xiabatian	15 (100%)	11 (73.3%)	12 (80%)	1 (6.7%)
	Yakou	8 (100%)	8 (100%)	8 (100%)	4 (50%)
Back-up resettlement	Lannitang	16 (100%)	15 (93.8%)	16 (100%)	3 (18.8%)
Non-agricultural settlement	Tianba	10 (100%)	10 (100%)	7 (70%)	–
On-site settlement	Pingzhang[a]	10 (100%)	10 (100%)	7 (70%)	3 (30%)
Control	Songshan	13 (100%)	13 (100%)	13 (100%)	3 (23.1%)
	Luqing	11 (100%)	10 (90.9%)	10 (90.9%)	4 (36.4%)
Total		**95 (100%)**	**89 (93.7%)**	**83 (87.4%)**	**21 (22.1%)**

[a]Pingzhang village is threatened by landslides. The government required the villagers to move to Yakou village and stopped investing in housing. Most former Pingzhang residents have a house or are planning to build a house in Yakou

Currently, access to energy has greatly improved. Firewood, electricity, solar energy, and methane are the main sources of energy used by resettlers (Table 8.7). All surveyed households have at least two types of energy, with 64% of households using three different sources. While reliance on wood has reduced, resettlers continue to collect firewood from the surrounding mountains to use for cooking and making feed for pigs in the winter. All households have access to electricity, and the price has decreased to 0.45–0.48 yuan per kWh, which is one third to half of the price prior to the construction of the dam. Electricity is now used for cooking, boiling water, washing machines, televisions, and other household activities. Solar energy is widely used for heating bath water. Although methane systems were subsidized and encouraged by local government, only 22% of households use methane systems due to lack of fuel availability due to the decline of livestock populations, high maintenance costs, and lack of suitable places for methane systems in mountainous areas.

As cooking is usually women's responsibility, improvements in access to energy greatly reduced women's workload. Moreover, the wide use of solar energy baths helped women ensure the hygiene of their families. In addition, air quality and women's health have improved with the shift from firewood to electricity for cooking. Lastly, conflict has reduced between the resettled communities and neighboring communities as the resettlers no longer need to take wood from the other communities.

Access to Water Resources

Prior to the construction of the dam, villagers along the Lancang River depended on springs and brooks for drinking water and irrigation. Some older villages developed canals and community-based irrigation management systems (Zheng, 2014). After resettlement, the villagers' access to water generally deteriorated (RGIMHSLR 2002). For example, the Hongyan village had water shortages due to the poor design of the infrastructure for water transportation. Many villagers had to collect water from rivers during the dry season. With the deterioration of water access, the newly reclaimed farmlands, which were usually located on high elevation, could not be irrigated.

Currently, access to drinking water has increased through improved infrastructure, although irrigation remains a problem (Table 8.8). All villages now have tap water systems and most villagers can get tap water for domestic use. During the dry season, villagers from five villages may not have tap water, up to a week at a time. Most families have water storage facilities as an adaptation to water shortages in the dry season. Through increased access to tap water, women's workload in collecting water has greatly reduced.

While access has increased, the safety of drinking water is not consistent. Most households use untreated water from brooks and small reservoirs for domestic use, except some households in Tianba village that use treated water from the Manwan hydropower project (Table 8.8). People from five villages reported that their water is not clear and there is sediment in the water during rainy seasons. Villagers further treat water for drinking or buy bottled water during this time. Upstream of the water source of Lannitang and Xiabatian, there are other villages whose livestock waste, fertilizer, pesticide, and domestic waste may affect water quality at the water source. These potential contamination sources concern the villagers, especially women given their responsibilities to care for the family.

Table 8.8 Summary of access and availability of different water resources in surveyed villages

Model of resettlement	Name of village	Source of drinking water	Clearness of drinking water	Abundance of drinking water	Drinking water fee	Irrigation conditions
Out-migration	Hongyan	Mengdi River	Unclear	Not abundant	15 yuan per capita per year	Lack of water source
Second-round out-migration	Xiabatian	Lvyintang Reservoir	Unclear	Not abundant	0	Lack of water source
	Yakou	Wanyou Reservoir	Fair	Not abundant	0	Lack of water source
Back-up resettlement	Lannitang	Mangshuai River	Unclear	Not abundant	30 yuan per capita per year	Lack of irrigation infrastructure
Non-agricultural settlement	Tianba	Hydropower station/small brooks	Clean/ unclear	Relatively abundant	0	Lack of water source
On-site settlement	Pingzhang[a]	Tiechang creek	Fair	Not abundant	60 yuan per household per year	Irrigation available
	Songshan	Bingbu River	Unclear	Abundant	10 yuan per capita per year	Irrigation available
Control	Luqing	Luqing River	Clear	Abundant	30 yuan per capita per year	Irrigation available

[a]Pingzhang village is threatened by landslides. The government required the villagers to move to Yakou village and stopped investing in houses in Pingzhang. Most former Pingzhang residents have a house or are planning to build a house in Yakou

Irrigation is also challenging, and proximity to water sources determines the feasibility of irrigating. As such, most villages have problems obtaining irrigation water with the exception of Luqing, Songshan, and Pingzhang. The out-migrated communities, including Hongyan, Yakou, and Xiabatian, lack irrigation sources. The back-up resettled Lannitang village also lacks irrigation due to a landslide that caused damage to their canals. As previously note, the lack of irrigation water is a contributor to the change in crop production in most villages.

Transportation

Although the study area is remote and isolated, prior to dam construction, a key road went along the Lancang River connecting the villages and a bridge connected the banks of the river before resettlement. Shipping was very active and transportation was very convenient at a local scale. After the construction of the dam, the road and bridge were submerged and the Lancang River widened. It became difficult for villagers to travel to markets and cities, resulting in an increased reliance on shipping

Table 8.9 Transportation conditions of surveyed villages and their distance to education, health-care resource, and markets

Model of resettlement	Name of village	Degree of transportation convenience[b]	Distance to nearest medical clinic (km)	Distance to nearest primary school (km)	Distance to nearest market (km)	Distance to nearest city[c] (km)
Out-migration	Hongyan	+	2.5	2.5	2.5	75
Second-round out-migration	Xiabatian	+++	0	0.5	3	3
	Yakou	+++	0.5	2	2	18
Back-up resettlement	Lannitang	++	10	10	10	65
Non-agricultural settlement	Tianba	++	4	4	4	60
On-site settlement	Pingzhang[a]	++	11	1	10	70
	Songshan	++	8	8	8	70
Control	Luqing	+++	3	3	3	57

[a]Degree of transportation convenience is shown by one or more plus signs (+), whereby an increasing number of plus signs denotes more convenient transportation
[b]Cities refer to the center of the county or municipality where there are better resources (i.e., health care and education) and more job opportunities
[c]Pingzhang village is threatened by landslides. The government required the villagers to move to Yakou village and stopped investing in houses in Pingzhang. Most former Pingzhang residents have a house or are planning to build a house in Yakou

(i.e., shipping supplies to the villages). In the out-migration village of Hongyan, the road conditions were much better than in the other villages around the reservoir.

By 2016, transportation in all villages has improved. Two main roads connect the villages along the reservoir to other towns and cities, and all villages are connected to main roads through paved roads. Villagers use motorcycles or minibuses run by individual businessmen to travel short distances (10–20 km). In general, there are more men than women who operate these vehicles. For longer distances, public buses are available. In emergency situations like medical emergencies, villagers typically hire cars to transport them to reach emergency services. Only a few households have their own cars, trucks, or minibuses.

Access to transportation varied among communities, depending on their distance to towns and main roads (Table 8.9). Hongyan village is far away from main roads and, thus, is not conveniently located to access transportation infrastructure. The villages along the reservoir have moderately convenient transportation. Some villages have very convenient transportation, including Luqing (close to the Linxiang main road), Tianba (located near Manwan town), and the out-migrated villages of Yakou and Xiabatian (close to the main road and the county center).

Improved transportation provides increased access to markets, education, and health care, thus providing more opportunities for villagers. This is especially ben-

Fig. 8.3 An example of a village clinic which benefited from the financial investment provided by the hydropower company. (Photo by Yanbo Li)

eficial to women as women's connection with the outside world was limited before dam construction and most women cannot drive. With improved road infrastructure and public transportation, women's connection with the outside world and markets increased leading to access to more information and job opportunities.

Access to Medical Care and Education

Following construction of the dam, women's access to medical care and education has improved. In the 1980s, there were very few schools and hospitals in the rural areas. Very few girls could participate in formal education or had access to medical care. However, in the past 20 years, these conditions have improved with the advancement of government policies. In 1994, the government began to provide free primary and middle school education, and since 2011, government has provided free school lunches. Girls and boys both have the same right to these benefits.

Medical care services have also improved. Clinics were built in every administrative village. The hydropower company also financially invested in a medical clinic (Fig. 8.3). Many villagers can reach the medical clinics within 20 min. Improved transportation also makes it easier for the resettlers to get treatment in the hospitals in towns and cities (Table 8.9). In 2009, both men and women resettlers began to receive medical care insurance that covers more than a half of the costs of medical services. Women, similar to men, have greatly benefited from the improved access to medical care.

Table 8.10 Summary of gender differences of interviewees in access to credit and proportion (%) of interviewees that have that type of access to credit

Gender	Formal (from banks)	Informal (from friends and relatives)
Male	15 (50.0%)	38 (44.2%)
Female	4 (13.3%)	22 (25.6%)
Both male and female	11 (36.7%)	26 (30.2%)
Total	**30 (100%)**	**86 (100%)**

Access to Financial Support

Several studies in developing countries suggest that rural women have very limited opportunities to access credit services. However, our study found that although men have advantages in obtaining credit from banks, women have access to some credit services. We found that in 50% of households, only men can obtain bank credit, while women can only acquire bank credit in 13.3% of households. However, in approximately one third of households, both men and women can obtain bank credit (Table 8.10). However of the 95 households surveyed, over 90% of households had informal credit arrangements, while only a third had a formal bank credit arrangement (Table 8.10). With informal loans from friends and relatives, the gender differences are not as significant. In almost half of the households with informal credit arrangements, the man was borrowing money from friends and relatives almost twice as often as a woman. However, over for many households (26) both women and men would both informally borrow funds from friends and relatives.

Participation in Collective Activities

Resettlement strongly influenced existing communities in many aspects, including how community members interacted with each other and the government. In the out-migrated villages, the communities were reorganized as the original structure of the community could not be maintained. The changes in community structure led to social impacts to the resettlers (community-wide and individual). For example, villagers in Hongyan village stopped conducing traditional festivals at the community level due to changes in the traditional governance structure of the community and poverty. This led to disappointment for resettlers who intensely missed their connection with their hometown and traditional celebrations (RGIMHSLR 2002). In contrast, interaction with outside organizations, such as the government, increased during the resettlement process (RGIMHSLR 2002). Vertical linkage (i.e., between government and villagers) displaced horizontal linkage (i.e., among villagers), and most of the collective activities that resettlers participated in were connected with the resettlement process and led by the government.

As the process of resettlement ended, the collective activities in which the resettlers engage in are now less connected with government and more related to the management of public affairs within the communities. The activities include election

Table 8.11 Frequency of surveyed households participating in collective activities and proportion (%) of each participating household who answered the question

Model of resettlement	Name of village	Frequency in participating collective activities				Total
		Always	Frequently	Occasionally	Never	
Out-migration	Hongyan	7 (63.6%)	2 (18.2%)	2 (18.2%)	–	11 (100%)
Second-out-migration	Xiabatian	7 (53.8%)	4 (30.8%)	–	2 (15.4%)	13 (100%)
	Yakou	3 (50.0%)	1 (16.7%)	2 (33.3%)	–	6 (100%)
Non-agricultural settlement	Tianba	3 (75%0.0)	–	1 (25%)	–	4 (100%)
Back-up resettlement	Lannitang	10 (100%)	–	–	–	10 (100%)
On-site settlement	Pingzhang[a]	2 (40.0%)	1 (20.0%)	2 (40.0%)	–	5 (100%)
	Songshan	11 (84.6%)	1 (7.7%)	1 (7.7%)	–	13 (100%)
Control	Luqing	10 (90.9%)	–	1 (9.09%)	–	11 (100%)
Total		**53 (72.6%)**	**9 (12.3%)**	**9 (12.3%)**	**2 (2.7%)**	**73 (100%)**

[a]Pingzhang village is threatened by landslides. The government required the villagers to move to Yakou village and stopped investing in houses in Pingzhang. Most former Pingzhang residents have a house or are planning to build a house in Yakou

of village cadres, cleaning public spaces, repair of roads and irrigation systems, and celebratory festivals.

Villagers' participation in such activities varied greatly among different communities (Table 8.11). Nevertheless, all the resettled communities participated less often in collective activities than Luqing village (control), except for Lannitang village. In the case of Yakou and Xiabatian, which are relatively new communities, reduced participation is due to two key factors. Firstly, livelihood is predominantly focused on migration labor. Secondly, government is responsible for most of the collective works and activities (e.g., cleaning and repairing public infrastructure). In Hongyan village, the degree of participation in public activities remains lower than in the villages along the reservoir, although the villagers have started to celebrate festivals again. In the villages along the reservoir, villagers' participation in collective activities is higher (Table 8.11).

When community-level challenges occurred, the power of collective action was not sufficient to address problems and challenges. For example, in Lannitang, the canals were damaged by a landslide, which seriously affected their agricultural production. The villagers could not repair the canals and needed support from the government. When comparing the different resettlement models, it appears in general that out-migration, especially earlier on, adversely affects community cohesiveness the most. However, community cohesiveness and participation do seem to recover over time.

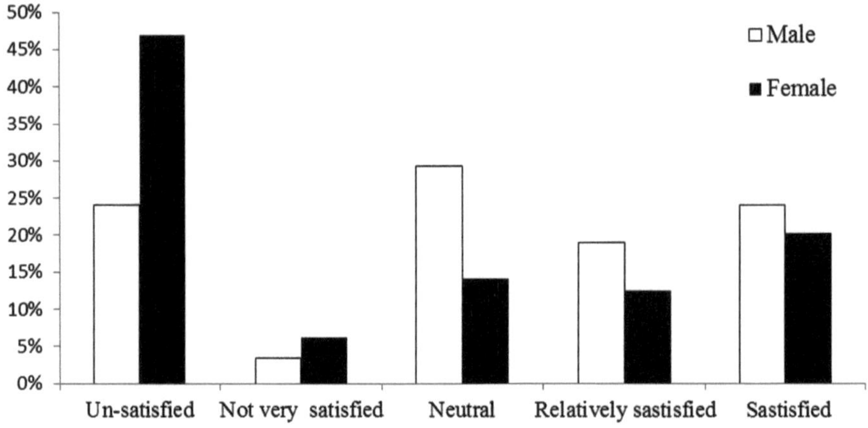

Fig. 8.4 Degree of satisfaction with current life conditions among surveyed households

Degree of Satisfaction About Current Life Conditions

Degree of satisfaction is a comprehensive indicator on resettlers' views about their life conditions. Results of the 2001 survey indicated that resettlers were strongly dissatisfied with their lives as they were facing many problems, especially in Tianba and Hongyan villages (RGIMHSLR 2002).

In 2003, disappointed resettlers gathered in Manwan town to protest the Manwan hydropower project and demanded actions of the Manwan town government. This event led to increased support to resettlers from the government to improve their living conditions. Through financial investments and additional construction projects, over 10 years, resettlers have adapted to new life and rebuilt their livelihoods. The degree of satisfaction about current life conditions increased, and the proportion of those strongly dissatisfied had decreased to 47% for females and 23% of males (Fig. 8.4).

Obvious gender differences exist in the level of satisfaction with regard to life conditions (Fig. 8.4). Most male interviewees did not have very strong views about life satisfaction, while more women expressed dissatisfaction. In general, men's opinions on life conditions are more evenly distributed than women. Over 20% of male interviewees indicated that they were dissatisfied with current life conditions, which is similar to those that replied that they were satisfied. In contrast, more than double of the women interviewed responded that they were dissatisfied with their current life condition than satisfied.

Among the surveyed communities, resettlers in Tianba, Pingzhang, and Lannitang have lower degrees of satisfaction about their lives, while villagers in Yakou and Luqing (control) are more satisfied (Fig. 8.5). Lack of income source, dependence on a single income source, and lack of irrigation water are the main reasons for the low degree of satisfaction in Tianba, Hongyan, and Lannitang. In Lannitang, the primary concerns are the damaged irrigation systems and challenges with access to

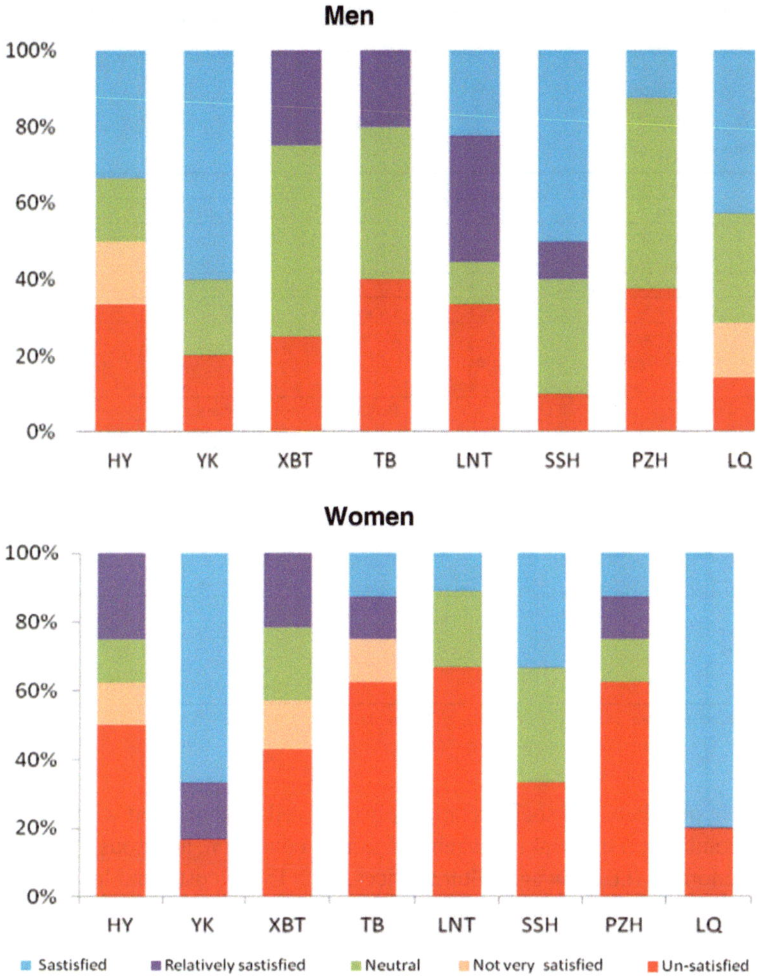

Fig. 8.5 Degree of current life satisfaction for men (top figure) and women (bottom figure) interviewed household in each surveyed village. (Village acronyms: HY(Hongyan), YK (Yakou), XBT (Xiabatian), TB (Tianba), LNT (Lannitang), SSH (Songshan), PZH (Pingzhang), and LQ (Luqing))

safe drinking water. Through the second-round out-migration, some households moved out of Pingzhang to Yakou. Those who remain in Pingzhang are relatively poor and living in poor housing and conditions; thus they feel dissatisfied about their current life.

Gender differences also existed in the rationale provided to why resettlers were dissatisfied (Table 8.12). Lack of income source, debt, and lack of land and labor were the top three issues that concerned men. Lack of development opportunities and agricultural and drinking water concerns were also a source of dissatisfaction to men. In contrast, lack of income, lack of land, and illness of family members are the top three issues that concern women. Women are also concerned about the welfare

Table 8.12 Summary of reasons why resettlers interviewed are not satisfied with current life by gender

Order of importance	Male	Female
1	Lack of income source	Lack of income source
2	Debt	Lack of land/paddy rice field
3	Lack of land Lack of labor	Illness of family member
4	Poor drinking water conditions	Difficulty balancing between income and expenditures
5	Lack of development opportunities	Lack of labor
6	Lazy son Poor agricultural infrastructure Low price of agricultural production	Debt
7		Poor drinking water conditions Separation of family members
8		Poor irrigation conditions Low price of agricultural production Unmarried adult son/lazy son and husband Poor housing conditions Unjust village leaders Limited government subsidies
9		High cost of living Poor transportation Poor education conditions Conflicts with other members of the community

Note: The reasons were reported by those who are not satisfied about current life. The factors were ordered by frequency. The factors in red are reported only by one gender.

of family members and social relationships, as well as concerned with the community's condition (i.e., drinking water, irrigation, education, and housing conditions) and economics (i.e., low agricultural prices, high living costs, and low government subsidies). Overall, women expressed more reasons than men for why they are dissatisfied.

The reasons the resettlers are concerned with life conditions has changed between 2001 and now. In 2001, resettlers were strongly dissatisfied with life and attributed problems to hydropower development that included submerged farmland, insufficient compensation to make a living, deterioration of housing conditions in some villages, and poor drinking water (RGIMHSLR 2002). Currently, although resettlers continue to be concerned about farmland and water-related issues linked to the hydropower project, resettlers are now more concerned with problems with their own financial situation, such as lack of income sources, opportunity, and labor. The livelihood of resettlers has improved; nevertheless dissatisfaction with life conditions continues to be an issue.

Table 8.13 Summary of gender roles in the different tasks associated with agricultural production and works, including the proportion (%) completed by each gender group based on interviewed households

Labor division	Heavy work	Light work	Irrigation	Animal husbandry	Sale of agricultural products
Hire workers	9 (9.8%)	–	1 (2.1%)	–	–
Male only	43 (46.7%)	19 (20.7%)	23 (47.9%)	16 (17.4%)	15 (23.4%)
Female only	18 (19.6%)	44 (47.8%)	11 (22.9%)	66 (71.7%)	38 (59.4%)
Men and women together	22 (23.9%)	29 (31.5%)	13 (27.1%)	10 (10.8%)	11 (17.2%)
Total	**92 (100%)**	**92 (100%)**	**48 (100%)**	**92 (100%)**	**64 (100%)**

8.2.3.2 Changes in Social Status of Women

Changes in women's social status due to hydropower development have been the focus of many studies. In summary, studies found that women are more negatively impacted by hydropower development due to disadvantages in access to resources, lack of participation opportunities in collective decisions, and their dual role in agricultural production and childbearing and child-rearing (WCD 2000; Scudder 2005; Simon 2013; Chen 2013b). In our study, we explored the labor division between males and females within families and gender differences in access to resources and participation in collective decisions. We then compared our results with the historic situation, to explore the impacts of hydropower development on women, and the changes in such impacts as the resettlement process concluded.

Gender Roles in Family

While men and women both participate in all family activities in all of the surveyed households, obvious gender differences exist in each task. The following section explores gender roles in the family in agricultural production, non-agricultural employment, and domestic work.

Agricultural Production

Notable gender differences occur in the division of work in agricultural production (Table 8.13). Based on the interviews, men are responsible for heavy work, such as plowing and transporting products (46.7%) and managing irrigation activities (47.9%). Women predominantly work in animal husbandry (71.7%), sale of agricultural products (59.4%), and lighter agricultural activities (such as weeding, fertilizing) (47.8%). In some households, men and women work together in agricultural tasks, while approximately 10% of households hire workers to do the heavy labor.

Table 8.14 Summary of gender roles in the different tasks associated with non-agricultural works, including the proportion (%) completed by each gender group based on interviewed households

Labor division	Migration labor	Small business
Male only	51 (63.0%)	7 (36.8%)
Female only	8 (9.9%)	5 (26.3%)
Shared by men and women	22 (27.2%)	7 (36.8%)
Total	**81 (100%)**	**19 (100%)**

Table 8.15 Summary of gender roles in domestic works, including the proportion (%) completed by each gender group based on interviewed households

Labor division	Collecting drinking water	Cooking and washing	Sending children to school
Male only	18 (48.6%)	12 (12.6%)	20 (40.8%)
Female only	14 (37.8%)	70 (73.7%)	22 (44.9%)
Shared by men and women	5 (13.5%)	13 (13.7%)	7 (14.3%)
Total	**37 (100%)**	**95 (100%)**	**49 (100%)**

As noted earlier, women are now the primary workforce in agriculture as men frequently work in migration labor. Although women have taken a larger role in agricultural production, their workload did not noticeably increase. This is because the area of farmland has decreased, crops have changed (e.g., annual to perennial crops), and men continue to contribute to agricultural production when they are available.

Non-agricultural Employment

In non-agricultural employment, some types of work have clear gender differences, while others do not. For example, based on the households surveyed, mostly men were involved with migration labor (Table 8.14). In contrast, little to no gender difference was detected for operating small businesses. The nature and context of the occupation likely explains the variation in scale of gender differences among forms of non-agricultural employment. It is easier for men to travel for migration labor, and these occupations tend to be in labor-intensive sectors which are less suitable for women given the differences in strength and family responsibilities. On the other hand, small businesses are usually located within local communities; therefore, women can conduct this work while still taking care of their families.

Domestic Work

Traditionally, it is women's responsibility to do domestic work, although men will also contribute (Table 8.15). In 73.7% of households, women take on the household tasks of washing and cooking. Gender differences are less distinct in other types of domestic work. Men are responsible for collecting drinking water in almost half of

Table 8.16 Summary of gender difference in domestic decisions, including the proportion (%) based on interviewed households

Who made decision	Domestic financial management	Agricultural production	Important domestic decision[a]
Male or female decide independently	3 (3.2%)		
Male only (all funds)	28 (29.5%)	27 (34.2%)	37 (38.9%)
Female only (all funds)	45 (47.4%)	30 (38.0%)	8 (8.4%)
Jointly between men and women	19 (20.0%)	22 (27.8%)	50 (52.6%)
Total	**95 (100%)**	**79 (100%)**	**95 (100%)**

[a]Important domestic decisions refer to decisions such as significant purchases, where to live, whether to support children in higher education, etc.

the households, while in over a third of households, the work is done by women. In the case of sending children to school, there is no distinct difference in gender with approximately the same amount of women or men completing the task.

Many studies on gender issues emphasize that decline in water resources, either by the dam or climate change, forces women to spend more time searching for water for domestic use leading to adverse impacts to women (WCD 2000; Scudder 2005; Simon 2013; Chen 2013b; Sun and Zhao 2017). In our study area, the villagers have access to tap water due to improvements in infrastructure, and they have storage facilities for times when water supply may not be guaranteed. Given these improvements, women's workload to search for water greatly declined. Nevertheless, 39% of households surveyed need to collect domestic water during water shortages, which is usually located within 2 km of their houses. We found that men have more responsibility for collecting water as men are more skilled at operating the motor cycles and tractors required to carry water. Our analysis demonstrates that the adverse impacts from dams on women regarding water collection may not be as significant as reported in other studies due to the improvements in infrastructure resulting in wide use of vehicles.

Influence in Domestic Decisions

We explored women's influences in domestic decisions with regard to domestic finances, agricultural production, and important domestic decisions (Table 8.16).

Regarding domestic financial control, in almost half of the households, women managed all domestic financial decisions, approximately 20% more often than men. Women have a more significant role in finances for two key reasons. Firstly, women traditionally have managed money and are viewed as proficient in accounting and to have a good attention to detail. Secondly, women are responsible for caring for the family and domestic needs. As such, it is logical for the women to make the necessary family purchases.

For agricultural decisions, the division of decision-making power between genders is relatively balanced, with 38% of decisions made by women, 34.2% by men, and 27.8% made jointly between women and men (Table 8.16). Women played and continue to play an important role in decisions with regard to agricultural production because in many households men are absent due to migration labor.

However, there are obvious gender differences regarding important household decisions, such as whether to move or relocate, to purchase expensive assets, or to support children in higher education. In approximately half of the household surveyed, men and women jointly made these decisions, while in 38.9% of households, important decisions were made by men. In only 8.4% of households are important decisions made by women (usually in households headed by women). Nevertheless, these results indicated that women have influence in important family affairs.

While both men and women have decision-making responsibilities, women remain disadvantaged in this regard. Women have more decision-making in financial management and agricultural production and also have a role in important decisions. However, it should be noted that women have less control over important decisions.

Gender Differences in Participating in Collective Activities

Participation in collective meetings is an important factor in generating societal influence. Our study found that there are no obvious gender differences in community meeting attendance. During the survey, we found that 28.8% and 38.4% of the men and women usually attended meetings on behalf of the household, respectively, while in the remaining household surveyed, whoever had spare time participated in community meetings. Women's participation in community meetings increased, as men were often working outside of the communities as migrant workers.

Although participation numbers are not significantly different between men and women, the interest and influence in participation has obvious gender differences. A large proportion of male interviewees (81.0%) are very willing to join collective activities, while only 75% of women are willing. Only a small percentage of male interviewees (9.5%) explicitly expressed no interest in collective activities, while a quarter of women interviewees (25%) have no interest. Regarding influence in these collective meetings (i.e., expression opinions), 13.3% of male participants will actively give their opinions, and 46.7% of male participants will speak in meetings occasionally (Fig. 8.6). In contrast, only 29.2% of women participants will speak in meetings, and 70.8% do not express their perspectives at all. Interviews with village heads (usually men) also demonstrated that opinions of men were more valued and respected by other participants than women's opinions. Reasons for the low degree of women's participation in collective activities include shyness, lack of self-confidence, and the impression that their opinions are not important or will not be taken seriously. For men who do not speak in meetings, it is usually because they

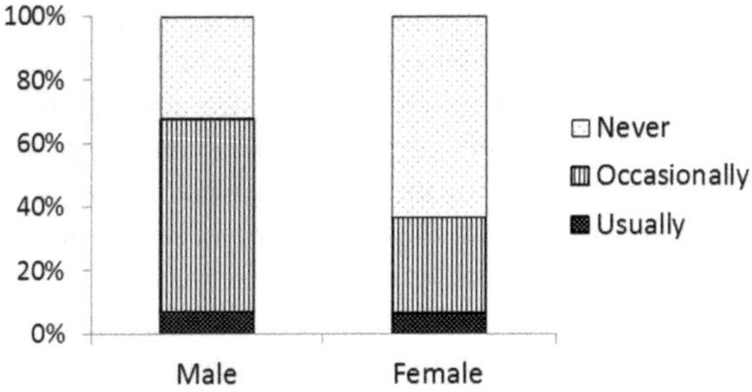

Fig. 8.6 Gender differences in level of expression of opinions at community meetings

feel their opinions will not be taken seriously and rarely due to shyness or lack of confidence.

Changes in Gender Equity

The study in 2001 found that the gender impacts of the construction of the Manwan hydropower project are complex and multifaceted (RGIMHSLR 2002). Overall, following the construction of the dam "men's labour intensity and women's labour time increased, with women suffering more adverse effects." "In the face of new and severe living environment…… men were traditionally expected to be responsible to make money, secure loans, do migrant work, understand new technology, culture, and build up the family fortunes amongst other duties..." (RGIMHSLR 2002).

In Hongyan village, the researchers' found that women actively and willingly recognized a man's strong societal position, which by extension weakened the status of women. On the other hand, the study also found that women's status improved in many other villages. Women were increasingly involved in agricultural production, such as plowing which was traditionally the work of men. Although generally important decisions were still made by men, women also had the right to speak and their words had considerable weight. Through increased participation in small business and petitioning government, women increased their participation in social affairs which in turn improved their social status in the family and public arena. Construction of the Manwan hydropower project increased women' work load but also indirectly helped to improve women's status in the family and society as women were given more responsibility and opportunities in domestic and public affairs (RGIMHSLR 2002).

In this study, we found that although men are still in a comparatively more advantageous position, the relationship between men and women has become more equal and the status of women has improved following development of the dam. Women have influence in domestic financial management and are involved in important domestic decision-making. Moreover, they are increasingly participating in village collective affairs. The gap between men and women in public service and financial support is narrowing. Given that many men are absent due to migrant labor, women have the opportunity and are required to make more family decisions and participate in collective affairs. Although women often are not willing to express their ideas as their ideas are thought to be less respected, participation nevertheless provides women with more knowledge about collective affairs and a formal platform to express their ideas and generate influence should they choose to do so. In one case, the wife of the village leader of Luqing performed her husband's role when he worked as a migrant labor and served and worked in organizing community meetings and collective works and passing government policies.

Nevertheless, women continue to be in an inferior position compared with men in obtaining non-agricultural employment opportunities. At present, only 31.6% of the households surveyed have female participation in non-agricultural employment, most of whom are unmarried young women. The main factors leading to this disparity are that non-agricultural employment opportunities exist mainly in heavy labor and because women have more responsibilities in caring for family members and thus must remain at home. For example, some women who work in urban areas leave their jobs and return to rural areas once they are married.

Some studies show that hydropower plant construction has caused a greater impact on the workload of women by forcing women to spend more time doing tasks such as collecting firewood and potable water (WCD 2000; Scudder 2005). This problem was not found in our study, mainly due to the improvement of rural infrastructure that reduced labor demand for such work. Use of electricity increased due to lowered prices, which in turn reduced dependence on firewood. Moreover, the construction of the water supply network and family water storage facilities made it unnecessary for women to spend much effort obtaining drinking water. Irrigation continues to be a common problem for resettled communities, and irrigation duties are mainly conducted by men in village households.

In summary, this study demonstrates that women's livelihood and social status are generally improved, due to diversification of livelihood, improved access to public services, and improved influences in domestic and collective decisions. The gap between women and men still exists but is narrowing. Nevertheless, there continues to be room for improved gender equality.

8.2.3.3 Resettlers' Livelihood and River Health

Ecosystem Services from Lancang River

The Lancang River (including the mainstem and tributaries) provides important ecosystem services for villagers along the river, including drinking and irrigation water, shipping routes, fishery resources, and hydropower. The construction of the Manwan hydropower project changed land utilization and caused resettlement away from the Lancang River. Consequently, over the past 20 years, the ecosystem services that the river provides and the beneficiaries of those ecosystem services have changed significantly (Table 8.17).

Before dam construction and resettlement, the cultivated land was dominated by paddy fields, and irrigation was the most important ecosystem service provided by

Table 8.17 Changes in ecosystem services obtained from the Lancang River and their beneficiaries

	Time period		
	Before 1993	1993 to 2010	2010s
Ecosystem services	Irrigation, fishing	Hydropower, irrigation, shipping, fishing, sand collection	Hydropower, irrigation, fishing, recreation
Beneficiaries of ecosystem services	Local communities	Hydropower plant, government, local communities	Hydropower plant, government, local communities

the tributaries of the Lancang River. Fishing provided an important source of protein for some villagers. The villagers who lived close to the river also collected floating wood from the mainstem Lancang River as one of their energy sources. Overall, prior to resettlement, the intensity of ecosystem services extraction from the river was not high. It was restricted to meeting the needs of daily life of the villagers.

The construction of the power station inundated a large number of cultivated lands, and some villagers were moved out of the reservoir area; thus, their dependence on the ecosystem services of the Lancang River declined between 1993 and 2010. The decrease in the area of paddy field and the deterioration of irrigation conditions reduced the irrigation services provided to villagers by the tributaries of the Lancang River. Conversely, the increase in water transportation has made shipping services more important. A few villagers engaged in fishing and river sand collection in the mainstem of the Lancang River during the early resettlement time period. During this stage, the hydropower development became the most important ecosystem service provided by the river, and the main beneficiary of this service was the power company. The government also benefited through tax revenue, and villagers benefited indirectly through government transfer payment, compensation, and agricultural production support.

At present, irrigation and fishing are still important ecosystem services obtained by local communities from the Lancang River, in addition to hydropower. Tourism services are also obtained from the Lancang River, including through government's support of tourism within the study area. With the improvement in road conditions, the importance of shipping has declined. River sand collecting activities were prohibited by local government to protect the Lancang River and associated ecosystems. Overall, out-migration combined with changes in local people's livelihood (i.e., from agricultural to non-agricultural employment and from paddy fields to dry land farming and orchards) reduced the dependency of local communities on the Lancang River.

Table 8.18 Summary of the types of treatment and disposal of domestic waste and sewage in each village in the reservoir area, including the proportion (%) based on interviewed households

Model of resettlement	Name of village	Treatment of sewage: Digestion tank	Treatment of sewage: Ditch	Treatment of sewage: No treatment	Treatment of garbage: Burn	Treatment of garbage: Store in fixed places/ tank	Treatment of garbage: No treatment
Back-up resettlement	Lannitang	2 (12.5%)	1 (6.3%)	13 (81.2%)	2 (13.3%)	1 (6. 7%)	12 (80%)
Non-agricultural settlement	Tianba	3 (33.3%)	1 (11.1%)	5 (55. 6%)	6 (66. 7%)	3 (33.3%)	0%
On-site settlement	Pingzhang[a]	–	1 (10%)	9 (90%)	2 (20%)	–	8 (80%)
	Songshan	1 (7.7%)	1 (7.7%)	11 (84.6%)	2 (15. 4%)	–	11 (84.6%)
Control	Luqing	1 (9.1%)	3 (27. 3%)	7 (63.6%)	3 (27. 3%)	3 (27. 3%)	5 (45.5%)
Total		**7 (11.9%)**	**7 (11.9%)**	**45 (76.3%)**	**15 (25.9%)**	**7 (12.1%)**	**36 (62. 1%)**

[a]Pingzhang village is threatened by landslides. The government required the villagers to move to Yakou village and stopped investing in houses in Pingzhang. Most former Pingzhang residents have a house or are planning to build a house in Yakou

Villagers' Impacts on River Health

While the dependence of local communities on the Lancang River declined, the impacts on river health have increased with the improvement of resettlers' living conditions. A modern lifestyle increases domestic sewage and garbage, including nondegradable components such as plastics. The decline in animal husbandry has made traditional methods of treating kitchen waste and crop straw no longer feasible and has led to an increased use of chemical fertilizer in the fields. Little to no treatment of sewage and garbage in the resettlement communities can also contribute to a decline in Lancang River health.

At present, in the communities around the reservoir, the vast majority of households do not treat domestic sewage, except for a few of households that have septic tanks (Table 8.18). Most families (76.3%) discharge domestic sewage without any treatment. In the study area, only 11.9% of the families treat sewage, mainly in Tianba and Lannitang. Similarly, 11.9% of households, mainly in the Luqing, discharge domestic sewage along the drainage ditch into farmlands.

Most families in the vicinity of the reservoir also do not treat household waste (Table 8.18). Most (62.1%) of families throw garbage outside without any treatment leading to garbage entering the river during rainfall events. In some villages, garbage is burned when the waste reaches a threshold amount. Only Tianba has a gar-

Table 8.19 Gender differences in environmental consciousness of resettlers

	Pollutant	Sewage			Garbage			Pesticide
Gender	Treatment	Digestion tank	Ditch	No treatment	Burn	Store in fixed places/ tank	No treatment	No treatment
Male	Harmful	2 (50%)	2 (50%)	16 (64%)	2 (66%)	7 (100%)	16 (89%)	34 (79%)
	No harm	2 (50%)	2 (50%)	9 (36%)	1 (33%)	–	2 (11%)	9 (21%)
Female	Harmful	1 (33%)	1 (33%)	9 (47%)	–	2 (100%)	11 (69%)	34 (85%)
	No harm	2 (67%)	2 (67%)	10 (53%)	1 (100%)	–	5 (31%)	6 (15%)

bage collection system, with other villages having no garbage collection or disposal services. A total of 12.1% of those interviewed send their garbage to a collection facility, mostly households in the villages of Luqing and Tianba, whereas 25.9% of families burn garbage.

Awareness regarding environmental impacts and protection varies among villagers. Most villagers, both men and women, are aware that the uncontrolled disposal of domestic waste to the environment and the use of pesticides may have adverse effects on human health and the environment. Additionally, villagers generally believe that burning garbage will cause adverse effects on the environment and human health. Villagers place more emphasis on odor when estimating the harm and lack understanding of the possible health risks and the impacts on ecosystems.

There are some gender differences in environmental awareness (Table 8.19). Most men (64%) think that sewage adversely affects the environment, while only 47% of women hold the same view. The majority of men (89%) think that untreated garbage would adversely affect the environment, but only 69% of women believe this to be true. However, with regard to pesticides, more women (85%) believe them to be harmful compared to men (79%). These results may imply that women are less environmentally conscious than men in the study area.

Another reason for lack of waste management in the study villages is poor collective action. In most villages, the only work in environmental sanitation management is to clean the public space once a month, with lower frequency in some villages. There are no regulations on daily storage and disposal of garbage. The lack of collective action is also a reflection of village's low social cohesion. The villagers think that environmental sanitation management of public space is the responsibility of the government. Many of the villagers, both men and women, believe that the government should invest to build necessary treatment facilities, such as sewers and garbage receptacles, to manage waste and protect the environment.

8.3 Jinghong Case Study

Jinghong hydropower development is a multipurpose project located in the lower reach of Lancang River, 4 km upstream of Jinghong City, which is the capital of Xishuangbanna. In total, there are eight hydropower projects planned in the upper Mekong River. Jinghong dam is the sixth cascade dam and currently the nearest hydropower project on the mainstem Lancang River to Laos. The project is a concrete gravity dam which began construction in 2006, began to store water in 2008, and was completed in 2009. The storage capacity is 11.39×10^9 m^3 in normal conditions. The 110 m dam inundated an area of 37 km^2 which resulted in the displacement and resettling of 5095 people. As the Lancang Valley within this area is deep with steep, high mountains and protected from floods by the Nuozhadu dam, the total population affected was small.

8.3.1 Background

Villagers in four areas were affected by the hydropower project. These were the Jinghong municipality, Menghai County, Simao District, and Lancang County (county, municipality, district are considered at the same administrative level). In 2013, the total population of these four areas was 1.45 million. Agriculture was the primary source of income for the majority of people (73% of people). Most of the population is comprised of minority groups (67%), such as Aini, Wa, Dai, and Lahu.

The total acquisition area for the project was 43.94 km^2 of which 30.84 km^2 (70%) is terrestrial and 13.1 km^2 (30%) is aquatic. The land area includes 833,444.67 mu farmland, 10,933.56 mu garden, 171,144.4 mu forest, 60.59 mu grassland, 84.68 mu other types of agricultural land, 790.48 mu construction land, and 8949.61 mu barren land.

During construction and dam operation, 5095 people have been resettled. In contrast to the Manwan case study, during the resettlement process, villagers could choose between a one-time financial compensation and receiving long-term rehabilitation support. Only 367 people chose one-time financial compensation. These individuals generally took the opportunity to move in with relatives in a city. Most (4728) people chose the long-term rehabilitation program. Of these, 2897 people moved to 16 new settlements that were built following the principle of maintaining original village structures. The remaining of the resettled people stayed in the same village but moved up to higher elevation, with protections put in place to ensure the safety of the affected village from flooding and landslides.

8.3.2 Methods

The Jinghong case study was used to analyze how hydropower development affected women during the resettlement through institutional changes at multiple scales (macro, meso, and micro). Analysis was conducted to understand what has changed for women and how women interacted with these changes. The case study relied on desktop research of secondary data and information regarding the resettlement policy, the resettlement report of the Jinghong dam, the Xishuangbanna annual evaluation report on woman development, legal action and complaints filed by villagers during resettlement, and minutes of county and village meetings. In addition, the analysis involved semi-structured interviews with government functionaries such as the migrants' bureau, Xishuangbanna women's federation, Jinghong women's federation, Menghai County women's federation, project authorities, sector experts, concerned individuals, and nongovernment organizations (Xishuangbanna Legal and Psychological Consultation Center for Woman and Children), as well as focus group discussions with people in project-affected areas.

Four focus groups discussions were conducted in the current settlements from January 9 through 11, 2017 and April 7 through 17, 2017 with communities affected by the Jinghong dam. Six sites were visited during the field work. Participants were encouraged to express any kind of opinion, regardless of whether their opinion is shared or not by the other participants. In order to facilitate open communication, focus groups were structured based on similar social status. The focus groups consisted of two resettled villagers meeting (one all-female and one all-male) and two resettled village leaders (one all-female and one all-male).

8.3.3 Results and Discussions

Interesting observations regarding women's rights emerged from the interviews with local officials and villagers. The Jinghong hydropower project had the benefit of lessons learned through other resettlement processes, which contributed to better design of the planning and policies to mitigate the adverse impacts of resettlement. As this was occurring simultaneously with other social changes in China, more emphasis was given to women in general and particularly in resettled communities. In the Jinghong case, we observed three levels of innovation that significantly promoted women's adaptability to the challenges of resettlement through the replacement of the traditional village social structure by government legislation, which required and resulted in macro, meso, and micro institutional changes.

8.3.3.1 Macro-level Institutional Changes

On the macro-level, China has always upheld the constitutional principle of the equality between men and women as the basic state policy for its laws and regulations. Since 1990, the State Council established the National Working Committee on Children and Women (NWCCW), which was commissioned with the responsibility to organize, coordinate, guide, supervise, and encourage government departments to promote gender equality and women's development. Over the last 20 years, Women's Federation, a division of NWCCW, has been working directly under the central government to support all administrative levels of government above the county level in 31 provinces, autonomous regions, and municipalities, forming a multidimensional and well-coordinated network for promoting gender equality and women's development.

The NWCCW has developed and implemented the gender statistics system for resettled villagers and a women's development plan, including:

- Regular monitoring and reporting of village-based women's socioeconomic status and providing women's empowerment training, including:
  - Gathering household- and community-level socioeconomic sex-disaggregated data, including income derived from the use of natural resources
  - Evaluating women's property right status, particularly during divorce and inheritance
  - Evaluating women's participation in household and community decision-making
  - Reporting evaluation results and making recommendations to promote women's empowerment in their health care
  - Developing programs to provide women's capacity building opportunities
- Implementation of gender development plans:
  - Developing a gender development plan and risk awareness strategy to ensure ongoing decisions, investments, payments, and activities incorporate the gender perspective
  - Supporting women involvement with monitoring, reporting, and program development process
  - Allocating sustainable funds, resources, and budgets annually through government funds
  - Encouraging hydropower developers to give priority to projects that support women's interests during the resettlement process

By expanding the scope of gender equality, women benefit not only from temporary financial support but through an emphasis on women's social well-being as a whole. Over time, women's interests were secured in a series of policies which granted them access to public resources and reduced their vulnerability. During the implementation of comprehensive poverty reduction strategies, China has focused on women by strengthening efforts to address women's poverty by providing access to numerous public welfare and charity programs. Examples of such programs include a program to support mothers suffering from breast or cervical cancer, a housing project for impoverished rural single mothers, and the Health Express for Mother's Program founded to improve access to medical care in poverty-stricken areas. Programs such as these support women suffering from illnesses, impoverished single mothers, and various other women in need.

During the resettlement process, the newly built villages were planned to include basic infrastructure (e.g., roads, clean drinking water, electricity, cell phone service, and TV signals) and services to enhance resettlers' prospects and education (e.g., 9-year compulsory education, skills training opportunities, cooperative medical services, minimum living allowances, and pensions). These infrastructural and capacity building supports were combined with production capital (e.g., housing, basic farmland construction, and land ownership) and access to public resources to support the resettlers (e.g., microfinancing, agriculture production technical support, and other temporary subsidies).

During the resettlement process, the local representatives of the NWCCW carefully reviewed the resettlement policies to assess whether they address gender equality, particularly for the land rights of rural women. Various systems have been established for managing rural collective funds, assets, and resources. When NWCCW finds that the village regulations and traditional conventions are in conflict with statutory regulations, they work at a local level to ensure the principle of gender equality is addressed. For example, during verification, registration, and certification of land contracts and land management rights, women's land-related rights and interests must be written in the registration book and on the land rights certificate, so that women in rural areas are ensured access to such resources.

Special projects were also implemented for women seeking employment and starting businesses when natural resources are depleted. Examples of these projects include small-sum guaranteed loans with financial discounts, capacity building for the rural female workforce, and support to the rural female workforce to enter non-agricultural sectors or urban areas.

8.3.3.2 Mesolevel Institutional Changes

On the mesoscale, women's participation is a component in the resettlement policy implementation, including through decentralized village development plans. In the Jinghong case, women collectively have a single vote to jointly decide which type of smaller-scale development project would improve the sustainability of community livelihoods and thus should be funded through the reservoir fund (which comes from the revenue of hydropower to support the development of affected communities). Examples of these types of development projects include improved irrigation and transportation systems. Previously, the allocation of the reservoir fund was determined by government only. During an interview in Jinghong, a fund manager made the following comment "If you really want to get things done, you have to involve wives, even those who seldom attend village meetings. Otherwise, even all husbands reach an agreement; there is still a big chance that they would come back to you on the next day saying they changed their mind because their wives disagreed."

Once the priority projects have been confirmed at the village level, the local resettlement bureau will apply the funds granted from a higher-level government. When the resettlement community supports a project through a willingness to contribute labor or other resources, those projects are prioritized for funding. For those projects that the resettlers requested but that were not funded, the local government maintains a database, so they can be considered for future funding opportunities. Villages also have the ability to revise the list of priority projects in discussion with the local government. If a project is cost-prohibitive, the government may seek to accumulate funds for a few years to fund the project providing they have full support from the villagers. For example, in Manmi village, the villagers requested the upgrading of the 17 km road connecting to the main townships. By 2017 the local government accumulated sufficient funds over a 5-year period to complete the 20 million yuan project.

8.3.3.3 Microlevel Institutional Changes

On the microscale, villagers interacted more with government and thus began to actively participate in rural governance. "Before dam construction, we rarely have officials visit us, now even the provincial governor has been in my house," one villager in Jiangbian village said. As they gained experience through these interactions, they became more willing and skilled at communicating with officials to obtain and secure more benefits.

Women are usually considered quick learners that have better skills to negotiate with government officials. During the resettlement process, the majority of women have learned how to speak Mandarin to interact with government, whereas men usually only speak their own ethnic language or local dialects. Over the past few decades, households have received social welfare subsidies and other forms of higher government support to assist them in securing support for basic needs from local government. Households are encouraged to participate in village meetings to elect trusted village leaders and to discuss the village's future development. Financial compensation provides basic income security and the opportunities for farmers to develop financial safety nets to manage risks.

It is fairly common in resettled villages for women to realize more significant positive impacts given they tend to spend much less time in agriculture and daily house work after resettlement. In the Jinghong case study, women in villages spend more time in formal education than men, which is quite different from the Manwan case study. With the reduction of cultivated land, fewer laborers are required in the village. A common saying about farming among the surveyed villagers is "Farming is too harsh for girls, if the girls go to school, they could have better chance to work outside the village, so daughters should go to school. Sons are usually naughty, and if they were not doing well in school, they could just return home to inherit lands and practice as farmers for the future." Conversely, women's capacity building could benefit from better access to sufficient quality and quantity of domestic water. Many studies have suggested that, with improved access to water, the time women devoted to non-agricultural income-generating activities would increase, which in turn would eventually increase women's capacity.

8.4 Conclusions and Recommendations

8.4.1 Conclusions

Through our case studies of the resettlers from the Manwan and Jinghong hydropower projects, we have explored the gender division relating to resettlement impacts, with a focus on impacts to women. When considering both case studies together, we arrive at the following conclusions:

1. The livelihoods of resettlers have clearly diversified. In the Manwan case, non-agricultural employment (especially migration work) became the primary source of income for the vast majority of resettled households although some households specialized in intensive agriculture. This is due in large part to the limited carrying capacity of nearby agricultural land. As a large number of men became migrant workers, women become the main labor force for agricultural production. Although women now perform more agricultural work than men, their workload has nevertheless decreased due to decline in area of agricultural land available, the shift from rice to perennial cash crops, and men continuing to conduct heavy agricultural labor. In the Jinghong case study, arable land remained relatively abundant. With effective resettlement planning, better access to markets provided more income to villagers, in particular through rubber tree planting, fish trade, and tourism activity.
2. In both cases, the living conditions of resettled communities and access to public services improved significantly, including housing, energy, transportation, education, health care, drinking water, and access to credit services. Although both men and women benefit from these improvements, women have benefited more considering their social roles and previously inferior social status. Women's health, workload, access to external information, and employment opportunities have also improved.
3. Women's influence in family decision-making and participation in collective activities has increased. However, men continue to have more influence in important family decisions, have access to more non-agricultural employment opportunities, and more actively participate in collective affairs than women in the Manwan case study.
4. The degree of satisfaction about life condition has improved for resettled families. In the Manwan case, women are less satisfied with life than men. Men are more concerned about production and income, while women are more concerned

about family life, social relations, social public service, and income in the Manwan case study.

5. Women's livelihood and social status have improved, due to the diversification of livelihood, improved access to public services, and improved influences in domestic and collective decisions. The gap between women and men still exists but is narrowing.
6. In Jinghong, hydropower development created new opportunities for the national gender equality institution to become deeply rooted in remote communities. During the resettlement process, women often benefited from the increased interaction with society and markets outside their villages with the dissolution of the self-sufficient small-scale peasant economy. Women earn more through diversified crops and nonfarm income, more actively participate in collective transactions, are active in decision-making, and gain access to more government services. As the resettlers gain more exposure to external societies with more gender equality, gender division is also reduced in the villages, leading to a better future for women. Women have embraced the changes resulting from resettlement and are actively gaining new knowledge to increase their capacity to adapt to the uncertainty of the future.

8.4.2 Policy Recommendations

Through the analysis conducted in this study, we developed the following key recommendations to improve the standard of living of women resettlers, to empower women, to promote gender equality, and to strengthen river health management:

1. Improve the infrastructure and public warfare policies to generate a favorable social environment. From the case studies, we found that the improvement in women's livelihood, social status, and influences is mainly due to the improvement in transportation, increased access to markets, and improved social services. Such improvements helped to diversify income sources, increased access to clean and convenient water sources and energy, reduced workload, and increased access to improved health care and education. In addition, these improvements increased women's participation in collective activities, job opportunities, and access to information from the world beyond their villages.
2. Provide more targeted training for women. This recommendation is anticipated to support women in transitioning to cash crop and non-agricultural employment, increasing their participation in collective affairs, and increasing their self-confidence. We found that although women's livelihood and social status have obviously improved, there are still gaps. This is due in part to long-term discrimination against women but also due to lack of skills training and self-confidence. Targeted interventions are recommended to support women in increasing their self-confidence, so they more actively participate in collective affairs and decision-making.

3. Improve resettlement policy and planning through adaptive learning from past resettlements. As shown in the Jinghong case, systematic and multilevel institutional arrangements that focus on gender issues could help improve women's abilities to adapt to resettlement.
4. Promote group and social cohesion within the village, including strengthening village governance and the management of rural household waste. In the Manwan case study, we found that river health management activities had not been addressed at the community level. As living standards of local people are modernized, their impacts on river health through domestic wastes have increase. Special efforts are needed to promote villagers to collectively foster and implement communal action with regard to domestic waste treatment. As men are increasingly working outside of the communities, the roles of women are very important in community environmental management. Targeted training programs on environmental impacts of domestic waste and its effective management would support women in reducing impacts on river health.

References

Cernea, M.M. 1998. *Involuntary resettlement in development projects: Policy guidelines in World Bank-financed projects*. Vol. 80. Washington, DC: World Bank Publications.

Chen, Q. 2007. Study on female resettlers' self-identify: Based on two case studies in the three gorges reservoir. *Academic Exploration* 3: 63–66 (in Chinese).

Chen, M. 2013a. Kinsfolk network reconstruction from the perspective of social gender roles: A case study in Y reservoir resettlement in Long County of Guangxi. *Nationality Forum* 5: 43–45 (in Chinese).

Chen, S. 2013b. Poverty and development of female resettlers of the three gorges hydropower project from a gender perspective. *Liaoning Agricultural Science* 5: 42–45 (in Chinese).

He, D., L. Chen, and Q. Li. 2001. Report of comprehensive impacts of Manwan Hydropower Project on local ecological environment and society and economy (in Chinese).

Koenig, D., and T. Diarra. 2002. The effects of resettlement on common property resources. In Scudder, T. 2005. *The future of large dams, dealing with social, environmental, institutional and political costs*. London: Earthscan.

Li, J. 2012. *Ecological changes and relevant impacts in Manwan reservoir in recent 20 years*. Master dissertations, .Yunnan University (in Chinese).

RGIMHSLR (Research Group of Impacts of Manwan Hydropower Station, Lancang River). 2002. *Social, economic and environmental impacts of Manwan Hydropower Station on Lancang River: Comprehensive study report*. .Unpublished (in Chinese).

Scudder, T. 2005. *The future of large dams: Dealing with social, environmental, institutional and political costs*. London: Earthscan.

Simon, M. 2013. *Balancing the scales – Using gender impact assessment in hydropower assessment*. Oxfam Australia: Melbourne.

Sun D., and Q. Zhao. 2017. *Social gender analysis of impacts of & adaptation to climate change*. Beijing: .Social Sciences Academic Press (in Chinese).

Wang, Z., X. Zhang, and Z. Jin. 2000. *The ecological environment and biological resources of Manwan hydropower station reservoir along Lancang River in Yunnan, China*. Kunming: Yunnan Science and Technology Press (in Chinese).

WCD (World Commission on Dams). 2000. *Dams and development: A new framework for decision making in water infrastructure*. London: Earthscan.

Zheng, H. 2014. *Nature, culture and power: An anthropological investigation on Manwan dam and dam debate*. Beijing: Intellectual Property Press (in Chinese).

Chapter 9
Case Study: Experience Sharing in Laos

9.1 Background

Laos, officially the Lao People's Democratic Republic (Lao PDR), is a country rich in hydropower resource potential, and the Laotian government is actively seeking to develop the hydropower industry with the goal of becoming the "Battery of Southeast Asia." The regulatory and technical capacities of environmental management of the hydropower sector in the Lao PDR are relatively weak but are steadily improving due to capacity building through national and international institutions. China has a rich history of experience and lessons learned over almost 40 years in the development and environmental management of hydropower that can be shared to facilitate sustainable hydropower development in the Lao PDR. The objective of sharing experiences and lessons in the Lancang River along with international best practices with the Lao PDR is to increase the level of knowledge and skills of the hydropower and regulatory agencies regarding environmental impacts and mitigation measures for hydropower development.

9.1.1 Current Status of Hydropower in the Lao PDR

The geography of the Lao PDR is ideally suited to the development of hydropower given its deep valleys, perennial flowing rivers, mountainous terrain, and low population density. The Mekong River in particular is considered a rich resource for potential hydropower in Lao PDR. The country contains 202,000 km^2 of basin area or 25% of the total catchment area of the Mekong River Basin, accounting for more basin area than contained in the other five countries of the Mekong Basin territory. Since 1986, the rise of regional banks and investors in Asia has increased outside investment into the economy of the Lao PDR and has generated interest from the Government of Laos (GoL) to develop the country's hydropower potential by

X. Yu et al., *Balancing River Health and Hydropower Requirements in the Lancang River Basin*, https://doi.org/10.1007/978-981-13-1565-7_9

building several new small- to large-sized hydropower projects to supply the increasing power demand of neighboring countries (Smits 2011).

Hydropower in the Lao PDR has been previously underutilized with only around 15% of hydropower potential developed over the past 30 years (ADB 2004, 2012). Prior to 2006, the Lao PDR contained nine hydropower dams, with a total capacity of 680 MW. Currently, in 2017, the country has 18 large and small hydropower plants in operation for a total capacity of 2800 MW (Fig. 8.4). The potential for hydropower development in the Lao PDR is theoretically estimated at 26,000 MW with approximately 18,000 MW technically exploitable in the Mekong River Basin (12,500 MW in major Mekong sub-basins and 5500 MW in minor sub-basins).

A major increase in hydropower development in the Lao PDR is currently underway (Fig. 9.1). Plans are in place for the construction of numerous dams in the Lower Mekong Basin tributaries by the year 2030 (MRC 2011). Seventeen hydropower plants are currently under construction (capacity of 3000 MW), including Xayaburi Dam in the Lower Mekong Basin (capacity of 1850 MW). Another 25 hydropower plants are at the planning stage (6500 MW), while 35 more are undergoing feasibility studies (approximately 10,600 MW) (EDL 2011; Mekong Institute 2012; MEM 2012). This surge in hydropower development could play an increasingly important role in the Lower Mekong Basin, serving as the answer to the rapidly growing demand for energy in countries within the Lower Mekong Basin while providing an alternative to fossil fuels. Considering the magnitude of the hydropower generating potential of the lower Mekong River, significant revenue and economic benefits can be expected from the export of electricity generated within the Lao PDR.

9.1.2 Regulations and Regulatory Institutions in the Lao PDR

The GoL has developed a comprehensive body of domestic legislation to deal with sustainability and environmental conservation issues in its policies and regulations to ensure sustainable development (MEM 2012). Key national laws and regulations are listed and described in Table 9.1.

The country's 1999 Environmental Protection Law was amended in 2012 (Law on Environmental Protection (Amended) No. 29/NA2012) and established a framework for the management of environmental resources with the objective of conserving and facilitating the sustainable use of natural resources. In December 2013, the Ministry of Natural Resources and Environment issued two ministerial instructions to implement provision of Articles 21 and 22 of the amended law. These included instructions for the initial Environmental Examination process required for these projects (No. 8029/MONRE 17) and the Environmental and Social Impact Assessment (ESIA) process required by investment projects and activities (No. 8030/MONRE 17). The specific objective of these ministerial instructions is to ensure the uniformity in conducting the ESIA and Environmental Examination by every investment project and activities of a public and private enterprise operation

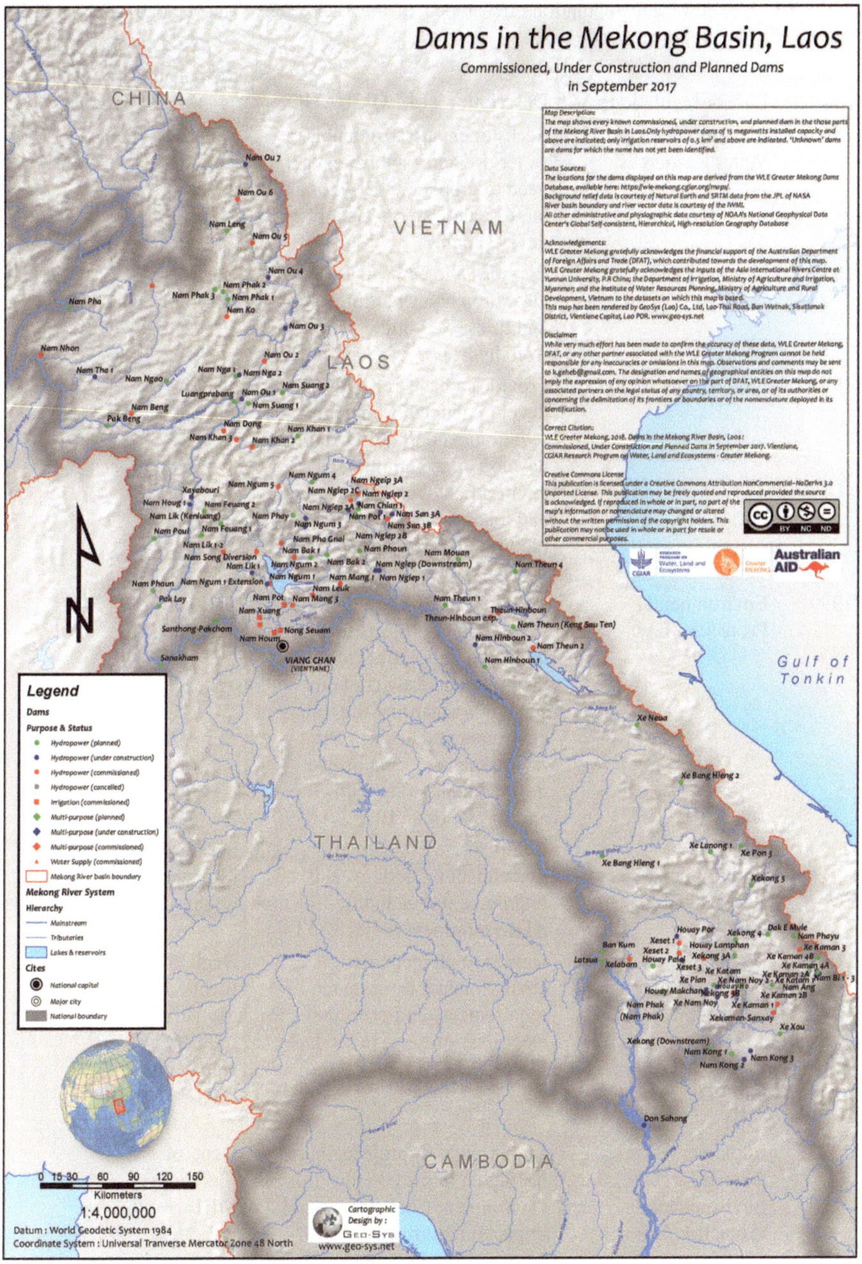

Fig. 9.1 Existing and planned hydropower projects in the Lao PDR (Figure by WLE Greater Mekong)

Table 9.1 Key laws and regulations developed in the Lao PDR to meet goals of sustainable hydropower development

Year enacted	Law or regulation	Description of LEGISLATION	Responsible agency	Year amended
1996	Law on Water and Water Resources	Seeks to ensure sustainable water use and establishes the need to prepare an EIA for any large-scale water project prior to development	Water Resource Committee under the Prime Minister's Office	2013
1997	Law on Land	Provides a legal basis for land allocation and awarding deeds and titles to resettled people from hydropower development	National Assembly	2003, under revision
1997	Law on Electricity	Provides a basis for hydropower development environmental requirements. Requires dam developers to submit environmental and social studies and management plans. EIA required for all hydropower projects	National Assembly	2008, 2012
1999	Environmental Protection Law	Requires an EIA for all development projects that have the potential to affect the environment. Law contains standards regarding the timing of environmental assessment requirements during project development along with content and format of an EIA	National Assembly	2012, 2013
2000	Regulation on Environmental Assessment No. 1770/ STEA	Refers to environmental assessment procedures and requirements (including EIA) for all development projects in Laos	STEA, Prime Minister's Office	
2001	Regulation on Implementing Environmental Assessment for Electricity Projects in Lao PDR No. 447/ MIH	Guides implementation of environmental assessment requirements and procedures for electricity projects in the Lao PDR	Dept. of Electricity, Ministry of Industry and Handicraft (MIH)	
2002	Implementation Decree for the Environmental Protection Law	Emphasizes all projects that having an impact on the environment (including social impacts) require an environmental assessment prior to approval	Dept. of Electricity, MIH	

(continued)

Table 9.1 (continued)

Year enacted	Law or regulation	Description of LEGISLATION	Responsible agency	Year amended
2003	Environmental Management Standards for Electricity Projects No. 0366/MIH.DOE	Provides minimum requirements for ESIA, EIA, EMP, and Resettlement Action Plans for electricity projects	Dept. of Electricity, MIH	
2005	National Policy on Environmental and Social Sustainability of the Hydropower Sector in the Lao PDR No. 561/CPI	Emphasizes the requirement of EIA reporting and EMPs for all large hydropower projects	CPI, Ministry of Planning and Investment (MPI)	2007
2010	Decree on the Environmental Impact Assessments No. 112/PM	Provides guidance on Initial Environmental Evaluation and EIAs and requires prevention and mitigation measures or management and monitoring plans for projects that may create adverse environmental and social impacts	WREA	

in the Lao PDR that causes or is likely to cause environmental and social impacts. Furthermore, the investment projects and activities shall conduct these assessments, shall contribute to the sustainable socioeconomic development of the country, and shall mitigate and enhance adaptation to global warming (LPDR 2013a, b). The specific objective of the ministerial instruction No. 8029 The National Policy on Environmental and Social Sustainability of the Hydropower Sector (NPSH) in the Lao PDR was built on the principles established during the development of the Nam Theun 2 (NT2) Hydropower Project (see Sect. 9.3.2.1). The NPSH is applied to the hydropower sector as a whole, particularly to large dams. In the Lao PDR, hydropower projects with capacity below 15 MW are classified as small-scale hydropower and large-scale hydropower has been defined as large dams with a capacity of more than 50 MW or inundating more than 10,000 hectares of land (GoL 2005; MEM 2011).

According to the Lao 1999 Environmental Protection Law, all large hydropower projects must develop a full environmental impact assessment (EIA) report and environmental management plan (EMP). Further, various governmental departments are responsible for the oversight of different components of project development. The Lao EIA process is largely compatible with international guidelines for conducting EIAs, and construction activities cannot commence until Water Resources and Environment Agency (WREA) approval is received. The ESIA department within the Ministry of Natural Resources and Environment (MONRE) is responsible for overseeing the implementation of the EIA process. MONRE issues environmental compliance certificates for the projects that have successfully completed the EIA process and coordinates with line agencies to carry out follow-up (compliance) monitoring and evaluation. Project proponents are required to submit regular monitoring reports to MONRE based on their environmental management and monitoring plans (EMMPs) (Wayakone and Makoto 2012).

9.1.3 Challenges of Hydropower Development in the Lao PDR

Currently, there are three main challenges to managing the environmental consequences of hydropower development in the Lao PDR. These are (1) limited enforcement of legislation and regulations currently in place, (2) limited institutional capacities, and (3) limited public engagement during the hydropower development process. These challenges are described in greater detail below.

9.1.3.1 Enforcement of Legislation and Regulations

Much of the existing legislation in the Lao PDR was issued under the framework of the country's Constitution adopted in 1991. The current legal system can be described as a hybrid of a civil code and common law, driven by the need to develop and adopt a large body of legislation to support the country's Constitution. However, capacity to implement the legislation remains challenging. The National Assembly, first elected in 1992 with 5-year terms, has been an active legislative branch, passing nearly 50 comprehensive laws, each requiring issuance of implementing legislation by the prime minister, ministries, and local authorities by way of decrees and regulations. Despite the rapid growth of the civil law issuance, the legal system is in the early stages of development with laws that are difficult to interpret, implement, and enforce, in particular with regard to the natural resources sector. Presently, few articles in the Environmental Protection Law (amended in 2012) have been actually implemented (e.g., Articles 10, 11, 21, 22, 47, and 48). Moreover, those articles that have been implemented have not yet been implemented for every hydropower project in the Lao PDR.

9.1.3.2 Institutional Challenges

The Lao PDR faces a large gap in governance and institutional quality for sound natural resource management. This institutional gap will continue to grow with increasing resource exploitation in the country. Over the past 10 years, the GoL has made important advances in the quality of institutional governance within the general public sector. For example, great strides have been made toward strengthening the public financial management system and improving internal oversight mechanisms. Yet, the GoL continues to face challenges in accountability, regulatory quality, and government effectiveness. This governance gap remains a crucial challenge that will increase over time if the government does not take strategic and continued action to enhance governance and institutional capacity. It is important to recognize that the governance gap is not simply related to national level policies and institutions, but given the extent of administrative decentralization in the Lao PDR, it is also related to the distributed implementation and monitoring capacity of local government in the areas where hydropower projects are developed.

9.1.3.3 Public Engagement

Public engagement is mandated in various components of the Lao PDR Environmental Protection Law (amended in 2012). Article 8 states that "EIAs shall include the participation of the local administration, mass organizations, and the population likely to be affected by the respective development project or activity." Furthermore, involvement of project affected persons and other stakeholders is outlined in the Ministry of Natural Resource and Environmental ESIA (Article 2.14 and 2.15) (LPDR 2013b). The provisions of these articles are fairly limited to providing information to the public about the project and allowing the public to participate in consultation meetings and monitoring activities. Further definition and clarification of the information to be made available to project affected people and other stakeholders by the project proponent is detailed in Ministry of Natural Resource and Environmental ESIA (Article 2.20) (LPDR 2013b) and includes information about the project proponent, information on social and environmental impacts of the project, ESIA reports and other relevant project reports, proposed mitigation measures, proposed budget for environmental and social management and monitoring plans, and any breaches of obligations of the project proponent.

Implementation of public engagement for hydropower projects in the Lao PDR is a key challenge as there remains limited public access to information on natural resource and environmental issues. Despite the National Assembly passing a press law in 2008 authorizing media access to government information, the government often is not cooperative or delays the release of certain information. Furthermore, there remain language and information format barriers to clearly communicate issues to local communities and other stakeholders to enable a full comprehension of project issues. These barriers hinder meaningful consultation with groups affected by a project.

9.1.4 The China Hydropower Experience

9.1.4.1 Overview of Hydropower in China

Large-scale hydropower development in China began in 1980s and has been ongoing for nearly 40 years. During the first half of that period, environmental protection was not a key consideration during hydropower development. Ecological degradation caused by hydropower projects was commonplace and eventually drew wide concern from the government and public. More recently, environmental management has improved significantly as a result of those concerns and lessons learned. The extensive experience gained in China can be shared with countries such as the Lao PDR, which are looking to significantly expand hydropower development, to facilitate early and effective environmental management. The central experiences gained through the development of hydropower in China are discussed below and include the establishment of relevant laws and policies, the development of sound technical standards, and integrating regulatory mechanisms and capacity.

9.1.4.2 Laws and Policies

China has enacted and implemented a series of sound laws and regulations relevant to the environmental management of hydropower development. Key laws and regulations include the Environmental Protection Law (enacted in 1979), the Water Law (enacted in 1988), the Water and Soil Conservation Law (enacted in 1991), the Environmental Impact Assessment Law (enacted in 2002), Regulation on Natural Reserves (enacted in 1994), and Regulation on Environmental Impact Assessment of Planning (enacted 2009). These laws and regulations established a legal basis for environmental management of hydropower development. For example, the Environmental Impact Assessment Law mandates that an EIA should be conducted for new projects that may cause significant environmental impacts. According to the Regulations on Environmental Impact Assessment Planning, a cumulative environmental assessment should be conducted in the case where multiple hydropower projects are being developed in a river basin due to the possibility of negative indirect and cumulative ecological impacts.

Based on relevant laws and regulations, the regulatory agencies including Ministry of Environmental Protection (MEP), National Development and Reform Commission (NDRC), and National Energy Administration (NEA) issued a number of policies for regulating environmental management of hydropower projects. Four policies issued in 2005, 2012, 2014, and 2016 guided the regulatory agencies to manage environmental issues stemming from hydropower project development and operations. The first of these policies issued in 2005 by the State Environmental Protection Administration (predecessor of the MEP) simply highlighted the importance of environment impact assessment, environmental mitigation, and the optimization of hydropower project operation. Subsequent policies were more clear and specific with regard to integrated planning, environmental assessment, mitigation, monitoring, and management.

9.1.4.3 Technical Standards for Hydropower

Government regulatory agencies in China established a series of technical standards relevant to environmental management of hydropower projects (briefly described below by stage and identified in Table 9.2). These standards cover almost the full

Table 9.2 Key technical standards for the various stages of hydropower development in China

Stage of hydropower	Relevant regulations or standards	
Planning	Regulations for environmental impact assessment for river basin planning	SL 45-92
	Specifications on compiling hydropower planning for rivers	DL/T 5042-2010
	Technical guidelines for environmental impact assessment and planning – general principals	HJ 130-2014
Design and Construction	Code for environmental impact assessment for water conservancy and hydropower projects	HJ/T99-203
	Guidelines for environmental flow assessment, release of water with low temperatures, and fish passage of hydropower and water projects	2006
	Specifications for environmental protection designs and mitigation for water conservancy and hydropower projects	DL/T 5042-2007
	Technical code for environmental protection and mitigation measures for hydropower and water conservancy during engineering and construction	DL/T 5260-2010
	Guidelines for fishways with regards to water conservancy and hydropower projects	SL 609-2013
	Guidelines for water resources assessment for construction	SL 322-2013
	Regulations for river and lake eco-environmental water demand computation (i.e., environmental flows)	SL/Z712-2014
	Technical guidelines for environmental impact assessment for construction projects – general program	HJ2.1-2016
Operation	Technical guidelines for confirmation that environmental protection for water conservancy for hydropower projects is complete	HJ 464-2009
	Guidelines for post-project environmental impact assessment for water projects	SL/Z 705-2015

scope of hydropower development including hydropower planning, design, construction, and operation; however, decommissioning has no current guideline since few hydropower dams in China currently need to be decommissioned.

Planning

Standards of hydropower planning relevant to environmental issues mainly specify the requirements of integrated hydropower planning with environmental considerations at river basin scale. As noted below, an EIA for river basin planning is a necessary condition prior to approval of individual hydropower projects.

Design and Construction

During the project preparation or design stage, an EIA should be conducted for the proposed hydropower project to assess the potential environmental consequences and recommend mitigation measures to avoid and minimize adverse impacts. Relevant standards provide guidelines for the EIA and the specifications and best management practices for the design of environmental protection. In the construction stage, the proponent should implement and monitor environmental mitigation. In addition to the standards presented in Table 9.2, environmental quality standards (e.g., water, air, and solid waste) are also used in preparation of the EIA and environmental protection design.

Operation

Once a hydropower project is in operation, government agencies will confirm that the environmental protection mitigation and offsetting measures, including offsetting facilities, are in place and are being implemented properly. In addition, environmental monitoring should be conducted throughout project operations to measure environmental impacts and manage them through the use of adaptive management.

9.1.4.4 Regulatory Mechanisms and Capacity

Regulatory mechanisms and capacity are fundamental to implementing environmental law and policy, which can safeguard the environment during hydropower development. China has established managerial procedures to environmentally manage hydropower development over time. These mechanisms and procedures have resulted in improved environmental assessments and more consistent implementation of mitigation measures.

MEP, MWR, NDRC, and NEA are the four main governmental agencies responsible for overseeing the environmental management of hydropower development in China. The MEP is the lead department that regulates environmental issues relevant to hydropower projects. Moreover, most of the policies and standards relevant to environmental management of hydropower projects are issued by the MEP, with a portion issued by the MWR and NEA. The EIA is reviewed by all of these environmental departments.

9.2 Methods

The current status and challenges of hydropower development in the Lao PDR were reviewed to determine ways that experience gained through the long history of hydropower development in China could be shared between the countries. Key lessons learned in China and recommendations were developed from this review to guide successful environmental management of the developing hydropower industry in the Lao PDR. The application of experience sharing and the use of international best practices were also explored through a hydropower case study in the Lao PDR and a training workshop held on November 3 and 4, 2016 in the Xayabury Province, where the first Mekong hydropower dam in the Lao PDR is being constructed. The training workshop provided the opportunity for 20 participants from the Lao PDR (mainly from Ministries of Energy and Mines, Natural Resources and Environment, Agriculture and Forestry), China (Asian International River Center), and Canada (Ecofish Research Ltd.) to share their experiences regarding environmental impact and protection of hydropower development.

9.3 Results and Discussions

9.3.1 Lessons from China Hydropower Development

The gradual advancement and enhancement of environmental management have improved the environmental assessment process, the implementation of mitigation and offsetting, and the overall oversight of hydropower projects in China. Through experience and lessons learned, China can share issues to be avoided and successes to leverage hydropower development in other countries. Three central lessons were identified: (1) implementing policy successfully, (2) strengthening regulatory requirements, and (3) instituting integrated planning. In China, the phenomena or mindset of "development given priority over protection" and "protection lagging behind development" in certain stages of hydropower development previously existed. This mindset was compounded with limited understanding of environmental issues and potential impacts and has caused significant adverse impacts on aquatic ecosystems and surrounding landscapes of some rivers (Wu 2011). To avoid reoccurrence of this issue, it is important to create awareness among policymakers and decision-makers across relevant regulatory departments on the establishment and application of good practices in law, policy, and managerial mechanisms and to ensure successful implementation of these laws and policies.

Management measures and requirements should be applied to all hydropower projects, irrelevant of scale and size, by strengthening regulatory requirements. In China, the management of small and large hydropower projects is different with the provincial and municipal government departments being responsible for the management of small hydropower projects, whereas central (federal) government ministries are responsible for managing large projects. Given different regulatory requirements, there is often less stringent environmental management and requirements for small hydropower projects, so the environmental consequences of small hydropower have been more prominent. The Chinese government is working on improving the environmental requirements for small hydropower developments, and the Lao PDR could learn from this process by strengthening regulatory requirements.

International good practices in hydropower planning and management such as integrated planning should be considered and applied to new hydropower development. Integrated planning is being adopted in China to reduce the unintended environmental consequences of the development of multiple dams within a watershed. Integrated planning is also effective in balancing competing interests, social, ecological, and industrial, within the watershed as a whole.

The three key lessons learned from hydropower development in China are discussed in detail below with recommendations for implementation.

9.3.1.1 Implementing Policy Successfully

The Lao PDR began developing policies for sustainable hydropower development in the mid-1990s (Table 9.1); however, successfully implementing the policies has lagged behind. Nevertheless, many of the laws, regulations, and institutional organizations have been updated to provide an improved framework for applying sustainable development policies with the aim of providing guidance and an indication that responsible agencies are overseeing investment projects in the hydropower sector. Responsible agencies must work effectively under the guidance of the interministerial committee formed by the Ministry of Energy and Mines to ensure effective communication of updated policies. Key components of updated sustainable hydropower development policy in the Lao PDR include:

- Successful planning and coordination of hydropower project development: Detailed data collection, planning, and successful policy application are required to achieve economic, social, and environmental sustainability in hydropower development. Collaboration with relevant stakeholders in the management and utilization of water resources is crucial to optimal beneficial use of water for all.
- Conduct adequate feasibility studies: Comprehensive feasibility studies need to be conducted before the hydropower project is approved to ensure economic, technical, and financial feasibility. A plan should be developed to prevent or mitigate potential adverse environmental and/or social impacts.
- Consider economic aspects: Special attention must be paid to economic aspects to ensure efficient, effective, and sustainable project development. Economic-technical feasibility studies should be conducted in parallel with environmental and social impact assessments and include appropriate measures to mitigate negative economic impacts. Economic studies in the Loa PDR must be implemented in accordance with the National Socio-Economic Development Plan and National Electricity Development Plan.
- Technical and engineering considerations: Developers must utilize the most advanced equipment when planning, constructing, and operating hydropower projects to ensure the safety of people and property and mitigate any potential risks to natural resources and the environment.

- Environmental and social impact assessment: A comprehensive ESIA must be conducted for all potential hydropower projects. Any project with large impacts or transboundary project must also undergo a cumulative impact assessment. Environmental and social management and monitoring plans (ESMMP) must also be developed prior to the construction and operation of a project.
- Social impact assessment on project affected people: The developer shall closely monitor and provide progress reports on social impacts to safeguard the statutory interest of project affected people due to resettlement and compensation. Resettlement and livelihood improvement plans, ethnicity development plans, and gender development plans should also be carried out to safeguard these interests and mitigate any potential adverse impacts prior to construction and operation of the project.
- Proper consultation: Reasonable, honest, accurate, and transparent public consultation needs to take place with adequate data and information provided for public review prior to decisions on hydropower project approval.
- Management and conservation of watersheds and water resources: Habitat losses due to the development of hydropower projects should be avoided. Where avoidance is not possible, the project developer must mitigate and compensate for losses and develop a sustainable biodiversity management plan. Additional funding must also be provided to help manage and effectively protect the watershed area, nearby watersheds, and other important conservation areas.
- Conduct monitoring compliance: Regular monitoring inspections and reporting by relevant government agencies or appropriate third parties should be completed to ensure that all hydropower projects are operated in accordance with relevant obligations required under laws of the Lao PDR, policies, strategies, contracts, or other implementation plans.

The implementation of the policy for sustainable hydropower development will need to be considered on a project-by-project basis as the technical requirements and environmental management will differ. Based on experience in the Lao PDR, the policy needed to be enhanced to consider economic and other technical aspects, as it originally only considered the environmental and social impacts associated with hydropower development.

9.3.1.2 Strengthening Regulatory Requirements

Enforcement of Legislation and Regulation

The Lao PDR has continued to enhance regulations relevant to environmental management of hydropower development over time, but application of the policy is in the early stages. Key legislation and regulations (see Table 9.1) provide a legal foundation for the consideration of how environmental management is to be planned and carried out (see Sect. 9.3.2.1). Standard steps in the EIA process are adopted in the Lao PDR based on practical experience and experience from other developing and industrialized countries. These steps include project screening, project scoping, ESIA report preparation, development of ESMMPs, assessment and review of the ESIA and ESMPP reports, project construction, project operation, project management, project monitoring, and compliance and enforcement. During each of these steps, the project proponent is required to undertake investigation and consultation activities to minimize the environmental and social impacts of the project. However, for smaller-scale hydropower projects, the implementation of environmental management is still not well developed, and enforcement of the EIA process is weak, especially for existing projects. Recommendations from hydropower experience in China for supporting the enforcement of legislation and regulation of sustainable hydropower development in the Lao PDR as well as the Mekong region as a whole include:

- Clearly defining governmental responsibilities so that the central and local levels of government and their respective agencies are fully aware of their mandates and responsibilities and are also able to effectively coordinate
- Developing clear separation of roles (e.g., implementation vs. monitoring) where governmental entities are engaged as owners or implementers of hydropower projects
- Establishing community developments, environmental monitoring, and remediation programs in areas affected by hydropower development

Institutional Capacity

Developments led by government institutions must consider the challenges of natural resource sector management in two crucial ways. The first is, at a national level, whether a country is "doing the right projects," and the second, at the project level, is whether a country is "doing the projects right." In the case of the Lao PDR, the GoL must focus on both these questions as it attempts to close the governance gap for sustainable hydropower development.

The national portfolio-level governance challenge, or "doing the right projects," is a central strategic issue facing a country rich in natural resources. The GoL must determine how it can assess resource utilization and development in a tactful manner, paying particular attention to ensuring that utilization does not proceed faster than the government's ability to manage the resource sector (i.e., hydropower sector). The Lao PDR still lacks a systematic regime for natural resource management. As resource development continues to grow and cumulative impacts from smaller investment projects proliferate, a project-by-project approach to governance will be ineffective and insufficient. A more strategic approach requires the ability to select the right projects in the context of the government's national socioeconomic development plan and institutional capacity. This approach requires an overarching natural resource policy framework along with a complementary sector-specific development policy. The project-level governance challenge, or "doing the projects right," requires concerted attention to the process of natural resource management itself, in order to implement the GoL's strategic vision.

Recommendations for improving natural resource management in the Lao PDR, based on China's experience, include:

- Establishing strategic oversight and coordination of institutional arrangements and capacities
- Enhancing transparency as a principal of natural resource management
- Standardizing the natural resource management process to increase predictability and efficiency

Public Engagement

The Lao PDR has a great need for more effective consultation and participation of stakeholders including local communities and affected people in the hydropower planning process. Careful and clear involvement of project affected persons and other stakeholders through the process of public consultation is a requirement under Article 8 of the Lao 2012 Environmental Protection Law. However, barriers remain that prevent effective communication with the public (see Sect. 9.1.3.3). It is recommended that the Lao PDR develop a more effective public engagement process to better address social and environmental issues associated with hydropower development.

9.3.1.3 Instituting Integrated Planning

Historically, electricity generation has been prioritized over environmental considerations for hydropower planning and development in many river basins in China with negative ecological consequences. For instance, six cascade dams on the middle and lower Lancang River were placed within close proximity causing backwatering of the downstream dam to reach the toe of the upstream dam. The effect of this cascade dam development resulted in the disappearance of the natural riverine habitat for fishes, thus fish diversity decreased significantly. Chinese government agencies have become aware of negative effects caused by cascade dam development and are using integrated planning when considering hydropower developments on the upper Lancang River. Through the use of integrated planning, the natural flows of a certain length of a river reach can be maintained by adjusting dam location and design (explained in Sect. 3.3.1). The principals of integrated planning are described below and could be used in the Lao PDR to plan for multiple hydropower projects within a river basin.

Integrated basin-wide hydropower development planning is the process by which decisions are made over different and competing demands for hydropower projects within a river basin. This type of plan takes into account individual project location, and design, while balancing these against the impacts of other proposed or existing hydropower projects within the river basin. Recently, cost-benefit trade-offs have been developed as an approach to assess economic benefits and social/environmental costs of multi-project planning using multiple criteria at the river basin scale to maximize hydropower's benefits while considering environmental and social sustainability (Hartmann et al. 2013; The Nature Conservancy et al. 2016).

Four principals are important to consider when implementing integrated planning for hydropower projects. First, hydropower planning should be consistent with the overall competing requirements of the river basin. Various domestic and industrial water uses should be considered while balancing the requirements of environmental protection, economic development, and social sustainability. Second, while planning hydropower development, it is important to conduct an environmental impact assessment of hydropower development at the basin scale. Third, it is important to consider geological activity when planning hydropower development. Dam safety is paramount. Thus, it is critical to avoid building dams in areas where significant geological instabilities may exist to prevent any disasters. Finally, it is critical that hydropower planning take into account the existing economic and social conditions of the river basin. Environmental capacity should also be examined with special attention paid to sensitive components of the ecosystem. If resettlement is necessary, a preliminary plan should be developed and instituted, in consultation with the settlers.

As noted above, integrated hydropower planning should be conducted at a basin scale. This level of planning involves five key procedural steps including:

1. Conduct field surveys. Surveys should be conducted to collect information on the natural river conditions, environmental setting, economic context, and social status of people living in the river basin. The baseline conditions of water resource utilization within the basin should also be considered to understand all of the factors that may be influenced by hydropower development.
2. Proposing preliminary development scenarios. Preliminary development scenarios (i.e., alternatives) should comprehensively consider geological conditions, distribution of existing hydropower developments, other water uses, sensitive social and environmental components, and joint operational requirements of multiple projects within a river basin.
3. Comparing preliminary scenarios. Analyze and compare preliminary scenarios in terms of effects such as water use, geological conditions, reservoir inundation, resettlement requirements, environmental protection, electricity transmission, and transportation.
4. Selecting a development scenario. Based on technical and economic comparison and analysis of preliminary scenarios, propose a development scenario with sound technical and economic indicators. This development scenario should result in development benefits without significant effects to the environment or people.
5. Proposing an implementation plan. The implementation plan should include a schedule detailing the sequence for developing cascade hydropower projects, an implementation process, an environmental protection and mitigation measures, a

resettlement plan, a transportation plan for the construction area, and a preliminary plan for transmission of electricity.

Integrated basin-wide hydropower planning will be particularly important in the Lao PDR for supporting sustainable hydropower and should be actively incorporated into government policies and best practices for hydropower developers. In general, governments have the greatest ability to implement integrated basin-wide hydropower planning, particularly through the approval and licensing of individual projects. It is important for the government to establish relevant technical guidelines and criteria for integrated basin-wide hydropower planning prior to approving individual projects within the basin. Through integrated planning each project can be assessed and effects can be determined in the context of all projects within the basin. Hydropower developers can also play an important role in promoting integrated basin-wide planning by supporting government and contributing to the adoption of the process where they can. For instance, hydropower developers can be commissioned by the government to conduct or fund hydropower planning. In these instances, the developer can commit to implementing integrated basin-wide planning.

9.3.2 *Application of Experience Sharing*

9.3.2.1 Case Study: Nam Theun 2 Hydropower Project

The manner in which the NT2 hydropower project was developed and operated is evidence that significant strides have been made to improve environmental management of hydropower developments in the Lao PDR through learning from and integrating experiences and best practices of other countries, including China. The following provides an overview of the project components that have incorporated those valuable lessons.

Project Background

The NT2 hydropower project is located in the Khammouane Province in central Laos, approximately 40 km upstream from the completed Theun-Hinboun Hydropower Project. The project began commercial operation in April 2010. The project consists of a 50 meter high dam on the Theun River (the fourth largest tributary of the Mekong River in the Lao PDR) that forms a 450 km^2 reservoir. Water from the Nakai Reservoir drops more than 350 m to the powerhouse with an installed capacity of 1075 MW. The dam provides 1000 MW of power for export to Thailand and an additional 75 MW for domestic consumption. The water discharged from the powerhouse flows to the Xe Bang Fai River through a waterway. Subsequently, the

Xe Bang Fai River flows into the Mekong River about 150 km south of the Nam Theun River.

Funding for the NT2 project was provided mainly through the World Bank and the Asian Development Bank which approved millions of dollars in guarantees and loans for the NT2 hydropower project in 2005. Requirements of the lenders include environmental evaluations of the various stages of the project. Independent assessments on the project were provided by a Panel of Environmental and Social Experts (POE), an International Advisory Group (IAG), and independent monitoring agencies including international experts in dams, environment, and social protection. Many environmental and social impacts from the development of the NT2 project were identified, and comprehensive measures were designed to mitigate the adverse impacts. According to a group of social and environmental experts advising the project, these mitigation measures could become a global model (The World Bank 2007). During the operational phase of the project, a number of monitoring programs were established and implemented (described below) (NTPC 2005).

Environmental Assessment and Mitigation

Potential impacts of the NT2 project on cultural resources and the physical, biological, and social environment were assessed. Additionally, cumulative impacts of the cascade of hydropower projects were assessed to analyze combined impacts in relation to the NT2 project. Based on the environmental assessment, a variety of measures was recommended for mitigating adverse impacts of the NT2 project.

Physical Environment

Key impacts of NT2 project operations on the physical environment are associated with changes to hydrology, water quality, erosion rates, and, to a lesser extent, climate and groundwater. The impacts of water diversion from the NT2 Project were quantitatively evaluated. Negative hydrological impacts are mitigated through operational management of the project, specifically regulating discharge releases from the dam. Changes in water quality during filling and the storage of water in the Nakai Reservoir could affect water quality in the Nam Theun River (downstream of the Nakai dam), Xe Bang Fai River, Nam Kathang River, and possibly the Mekong River. The identified impacts on water quality in the Nakai Reservoir and downstream in the Nam Theun River and Xe Bang Fai River are mitigated by reservoir operation and control of pollutant sources. The EIA indicated that sediment accumulation in the Nakai Reservoir and river channel erosion below the dam were expected to occur during project operations. Several project components have been designed to minimize erosion including controlled and consistent release of water from the dam, limiting the rate of increasing and decreasing discharge into the Xe Bang Fai River, and installing erosion protection along the Xe Bang Fai River downstream of the project.

Biological Environment

Potential impacts on aquatic habitat, fish diversity, terrestrial biodiversity, and threatened species were evaluated in the NT2 project EIA, and mitigation measures were recommended to avoid and minimize these impacts.

The EIA indicated that aquatic habitats were likely to be impacted by activities during NT2 project construction and operation. Predicted impacts during construction included sedimentation caused by work in the riverbed; clearing of vegetation in the inundation area; erosion at construction sites; water pollution caused by oils, fuels, and chemicals used; and the use of explosives. Long-term impacts on aquatic organisms, aquatic habitats, and fish biodiversity in the project area were predicted to occur after completion of construction. The impoundment of 195 km of the Nam Theun River and the creation of the Nakai Reservoir were assessed to have the potential to reduce habitat and adversely affect water quality resulting in the displacement of many aquatic species that cannot adapt to the new environmental conditions. Sedimentation resulting from reservoir infilling was assessed to have the potential to contribute to habitat loss by removing distinct habitats favored by some species. The construction of the Nakai dam blocks possible fish migration routes between the plateau and downstream areas. Downstream of the Nakai dam, reductions in water flow were assessed in the EIA to reduce the carrying capacity of the river, both in terms of fish diversity and abundance.

Mitigation measures to reduce the assessed impacts to the biological environment were developed during the EIA. During construction, strict management and regulation of construction activities were recommended to mitigate construction-related impacts. A number of the operational impacts (e.g., impacts caused by impoundment and creation of the reservoir which prevents fish migration) could not be mitigated. For example, there are no plans to provide fish ladders as these have proven to be ineffective elsewhere in the tropics. Management of the Nakai Nam Theun 2 National Protected Area (NNT NPA) should, however, ensure the survival of most fish species that are present elsewhere in the river system, hence compensating for this impact. It was recommended that populations of critical species be monitored to detect possible declines and to provide recommendations for appropriate supportive measures (e.g., restricting or banning fishing of species of concern, establishing and using captive stock to supplement the wild population, and increasing the amount of protected pool areas). No new fish species will be considered for potential introduction into the reservoir for 10–15 years to allow time for the existing species to acclimatize to the new aquatic conditions. To mitigate the impact of reduced flows in the Nam Theun River downstream of the dam, adjustments to the river morphology were considered for purposes of maintaining water flows and depths in critical areas.

The EIA predicted that direct impacts on terrestrial biodiversity would occur as a result of land clearing for construction works and reservoir inundation and as a result of degradation and disturbance to the ecosystem. Indirect impacts on terrestrial biodiversity were expected to occur as a result of increased human population density and improved access to the area. Mitigation of impacts on terrestrial biodiversity identified in the EIA included a wildlife management and protection program in the NNT NPA to compensate for the losses caused by the NT2 project and the development of a plan for the management of animal relocation from the Nakai plateau to reduce potential conflicts with the local population and reduce hunting pressure.

The EIA predicted that threatened species may also be affected as a result of increased human populations on the Nakai plateau, destruction of habitats as a result of construction and reservoir inundation, and increased access to habitats where threatened species occur. Specific conservation programs for the Asian elephant and white-winged duck have been established. A survey of wildlife on the Nakai plateau was conducted during the construction phase of the NT2 project. Wildlife (particularly indicator species such as elephants, primates, and hornbills) will be monitored to detect trends in population sizes and enable further management. Minimization of construction impacts on threatened species was identified in the EIA and included careful location of work camps and the implementation of construction environmental management and monitoring plans.

Social Environment

Social impacts were anticipated in the EIA in five key locations, namely, the Nakai plateau, the downstream Nam Theun River below the Nakai dam, in the NNT NPA, along the Xe Bang Fai River, and in the surrounding area of the NT2 project. The key social impact identified was the relocation of households. Additionally, local residents would also be affected by construction activities including disturbance, increased pressure on resources and services, possible economic inflation due to increased demand, health risks, and human trafficking risks associated with an influx of workers and camp followers. The Resettlement Action Plan was designed to ensure that all resettled families are significantly better off after relocation. Efforts were made to select resettlement sites within existing traditional and spiritual territories to ensure cultural continuity and familiarity. The NT2 project has committed to provide $31.5 million of financial assistance and management support for the conservation of biodiversity and improvement of livelihoods of the communities residing in the NNT NPA. A Public Health Action Plan was prepared and includes two health programs responsible for preventing and mitigating adverse health effects due to the NT2 project.

Cultural Resources

The Nakai Reservoir was identified in the EIA to have the potential to affect a number of physical cultural resources within the inundation area, including sites of spiritual significance and 26 cemeteries. These resources are part of the cultural traditions of the villages in this location and required the performance of appropriate spiritual ceremonies prior to imposing any impact. NT2 project construction activities were also identified to have the potential to impact a number of physical cultural resources throughout the project area. The results from the most recent physical cultural resources survey (carried out in the mid-2004) were used to develop the management plan that included community awareness programs, relocation of physical cultural resources, appeasement ceremonies, securing moveable physical cultural resources, archaeological salvage operations, additional risk assessments, and a procedure for dealing with chance finds. Survey results and management measures were also integrated into resettlement plans.

Cumulative Impacts

The scope of the cumulative impact assessment included the effects that other (future) developments may have on the type and magnitude of NT2 impacts (added impacts) and the impacts of development in other sectors that are induced by the NT2 project (induced impacts). Two development scenarios were assessed based on 5-year and 20-year planning horizons.

The cumulative impact assessment concluded that the NT2 project alone will have an insignificant negative impact on the Mekong floodplain and on all aspects of the Tonle Sap including fish production. The assessment recommended several best practice actions to mitigate and compensate for impacts of all developments in the basin and predicted the results of these actions for 5-year and 20-year scenarios.

Environmental Management Plan and Monitoring

A variety of environmental management plans and monitoring activities have been implemented for the NT2 hydropower project and are described below.

Water Quality Management and Monitoring

Risks for poor water quality in the Nam Theun River arising from discharge of high organic loads or agricultural chemicals resulting from agricultural land use by resettled villagers have been taken into account through management planning. Mitigation measures include early partial clearing of the reservoir to maintain soil stability as it fills and an aeration weir and riffle structures in the downstream channel to redress potentially low levels of dissolved oxygen in reservoir water. There is also a plan to fund the construction of tube wells for drinking water if the river water becomes contaminated. Arsenic is known to be widespread in groundwater in the Lao PDR, and all developers should ensure that tube well construction be implemented only after conducting a water quality risk assessment to ensure that appropriate drinking water quality guidelines would be maintained (e.g., World Health Organization, AusAID).

Water quality monitoring programs for the NT2 project include the establishment of an on-site laboratory that is capable of monitoring water chemistry, hydrobiology, and greenhouse gasses. Additionally, a network of automated water quality monitoring stations with satellite telemetry for "real-time" monitoring is in operation. Water quality sampling effort is comprehensive with over 20 parameters being sampled weekly to bi-weekly over 22 downstream sampling sites.

Discharge and Water Level Monitoring

Diversions through the NT2 power station were predicted to almost double the annual discharge in the upper reaches of the Xe Bang Fai River and result in an increased water level of about 5 meters in the dry season and 1.5 meters in the wet season. The impact in terms of flooding was not studied in great detail (due to uncertainties with the relatively short-term dataset), but it was predicted that overall flood levels in the Xe Bang Fai River and its floodplain could increase by approximately 0.5, 0.4, and 0.2 meters in the upper, middle, and lower reaches, respectively, with an estimated 3.75% increase in flooded areas. Programs taking place to monitor the effects on flow from NT2 infrastructure and operations include flow monitoring by a flow meter installed at each discharge point, "real-time" monitoring of water level and rainfall through a network of automated stations, and data feed into a supervisory control and data acquisition (SCADA) system to optimize project operations.

Erosion Monitoring

Erosion within the watershed was a possible effect described in the NT2 project EIA. The extent of erosion is dependent on land use resulting from resettlement on the Nakai plateau. The major impact of the NT2 project is the doubling of discharge in the Xe Bang Fai River downstream of the dam, resulting in alluvial channel adjustment. The increase in discharge in the Xe Bang Fai River was investigated through hydrological studies, the EIA, and is addressed in the EMP and social safeguard documents, but the hydrological predictions remain uncertain because of the limited duration of the recorded flow data for the system. Studies predicted increased bank erosion in the Xe Bang Fai River due to increased flows, low sediment load in the diverted water, and daily variation in discharge (especially on weekends). It was estimated that the river would widen by up to 20 meters, with impacts most marked in the upper reaches.

Erosion management plans were developed for the Xe Bang Fai River at key points, particularly the point where the diverted water enters the mainstem. Compensation and mitigation measures are proposed to address the social impacts of riverbank erosion. In the entrenched reaches in the lower Xe Bang Fai River below the Mahaxay District, very little additional sedimentation after the first flood season was predicted due to widening of the alluvial reaches of the river. Monitoring measures to assess riverbank erosion on the Xe Bang Fai River include river cross section surveys, suspended solids monitoring, film and photographic documentation of erosion, and recurring visual surveys by boat.

Fish Monitoring

Effects from the NT2 project include major hydrological changes that affect existing patterns of fish species distribution and abundance; however, the limited data available indicate that fish extinctions are unlikely and that a viable fishery can be reestablished. Current data on fisheries are abundant, but not systematic. Various studies (ESIAs, social safeguard documents, and the EMP) have documented a wide variety of fish species present in the NT2 impact area and have indicated the distribution of some species; however, there is no reliable data on species abundance. Comparative data are available for similar areas in the Mekong River Basin and have been used by both proponents and opponents of the NT2 project to predict project effects. The social implications of effects on fish are currently unknown because few studies on fish consumption in the impacted area exist. The first fisheries studies in the impacted areas began in the late 1990s and intensive fish catch monitoring studies began 2005. The baseline fish, aquatic species, and habitat inventory report was finalized in September 2006.

A fisheries monitoring program is currently in place that utilizes adaptive management techniques for managing fisheries. To date, no fisheries collapses have occurred since the NT2 project began operations. Long-term effects such as changes in migration, spawning, habitat changes, and food availability are still uncertain and will continue to be monitored. Current fisheries studies and monitoring in the NT2 affected area include impacts to upstream and downstream river fish species, fish species and habitat inventory in the Nam Theun River and the Xe Bang Fai River, fish productivity monitoring, fish migration studies in the Nam Theun River, fish habitat and flow studies in the NPA and Xe Bang Fai River, downstream Nam Theun River fish habitat modification monitoring, and the implementation of a flow adaptive management program. In addition to the monitoring programs, river protection was incorporated into the management of the NPA and its corridors.

Contribution to Climate Change

The NT2 project was not predicted to make a significant contribution to greenhouse gases, particularly when compared to forms of power such as coal, oil, and gas-fueled generators. In fact, there were very limited efforts to protect watersheds in the area until the planning of the NT2 project. Hence, the current scenario is presumably more sustainable at least from an environmental point of view than the "no project" scenario. It cannot be said with certainty if this is the case given that there are factors that are unpredictable with regard to degrading and exploiting natural resources. Moreover, sustainability is a subjective concept. However, there is no current environmental management plan or specific climate change monitoring activities for the NT2 project. However, greenhouse gasses are being monitored as part of water quality monitoring (see above).

9.3.2.2 Training Workshop

The training workshop provided the opportunity for participants to share firsthand experiences in China, the Lao PDR, and internationally to address environmental challenges in hydropower development, including technical and policy issues through technical presentations, discussions, and a field excursion.

The workshop was a success, and the participants gained valuable insight from examining and discussing commonalities and themes of hydropower project environmental impacts and mitigation in the Lancang River in China and the Columbia River in the USA. The participants from Laos learned that the most effective measure for reducing the adverse impacts of hydropower project development is avoiding the impacts to the extent possible by careful planning in the design phase. The mitigation measures implemented at hydropower projects on the Lancang River also provided insight to participants from Laos on the design of environmental mitigation measures for their projects. There were also lessons learned regarding how to address transboundary environmental effects of hydropower projects, an issue facing participants from China and Laos. A case study in the Lao PDR was also presented on hydropower impacts to fisheries and the mitigation measures used to minimize the predicted impacts. Participants were trained on how to prepare and implement an EMP for a hydropower project. The field trip to the construction area of the Mekong hydropower dam in Xayabury Province provided the opportunity for participants to exchange information on environmental impacts and the best management practices that can be employed by hydropower developers. The knowledge and skills gained by participants during the workshop can be utilized to further improve the management of hydropower development in the Lao PDR.

9.4 Conclusions

The development of the hydropower sector in China has been ongoing for nearly 40 years with numerous lessons learned. China's experience in developing and shaping environmental management of hydropower development can be leverage by the GoL to implement good environmental management of the rapidly developing hydropower industry in the Lao PDR. Herein, we have discussed the current status of hydropower development in the Lao PDR, the regulations and institutions that oversee the hydropower industry, and the challenges of hydropower development in the Lao PDR. We also discussed the experience of hydropower development in China and presented lessons and key recommendations for the Lao PDR based on these experiences. We provided a case study of the NT2 hydropower project as an example of where lessons and best environmental management practices from China and other countries have been instituted in the Lao PDR. Finally, we referenced a training workshop held in the Lao PDR where further experience sharing by presenters and attendees from China, Canada, and the Lao PDR took place to develop enhanced understanding of the issues surrounding good environmental management of hydropower projects. By learning from these experiences shared, the Lao PDR can plan for more effective environmental management of their rapidly developing hydropower industry.

References

ADB (Asian Development Bank). 2004. Summary environment and social impact assessment Nam Theun 2 Hydroelectric Project in Lao PDR. Vientiane, Lao PDR. Vientiane, Laos.

———. 2012. *Greater Mekong subregion ATLAS of the environment*. 2nd ed. Manila: Asian Development Bank.

EDL (Electricite Du Laos). 2011. *Electricity statistics: Statistical yearbook 2011*. Vientiane: Vientiane, Statistic-Planning Office-Business – Finance Department of Electricité Du Laos.

GoL (Government of Laos). 2005. National policy on the environmental and social sustainability of the hydropower sector in Laos. No. 561/CPI Stat. Vientiane, Laos.

Hartmann, J., D. Harrison, J. Opperman, and R. Gill. 2013. *The next frontier of hydropower sustainability: Planning at the system scale*. Washington, DC: Inter-American Development Bank (IDB).

LPDR (Lao People's Democratic Republic). 2013a. Ministry of Natural Resources and Environment Ministerial Instructions on the Process of Initial Environmental Examination of the Investment Projects and Activities. No. 8029/MONRE. December 17, 2013. Available online at: https://www.investlaos.gov.la/images/sampledata/pdf_sample/ESIA-IEE/IEE_Ministerial_Instruction_Eng.pdf.

———. 2013b. Ministry of Natural Resources and Environment Ministerial Instructions on the Process on Environmental and Social Impact Assessment of the Investment Projects and Activities. No. 8030/MONRE. December 17, 2013.

Mekong Institute. 2012. *The proceeding of the Water Energy Development and Environmental Protection in the Greater Mekong Subregion*. A Regional Seminar "Meeting the Needs and Keeping the Ecological Balance" March 21–23, Phnom Penh, Cambodia.

MEM (The Ministry of Energy and Mines of Laos). 2011. *Renewable energy development strategy in Lao PDR*. Vientiane: The Ministry of Energy and Mines of Laos.

———. 2012. *Power sector development in Lao PDR Policy Dialogue*. Viantiane: The Ministry of Energy and Mines of Laos.

MRC (Mekong River Commission). 2011. *Planning Atlas of the lower Mekong River. Mekong River Commission-Basin development plan programme*. Vientiane: Mekong River Commission.

NTPC (Nam Theun Power Company). 2005. Social and environment management framework and operational plan (SEMFOP) [1st April 2005 to 30th September 2011]. Volume 1 of 2: management framework and operational plan.

Smits M. 2011. Hydropower and the green economy in Laos: Sustainable developments? Book chapter (8) in Hezri, A., Hofmeister, W. (Eds.), 2012, Towards a green economy: In search of sustainable energy policies for the future. Singapore: Konrad Adenauer Stiftung.

The Nature Conservancy, WWF, and the University of Manchester. 2016. *Improving hydropower outcomes through system-scale planning: an example from Myanmar*. Prepared for the United Kingdom's Department for International Development. Arlington, Virginia, USA.

The World Bank. 2007. *Nam Theun 2: A way to better hydro projects*. September 19, 2007. Available online at: http://www.worldbank.org/en/news/feature/2007/09/17/nam-theun-2-a-way-to-better-hydro-projects.

Wayakone, S., and I. Makoto. 2012. Evaluation of the environmental impacts assessment (EIA) system in Lao PDR. *Journal of Environmental Protection* 3: 1655–1670.

Wu, X. 2011. *Development of hydropower and protection of eco environment in China*. http://english.mep.gov.cn/Ministers/Speeches/201109/t20110907_217061.shtml, accessed April 17, 2017.

Chapter 10
Closure

This book has presented key components of a Project funded by the CGIAR Research Program on WLE focused on ecological and social effects of hydropower development on the Lancang River. Objectives of this Project were to investigate and address key ecological and social issues, within China and in countries downstream, related to hydropower development on the upper Lancang-Mekong River in China, and to make recommendations for improvement. In accordance with the broad scope of this Project, there were many goals, some of which were the acquisition and compilation of information, whereas others were applied, with a focus on facilitating positive change. The investigations and analyses presented in this book are an important component of the tasks required to meet the Project's goals by investigating and identifying key issues that communication and collaboration can then further address.

Five key topics which had been identified as important gaps in knowledge essential to the understanding of ecological and social consequences of the Lancang hydropower projects were investigated in this book, with each presented in one of eight main chapters. In the first of these chapters (Chap. 2), the assessment of river health and hydropower project impacts on river health was conducted by developing a framework that considered both ecological (physical, chemical, and biological) and human values in the definition of river health. Through the identification of appropriate physical, chemical, biological, and social indicators of river health, as well as reference values to score indicators, river health and the positive and negative impacts of hydropower projects were evaluated. In the next chapter (Chap. 3), the existing mitigation, management, and monitoring activities that are employed on the Lancang River to improve river health, given the potential for adverse effects of the hydropower projects, were reviewed and recommendations were made for improvement. Shortcomings were identified, and recommendations were provided following examination of the effectiveness of existing mitigation measures and the evaluation of existing environmental management framework and monitoring programs.

X. Yu et al., *Balancing River Health and Hydropower Requirements in the Lancang River Basin*, https://doi.org/10.1007/978-981-13-1565-7_10

Chapters 4, 5, 6, and 7 investigated four key potential transboundary environmental effects of hydropower: hydrology, water temperature, sediment transport and geomorphology, and fish community. For each of these potential effects, assessment of transboundary effects was conducted through a number of approaches, including pre- and post-development comparisons and assessment of the changes in detectability of effects with distance downstream (including south of the Chinese border). Chapter 8 investigated the effect of the Lancang hydropower projects on the livelihood of women, which was a key Project objective given that impacts on women had not been considered in China, and women are generally considered more adversely impacted than men. Effects on women, particularly in relation to forced displacement and resettlement, were investigated through two case studies in which research, interviews, and/or focus group discussions were used to assess the gender division in resettlement impacts and to make policy recommendations. The last key topic investigated in this book addressed the Project objective of facilitating and documenting cross-border experience sharing. This was conducted in Chap. 9 through a case study that documented how knowledge and skills related to environmental impacts and mitigation measures were improved in Laos through the sharing of China's experiences, best management practices, and lessons learned.

Hydropower is always a hot topic in the Greater Mekong countries, the authors tried to explain whether a river is healthy under the impacts of hydropower. In essence, the debates on hydropower are rooted from the anthropocentrism and naturalism in philosophy. Anthropocentrism considers human beings to be the most significant entity of the universe and interprets or regards the world in terms of human values and experiences. Naturalism is the belief that only natural laws and forces operate in the world. Since the industrial revolution in the 1800s, the world has been realized that exploitation of natural resources with industrial technologies may increase society's living standard. Meanwhile, using our huge natural resources wisely and leaving a fair share for future generation (environmental sustainability) are crucial for both mankind and nature. More and more people believe that mankind and nature are an inseparable and interactive whole. Similar to the relationship of mankind and nature, both ecological integrity (maintaining structure and function) and human values (providing goods and services) were explicitly incorporated into the definition of river health in this book. Inclusion of human values doesn't only mean human can benefit from rivers but also emphasizes the balance between ecological integrity and human values. Often human uses of rivers conflict with ecological integrity or within each other, and a river is unhealthy if only a few of components or uses can be satisfied. Furthermore, the status-impact matrix developed in this study quantitatively illustrates river health status and the impact of hydropower. Key impact pathways that have significant impacts on river health indicators in poor/critical state can be identified in the matrix and may help hydropower developers to implement improvement measures.

The research on transboundary environmental effect of hydropower projects on the Lancang River is still limited due to the lack of monitoring data on the Mekong River. The limitations of data availability include monitoring frequency, access of data, and monitoring methods. The frequency of hydrological monitoring of gauge stations on the Mekong River is not enough to support the analysis of hydrological effects. The Chinese researchers are not able to have access to the monitoring data for their research works. The monitoring method of sediment transport is different from that in China, so the data cannot be compared directly with the data in China.

In the future, the study on river health and hydropower needs to be constantly improved in the Lancang-Mekong River Basin. In November 2015, six countries reached a consensus that priorities should be given in five directions including connectivity, production capacity, transborder economy, water resources, agriculture, and poverty reduction. Among the five directions, water resources collaboration is expected to be built into a flagship field in Lancang-Mekong River cooperation. At present, only around 10% of the estimated hydroelectric potential in the lower Mekong River Basin is developed. Hydropower development in the Lancang-Mekong Basin has long been debated and is expected to last in the future decades because new hydropower project is still under construction. The transboundary effect of hydroelectric project is a major topic of conflicts in the Lancang-Mekong region. A variety of environmental, cultural, economic, and social components are involved with hydropower development, so hydropower will be the core of water resources collaboration in this region. In the past, most of the concerns over hydropower's impacts on river health are focused on Chinese hydropower projects on the Lancang River. In the near future, the studies on river health and hydropower will include hydropower projects of Laos on the mainstem Mekong.

MIX
Papier aus verantwortungsvollen Quellen
Paper from responsible sources
FSC® C105338

If you have any concerns about our products, you can contact us on
ProductSafety@springernature.com

In case Publisher is established outside the EU, the EU authorized representative is:
Springer Nature Customer Service Center GmbH
Europaplatz 3, 69115 Heidelberg, Germany

Printed by Libri Plureos GmbH
in Hamburg, Germany